Proceedings of the Royal Institution of Great Britain

Proceedings of the Royal Institution of Great Britain

Volume 69

Edited by
P. DAY

Oxford New York Tokyo
OXFORD UNIVERSITY PRESS
THE ROYAL INSTITUTION
1998

Oxford University Press, Great Clarendon Street, Oxford OX2 6DP

Oxford New York
Athens Auckland Bangkok Bogota Bombay
Buenos Aires Calcutta Cape Town Dar es Salaam
Delhi Florence Hong Kong Istanbul Karachi
Kuala Lumpur Madras Madrid Melbourne
Mexico City Nairobi Paris Singapore
Taipei Tokyo Toronto Warsaw

and associated companies in
Berlin Ibadan

Oxford is a trade mark of Oxford University Press

Published in the United States
by Oxford University Press Inc., New York

A catalogue record for this book is available from the British Library

Library of Congress Cataloging in Publication Data
(Data applied for)

ISBN 0 19 850366 0

Typeset by EXPO Holdings, Malaysia

Printed in Great Britain by
Bookcraft Ltd., Midsomer Norton, Avon

PREFACE

This annual volume, prepared for the Members of the Royal Institution, brings together the texts of Friday Evening Discourses delivered over the last year of so in the historic Lecture Theatre at 21 Albemarle Street. As in earlier volumes, the topics covered range very widely over many aspects of contemporary science and technology. In addition, following the tradition of the Royal Institution, others also treat historical or biographical subjects relating to science and its impact on society. It is to be hoped that this collection gives an impression of the continuing vitality of Friday evenings at the Royal Institution, as a medium through which scientists and engineers expound their work to lively general audiences.

As Editor, my warm thanks are due to Sarah Cripps and Evie Jamieson for keeping track of the manuscripts and the publishing process, and finally, of course, to the authors for putting their Discourses into printed words.

London P.D.
May 1998

CONTENTS

PLATES

The plates section falls between pages 68 and 69.

1. The sun in both visible light (*right*) and in X-rays (*left*). The dark areas visible (sunspots) are seen to be the seat of energy which manifests itself as X-rays in the solar corona. (Yokoh satellite.)
2. Patches of orange carbonates on the surface of martian meteorite ALH 84001. The field of view is about 0.5 mm across. The grains were produced below the surface of Mars when water circulated through it.
3. The historiated letter 'I' from *Genesis* in the Lucka bible.
4. Magnified (× 1000) portion of the dark grey column depicted in a sixteenth century German manuscript showing the presence of at least seven pigments.
5. Photographs and microphotographs of elaborately decorated initials on a fifteenth century German manuscript MS Ger 4 (f 2r and f 28v) from the DMS Watson library.
6. Faces on the Byzantine/Syriac lectionary (upper f 188v, lower f 67v) blackened by degradation of white lead to black lead sulphide.
7. Partially reconstructed glazed bowl painted in black on a blue background (CF 1412). The scale section on the left is 2.5 cm long.
8. Polychromatic Egyptian faience samples from the Petrie Museum, (a) UC. 686 red pigment identified to be red ochre and (b) UC. 888 (lotus) yellow pigment identified to be lead(II) antimonate, $Pb_2Sb_2O_7$ (scale in mm).

CONTRIBUTORS

Roy F.V. Aylott
City Engineer,
Corporation of London,
PO Box 270,
Guildhall,
London EC2P 2EJ

John Burland
Professor of Soil Mechanics,
Department of Civil Engineering,
Imperial College,
London SW7 2BU

Robert W. Cahn
Department of Materials
Science and Metallurgy,
University of Cambridge,
Pembroke Street,
Cambridge CB2 3QZ

Robin J.H. Clark
Professor of Inorganic Chemistry,
University College,
20 Gordon Street,
London WC1H 0AJ

Peter Day
Director and Fullerian Professor
of Chemistry,
The Royal Institution of
Great Britain,
21 Albermarle Street,
London W1X 4BS

Mike Garrett
Director of Innovation,
BOC Gases, 10 Priestley Road,
The Surrey Research Park,
Guildford,
Surrey, GU2 5XY

Monica M. Grady
Department of Mineralogy,
The Natural History Museum,
Cromwell Road,
London SW7 5BD

Richard Holmes
42 Dartmouth Road,
London NW5 1SX

Dan McKenzie
Department of Earth Science,
University of Cambridge,
Madingley Road,
Cambridge CB3 0EZ

Norman Myers
Consultant in Environment and
Development,
Upper Meadow,
Old Road,
Headington OX3 8SZ

Kuniaki Nagayama
Laboratory of Ultra structure Research,
Department of Molecular Physiology,
National Institute for Physiological Sciences,
Myodaiji-cho, Okazaki,
444–8585 Japan

John Pickett
Department of Biological and Ecological Chemistry,
IARC-Rothamsted,
Harpenden,
Hertfordshire AL5 2JQ

Sir Brian Pippard
Cavendish Laboratory,
Madingley Road,
Cambridge CB3 0HE

Andrew Wallard
Deputy Director, National Physical Laboratory,
Teddington,
Middlesex TW11 0LW

Sir Arnold Wolfendale
Department of Physics,
University of Durham,
South Road,
Durham DH1 3LE

Will Wyatt
Chief Executive,
BBC Broadcast,
Broadcasting House,
Portland Place,
London W1A 1AA

The search for extraterrestrial life—and the future of life on Earth

ARNOLD WOLFENDALE

Introduction

One of the great questions for humanity is 'Are we alone?' The question is not a new one, of course, but has come to the fore over the last year or two for a number of reasons. First, radio astronomy has achieved such high sensitivities that to look for signs of life from the directions of other stars, by way of 'intelligent' radio signals, is becoming practicable. Secondly, recent claims for the detection of past (and present?) 'life' on Mars has raised the possibility of the actual formation of 'elementary' life being very common. Finally, the observations of 'Earth-grazing' comets and asteroids has pointed out the threat to life on Earth from the impacts of these bodies. Similar impacts on other (potential) life-bearing planets is one of the factors which must be considered when estimating how many 'civilizations' there are likely to be in the Galaxy.

Historical attitudes

The idea, commonly held nowadays, that 'intelligent life' is quite common in the Galaxy is not a new one. From many examples, one can quote just two: by the Greek Philosopher, Lucretius and by the Chinese Philosopher, Teng Mu.

The universe is infinitely wide. It's vastness holds innumerable atoms so it must be unthinkable that our sky and our round world are precious and unique ......
Out beyond our world there are, elsewhere, other assemblages of matter making other worlds.
Ours is not the only one in airs embrace.

Lucretius First century BC
(after Drake and Sobel[1])

How unreasonable it would be to suppose that, besides the heaven and earth which we can see there are no other heavens and no other earths.

Teng Mu AD Thirteenth century[1]

However, by the sixteenth and seventeenth centuries such views were anathema to the Christian Church and highly dangerous ones to hold. The Church's teaching was, essentially, that the Universe was perfect, with the exception of the earth on which resided those unique, generally sinful, human beings whose redemption could only be arranged via the Church's intervention. It was natural to teach, therefore, that the Earth was at the centre of all things and woe betide anyone who thought otherwise. As is well known, Giordano Bruno (1548–1600), the Italian monk, paid the price for a twofold heresey: that it was the sun, not the earth, that was at 'the centre', and that we are not alone. He was burned at the stake in 1600.

'Innumerable suns exist; innumerable earths revolve about these suns in a manner similar to the way the seven planets revolve around our sun. Living beings inhabit these worlds'.

Giordano Bruno Sixteenth century[1]

Bruno had been influenced by Nicholas Copernicus (1473–1543) whose famous book in the year of his death put forward the idea of a sun-centred planetary system.

It was Galileo Galilei (1564–1642) who really clinched the idea, thereby demoting the Earth from its pride of place. This brilliant physicist-cum-astronomer used the newly invented telescope to observe the heavens and discovered the now-called Galilean satellites orbiting Jupiter (Fig. 1). Galileo's view was that since the Moon orbits the Earth and the satellites orbit Jupiter so will the planets orbit the Sun—their

Fig. 1 Jupiter and a satellite—a modern Hubble Space Telescope photograph. Galileo would also have seen the satellites away from the planet's disc; he observed that they were rotating round the planet.

nearest big mass. Galileo, too, received the attention of the Inquisition and was lucky to escape with his life (see, for example, Sharrat[2]); instead of death he suffered a form of house-arrest.

In many ways, Galileo was the first mathematical physicist and his work on inertia, falling bodies, the orbits of projectiles and on pendulums, was ground-breaking; this work was built on by Isaac Newton, who was born the year that Galileo died.

The Drake equation

Having established that the Earth is not at the centre of everything and knowing, as we do, that the sun appears to be a rather typical star, the stage is set for a serious look at the possibilities for life elsewhere.

It can be said, immediately, that there is, as yet, no evidence at all for intelligent life elsewhere. Serious attempts are being made with large radiotelescopes (such is that at Arecibo, Fig. 2) to detect intelligent

Fig. 2 The Arecibo radio telescope in Puerto Rico. This telescope is used from time to time to search for radio signals from extraterrestrial intelligent life.

signals but without success so far; this SETI project (see NASA brochure, 1990;[3] Project Phoenix, 1994[4]) was funded initially by NASA but after support was withdrawn the research groups have had to rely on private donations. The present null results show that there is no detectable intelligent life out to a few tens of light years from the Sun—not a negligible distance, but one small compared with the dimensions of the Galaxy (distance to the Galactic Centre $\simeq$ 25 000 light years).

Because of the lack of evidence we are reduced to hypothesis and thus to an estimate of how probable extraterrestrial (ET) intelligent life is, i.e. how many such locations there might be. It is usual, in this context, to use the so-called Drake equation, named after Frank Drake, the American scientist, who has done so much work in this field. Before giving the equation it is important to stress just what it is that it gives the rule for estimating. It relates to the number of *detectable* civilizations and it assumes that the intelligent life is *similar* to ours in that it resides on a planet orbiting a star. It also assumes that the message is recognizable as such. Thus life-forms very different from ours are not included (we have enough problems trying to work out the numbers for earth-types without spreading the net wider!).

The Drake equation runs as follows:

The Drake equation

$$N = Rf_1\, nf_2f_3f_4T$$

R = rate of star formation
f_1 = fraction of stars with planets
n = number of planets hospitable to life per star
f_2 = fraction where life emerges
f_3 = fraction where intelligent life appears
f_4 = fraction capable of communication
T = length of time for which life remains detectable

Although one can make criticisms of the equation from the standpoint of the interdependence of some of the terms, and so on, it does give a list of the obviously relevant parameters. It is clear that, although some of the parameters are known, others are not, and this relates primarily to the biological ones. As an astrophysicist, however, I will have to concentrate mainly on the astronomical factors, although the recent work on the Martian meteorite will also receive attention.

Before continuing, it is necessary to make the rather obvious remark that the uncertainties in our knowledge of many of the parameters are so great that no credence can, yet, be put on a derived value for 'the number of ...'. Nevertheless, progress is being made all the time with the individual terms and it is useful to make an assessment. The value of 'T', the length of time for which intelligent life remains detectable, is one that will receive particular attention insofar as it has great relevance to the human condition, namely the 'future of life on earth'.

Life on Mars?

It has long been the stuff of science fiction that there is life on Mars—'little green men' and that sort of thing. Although in recent times such ideas have been dismissed as fanciful, the view that a very primitive form of life may have existed at very early epochs (say some 4 billion years ago), when the necessary free water was available, has remained respectable. The idea took something of a knock when the American Viking Orbiter/Landers (1976–82) sent back some 50 000 photographs of the Martian surface, and other data, which gave no signs of life of any form. The Martian surface appears to be a most inhospitable place.

Potential help was at hand, however, with the demonstration that a certain class of meteorites found on earth were, in fact, of Martian origin. Excitement reached fever pitch with the claim by NASA that one of these meteorites contained evidence of 'life' in the form of fossils. Figure 3 summarizes the life history of the rock chipped out of Mars by the impact of an asteroid some 16 million years ago. The techniques involved in determining the chronology involve the use of radioactive nuclei, some from the initial supernova, others induced by cosmic rays at different times. Understandably, the media reacted with great interest and much merriment. One of the ensuing newspaper cartoons is illustrated in Fig. 4. The more sober-sided scientific community reacted with great caution, the role of dramatic claims in releasing much-needed research funds not being lost on them! The scientific problems associated with the discovery were—and still are—mainly associated with the claimed fossils—they are almost entirely much smaller than those on earth—and there is a lack of evidence for cell-walls. Figure 5 shows the biggest of the small fossils claimed by the NASA group.

It is appropriate to point out that some meteorites have been known for some time to contain quite complicated biological-style molecules

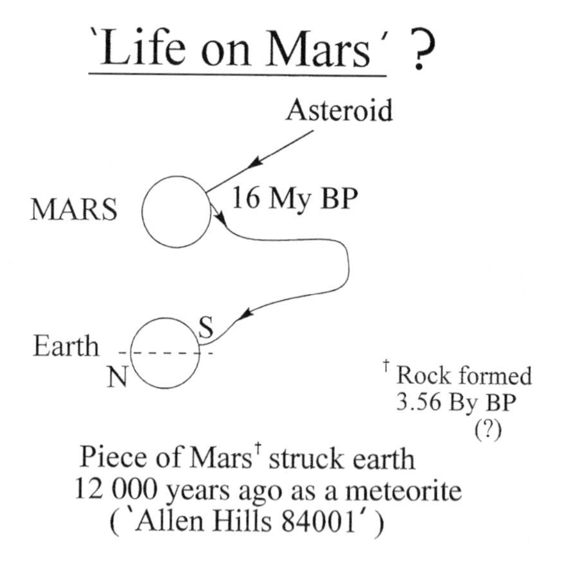

Fig. 3 Life history of the Martian meteorite 'Allen Hills 84001'.

Fig. 4 Cartoon from the *Daily Mail*, 8 August 1996.

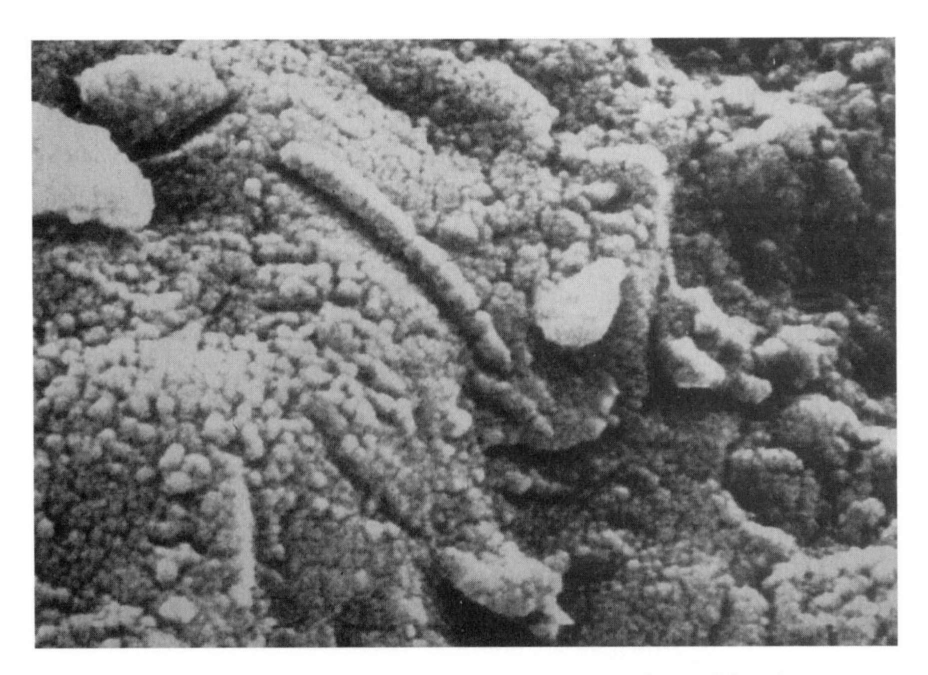

Fig. 5 An example of the sort of structure claimed by the NASA group in 1996 to be a fossil.

(amino acids); the problem has been the lack, until now(?), of fossils of true replicating systems, namely systems that could, in principle, join the life-sequence.

The jury is still out here, but it is interesting, and important, to note that recent work at The Open University and the Natural History Museum (described at a recent meeting at The Royal Society) has provided evidence by way of isotopic ratios in Martian meteoritic material which gives support to the idea of very primitive life-forms. We must conclude, then, that one of the important terms in the Drake equation ('f'_2—the fraction where life emerges) *may* be quite appreciable.

The search for planets round other stars

Staying with the Drake equation we turn our attention to f_1, the fraction of stars with planets. It might have been thought that this number was known with reasonable accuracy but this is not the case. The parent stars are so bright compared with their (possible) planets that the latter cannot be seen directly. It is true that accretion disks of solid material have been seen round a number of stars—with some nice recent examples from the Hubble Space Telescope—and these are generally regarded as the source of planets-to-be—but that is not the same thing as detecting actual planets.

Help is at hand, however, from neutron stars, the incredibly condensed stars which form when a more normal star collapses. Essentially, the protons in the hydrogen atoms eat up the electrons to form neutrons and neutrinos, the latter, having phenomenal penetration, immediately escape. The stars thus contracts by a factor of about 100 000—this being the ratio of atomic to nuclear dimensions—and the gentle rotation of the original star speeds up accordingly. For some reason, not yet fully understood, the neutron star has a magnetic field which is not aligned with the axis of rotation and this rotating field causes effects in the surrounding plasma which generate a rotating beam of radio waves (and other forms of electromagnetic radiation and fast atomic particles). The result is the famous 'pulsar', or pulsating radio star, discovered in 1967 by Jocelyn Bell and Tony Hewish.

The pulsars have, understandably, remarkably good time-keeping properties and it is this feature which has allowed extra-solar system planets to be inferred. If we imagine two skaters on an ice rink joined together by a taut rope going round one another in circles then, if one is much more massive than the other, the lighter one will go around in a much bigger circle than the heavy one. Replace the heavy skater by a pulsar, the light one by a planet and the rope by their gravitational attraction and we have a good analogy. In fact, because the ratio of the masses is very great ($\approx 10^6$) the orbit of the pulsar is very small but it is just sufficient for the radio beam to be 'modulated' in frequency because of the motion of the source and for this modulation to be detected.

The present situation is that one pulsar (PSR 1257 +12) with a millisecond period has at least three planets circling it, with masses from 0.015 to 2.8 times the mass of the Earth. There are also claims for three other pulsars having associated pulsars.

Now these planets will not be inhabited—the radiation level from the pulsar's emissions will be intolerable—but at least *they are there.* Indeed, one can argue that if pulsars can have planets then stars of a much gentler disposition should also have them—the point being that pulsars are born in the most dramatic circumstances following stellar collapse.

Turning to more normal stars the difficulties of detection are very great indeed but a start has been made at inferring the presence of planets much more massive than the Earth, specifically of mass similar to that of Jupiter (M_J). To date there have been seven good detections for stars out to 55 light years from the Sun, and five possible detections. The planets have masses in the range 0.5–5 M_J.

It would not surprise me at all if Drake's f_1 were quite high, say 0.01 or so.

T—the time for which life remains detectable

The parameter, *T*, is of course, one that has great relevance to the future of our own life as well as being of importance for life on other planets. Attention devoted to *T*, therefore, is time well spent whether there is intelligent life out there, or not!

Concerning the continuation of life on earth what are the mechanisms to be considered? First, there is the effect of humankind's own activities—the possible reduction of fertility, pollution of the environment, etc. This important aspect will not be considered here, again, we will stick with the astrophysics.

The astronomical hazards can be listed as: comets and asteroids, local stars exploding, and mischief with the Sun.

Starting with comets and asteroids—and large meteorites, too—their potential hazards for the earth are well known, although the actual frequency of impacts is rather uncertain. A fairly recent summary of likely rates is given in Box 4.

Likely rates of impact of 'bolides' (meteorites/asteroids/comets) with earth—and their effect.

(after Ahrens and Harris, 1992)[5]

Average time between impacts	Bolide diameter (km)	Effect
1000 y	0.1	Local destruction (e.g. Tunguska, 1908)
100 000 y	1	Would kill ~ 25% of the population
100 My	10	Would probably kill everyone

At first sight, we can put $T = 100$ million years (My) for life on Earth but there are some complications. For a start, the rates (or, more precisely, intervals, T) may be variable. Specifically, the rate of cometary impacts may increase over the next 20 000–30 000 years as we approach another star (α-Centaui). This star will, by gravitational attraction, so perturb the great Oort Cloud of Comets which extends out to nearly halfway to the next star, that more comets will flood into the Inner Solar System. Very recent work[6] has led to the claim that for other reasons, too, we might be in for an enhanced rate of impacts of cometary debris in forthcoming centuries. Indeed, it seems that the present period may be a comparatively quiet one from the standpoint of impacts of bodies of all sizes.

Returning to Box 4, we have a very rough and ready confirmation that $T \approx 100$ My for 10 km diameter comets from the fact that it is very likely indeed that the dinosaurs died out 65 My ago because of the impact of such a comet. The confirmation comes from a variety of sources, including the location—in the Gulf of Mexico—of the likely impact zone.

Other things being equal, it seems that T for the Earth could well be of order 100 My from the standpoint of the impact of large comets. However, help may be at hand, in view of the distinct possibility of firing a nuclear weapon at the comet and causing partial disruption and thereby a near-miss of the comet with the Earth, instead of a direct hit. The later such a civilization-affecting comet appears the more advanced will be our technology and thus the better the chance of negating its effect.

Allowing for all the above: increased comet rates, destruction possibilities and so on, it looks as though *T* is much bigger than 100 My for us.

What *T* might be for other planetary systems is, at present, anyone's guess. No doubt some planetary systems will have more comets and smaller *T* and others the reverse but we cannot yet be quantitative. It would surely be unlikely if all systems had *T* much less than ours, although it may be that possession of a large planet, such as our Jupiter, is a prerequisite for 'large *T*' and most planetary system may not have one. Jupiter's role for us is very interesting—it deflects many comets that venture into the Inner Solar System, where we are located, out of the Solar System all together; thus, a comet does not often have repeated chances of hitting the earth.

T—the lifetime of the Sun, and other stars

As mentioned already, 'mischief with the sun' is potentially a life-threatening phenomenon. At first sight, the sun is benign but in fact there *are* excitements from time to time in the form of solar flares (Plate 1). It is well known to cosmic ray workers that these flares can give rise to quite significant fluxes of energetic particles at the earth. The late George Wdowczyk and I[7] some years ago put forward evidence favouring rather dramatic effects on earth—on a statistical basis—every hundred thousand years or so. Proof positive is hard to come by but it does seem that, although there may be *some* effect, it is unlikely that life on Earth will be terminated by such phenomena.

Cometary impact, or more likely, still, our own incompetence, is a better bet if it is the demise of civilization that is required!

Similar remarks probably apply to a significant extent to other stars for most of their lifetimes. 'Lifetime' is, indeed, the parameter which can be guaranteed to be the maximum possible value of *T*. The Sun was born about 4.5 By ago and is about half-way through it's life. After about 3 By it will start to swell, eventually engulfing the Earth and, clearly, we will have had to leave a long time before that happens. This point has great significance for the whole 'life elsewhere?' problem, as will be demonstrated later.

The situation with many other stars is very similar. Admittedly, some stars explode dramatically and without much warning (supernovae) but they are few—only about one per 30 years in the Galaxy, and the Sun is not one of these. Furthermore, as Wdowczyk and I showed, the chance of another star sufficiently near us to be civilization-threatening is about once per 10 By, i.e. longer than the remaining solar life.[7]

The astro terms in the Drake equation

Taking everything together, it would be no surprise if, for a few per cent of the stars in the Galaxy, there were planets not too different from the earth. Comets, nearby stellar explosions and such like might limit the average value of T to, say 1 By, i.e. about one-tenth the age of the Galaxy. If the Mars detection of 'life' is correct then one finds that there could easily be several million sites where life exits. The big questions now are: How much of it is 'intelligent'? What fraction can communicate?

Perhaps 'absence of evidence' *is* 'evidence of absence'

The main argument against intelligent life being common is the fact that we have not seen any. At first sight this is bad science but in fact there is a good reason for saying it, for the following reason. As remarked earlier, when one's star is about to run out of fuel and become a giant it is 'time to go'. Now, very many stars in the Galaxy have been and gone, so that intelligent life on surrounding planets will have left. Thus, there should be very many space travellers visiting or passing by. The degree of intelligence—or technical mastery—does not need to be much superior to ours for this to happen. Despite many claims for the detection of UFOs none has been substantiated and, in my view, there is no firm evidence for visitors at all.

Thus, we may, in fact 'be alone'. [At this stage in the lecture, the lecturer's arguments appeared to be negated by the sudden appearance of a Martian, suitably clad, and with lights flashing. However, in view of 'its' inability to answer an elementary question about the mean mass, lifetime, and decay scheme of the muon, the lecturer categorized the visitor as 'unintelligent life' and argued that his thesis was still valid.)]

References

1. Drake, F. and Sobel, D., 1992, *'Is anyone out there?'—The scientific search for extraterrestrial intelligence'*, Souvenir Press.
2. Sharratt, M., 1994, *Galileo—decisive innovator*, Cambridge University Press.
3. SETI, 1990, Jill Tarter *et al.*, *NASA*, Ames Research Center, SETI Office, Moffett Field, California 94035-1000, USA.
4. Project Phoenix, 1994, *SETI Institute*, 2035 Landings Drive, Mountain View, California 94043, USA.
5. Ahrens, T.J. and Harris, A.W., 1992, *Nature*, **360**, 429.
6. Bailey, M.E., 1995, *Vistas in Astronomy*, **39**, 647.
7. Wdowczyk, J. and Wolfendale, A.W., 1977, *Nature*, **268**, 510.

SIR ARNOLD WOLFENDALE

Born 1927 in Rugby but brought up in Lancashire. He was educated at Stretford Grammar School and Manchester University where he joined the staff under the legendary P.M.S. Blackett. Moving with his family to Durham University in 1956 he built up a group working in Cosmic Ray Physics and later led a transition of much of the Physics Department's work into Astrophysics and Astronomy. He was elected to the Royal Society in 1977, was President of the Royal Astronomical Society, 1981–83, Astronomer Royal from 1991–95 and President of the Institute of Physics from 1994–96. Sir Arnold Wolfendale was Knighted in 1995. He has received a number of Honorary Degrees and Medals and has been honoured by a number of foreign Academies and Institutions. Sir Arnold has been giving Public Lectures for many years; he recently chaired a Government Committee looking into the Public Understanding of Science and several of the committee's recommendations have already been taken up. He has published some 500 papers. He is still trying to find out the origin of the cosmic radiation and feels that he is now very close—a feeling that he has had before!

Magellan looks at Venus

DAN MCKENZIE

The beginning

Unlike astrophysicists, earth scientists have until recently had only one object on which to test their ideas and theories about how the Earth formed and evolved. Furthermore, and perhaps surprisingly, the large-scale structure and dynamics of the Earth is in some ways harder to model than is that of stars. The problem is that nucleosynthesis inside starts is a very energetic process, and is controlled by nuclear reactions that have been studied in exquisite detail in the laboratory. Some of the most successful models of complicated natural processes have been developed to account for the observed abundances of elements and isotopes. The pressure and temperature within stars are also principally controlled by the interactions between photons, electrons, and ions, which can be studied in the laboratory. The principles governing the structure of stars were therefore understood before we had any comparable understanding of the Earth's structure and dynamics, for which we have had to depend on images produced by sound, rather than light, waves. Although we now think we understand, at least in outline, the processes involved, most of our ideas have come from observing what happens, and then arguing that we can understand what we see using simple physical principles based on laboratory experiments. Although such an approach is more sensible than the speculations it has replaced, and is obviously a useful way to start, it is possible that completely different processes may be important on other planets, and perhaps also early in the Earth's history. We would like to be in the same position as those who work on stellar structure. They can calculate the evolutionary path of a star as it grows older and compare it with a variety of observations from stars of different ages. One of the reasons astrophysicists have such confidence in their models is that they have a huge number of stars that they can observe, whereas the geologist has only one Earth. Hence the importance of looking at other planets and understanding how they work. As I have spent much of my scientific life thinking about the Earth, starting with the ideas of plate tectonics to describe surface deformation and then studying how mantle convection can maintain these motions, I was interested in planets that were likely still to be tectonically active. From this point of view the Moon was not a good bet. Although it is not completely inactive, the few moonquakes that take place are very small and occur at depths of about 700 km. Furthermore, they appear to be produced by the tides that are generated in the solid Moon by the Earth, rather than being an expression of large-scale tectonic deformation. Mars is bigger and is more likely to be tectonically active. However, much of its surface is old and probably not now being deformed. It does, however, have the largest volcanoes yet found in the solar system. With some luck, the spacecraft that are planned over the next few years will allow us to understand more about its interior. The experiments carried out by the earlier US landers were dominated by the desire to find life, and therefore were not much help in studying geological problems. But Venus was always much the most promising planet from my point of view, and I jumped at the chance to study it when it came, in about 1980, in the form of a boring technical document in a cheap envelope entitled 'Announcement of Opportunity: Venus Orbiting Imaging Radar'. At that time NASA sent out hundreds of such documents (they are now on the web), and you could send in a proposal to work on the data, which I did. And so began the project that has lasted the longest of any that I have ever undertaken.

Before Magellan I had never had anything to do with space, and knew nothing about what was involved. I had not thought about the difference between putting a satellite into orbit round the Earth, which is easy, and into orbit round Venus, which is very difficult indeed, because you must communicate with a small object at a great distance from the Earth, which is travelling very fast, and make it go into orbit as it passes the planet. The spacecraft that was finally sent to Venus had a transmitter whose power was about 300 W, and which could transmit about 200 k bits a second from the other side of the Solar System. At the beginning, the spacecraft was going to carry many different experiments. I went to my first meeting, at the Jet Propulsion Laboratory (JPL) in Pasadena, and was astonished when I entered the lecture room to find about 300 other scientists, all of whom had experiments they wanted to fly or who wanted to work on the data the spacecraft would return. Over the next few days they explained what they wanted to do, and I had my first experience of the opaque language that surrounds every aspect of JPL. Everything is referred to by a string of capital letters. I slowly got used to TBD (to be determined), BIDRS (basic image data records), MIDRS (mosaiced image data records), and so on. I still need a sheet which tells me what the many hundreds of such terms mean before I can understand the simplest talk. However, one thing I had understood before I became involved, which was that everything anyone was going to do *had* to be done with an instrument on the spacecraft, and so it was essential to make sure that *your* instrument was kept on and not other peoples'. I am not a builder of instruments, and the data I really wanted was going to come from the carrier frequency of the communications link, which *had* to be on board. So I was quite relaxed about this process, once NASA had agreed that I should be part of the project. But the same was not true of most of the people in the room, who were at their most persuasive. 'The Project', as it was called, gave some limited support to help people design their instruments, and we all went home to return in a year's time. When we did so, we were told that the estimated cost of the spacecraft was now $100 million more than it was last year! After the same thing happened for several years in a row, and the price tag reached about a billion dollars, NASA announced that the spacecraft was now too expensive, and would have to be completely redesigned to make it cheaper. 'The Project' then did two astonishing things: it threw off almost all the experiments that had been planned, on which many people had spent years of their lives, and reduced the mission to the task of making images of the surface of Venus. 'The Project' also collected a large number of existing bits and pieces of previous spacecraft to construct a remarkable object out of what was essentially NASA's version of the 'might come in handy' box of parts that I keep under the stairs.

The fundamental feature of Venus that dominates any attempt to map its surface is that is is always covered by clouds. They are made of droplets of sulphuric acid, and they never disappear. So photographs show nothing of the planetary surface. The only way forward is to use the spacecraft to illuminate the surface, in a waveband where the sulphuric acid is transparent, and in which the spacecraft can produce enough power to get a clear return. The spacecraft therefore must have a radar transmitter and receiver, so that it can illuminate the surface itself. But it cannot fly close to the surface, because it would burn up in the atmosphere, and must be powered by converting sunlight to electricity using solar panels. These problems dominated the final design (Fig. 1), which consisted on a big radar dish and two big flat solar panels to generate the power. The other features you can see are a long tapered horn at one side of the big dish, which is the altimeter. It works by sending the radar pulse straight downwards, to measure the height of

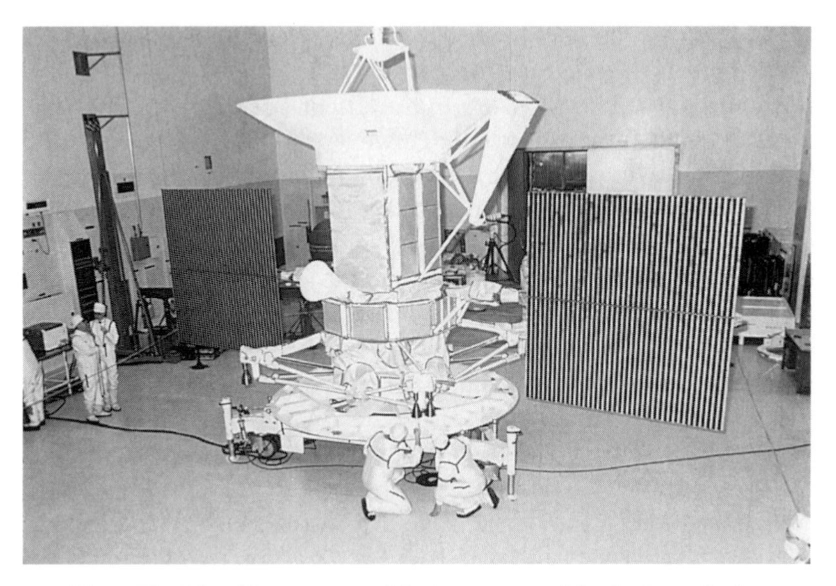

Fig. 1 The Magellan spacecraft being prepared for its launch. It is powered by electricity from solar panels, and uses the high gain antenna for mapping and communication with Earth. Rocket engine modules are used to control its orientation, and a star scanner to find its orientation.

the spacecraft above the surface and hence the shape of the planet. You can also see some small rocket motors on long arms sticking out from the bottom of the spacecraft. These are used to control its orientation. The main body of the spacecraft consists of a squat many-sided box, which contains the control bus, and a rectangular box beneath the big dish, which contains the radar. The big dish and the bus were spares left over from the Voyager mission, but the radar was new. The idea that saved a great deal of money was to use the *same* big dish to direct the radar on to a patch of the surface of the planet, to record the returned signals on tape recorders. Then the whole spacecraft rotates, to point the dish at the Earth, and the tape recorders play back the data, which are received by the Deep Space Network (needless to say, always known as the DSN). The DSN consists of three collections of antennae, in Owens Valley in California, near Madrid and near Canberra. This turn had to be very accurate, because the spacecraft had to point exactly at the Earth. And it had to do this twice every orbit, or twice every 3 hours, for at least 1 Venusian days of 243 Earth days. Rotating the spacecraft with rockets would use too much fuel, so it was done by spinning inertial wheels, in the same way as a cat can rotate to land on its feet when dropped from a high building. When I heard all this for the first time I thought 'Well, that's that. They will *never* get such a complicated system to work for so long'. I was wrong, but only just!

The next event was the Challenger explosion, which occurred after the spacecraft, by this time known as 'Magellan' (with, for some reason, a soft 'g') had been built. Like everyone, I was amazed, and horrified and fascinated by the stories that came out in the resulting enquiry. But I also had a personal interest in the outcome! The major effect on 'The Project' was that the large liquid hydrogen and oxygen rocket, that was to be used to propel Magellan out of the Earth's gravity field, had to be replaced by a solid fuelled rocket that was not so powerful. This in turn meant that there was less weight (and so less power) available for the rocket that was needed to insert Magellan into orbit round Venus, and the orbit was therefore going to be more elliptical. These problems were caused by NASA's unwillingness to fly liquid oxygen and hydrogen rockets in the Shuttle, but I could scarcely condemn them for their caution! So, finally, all was ready and the launch date was set, for the spring of 1989. As my teaching for the year was over, I thought I would go to Florida and watch the launch. I was astonished by the number of retired people who had come to watch the Shuttle in their campers. 'Oh yes', I was told, 'the number went up by 100 000 after Challenger: they are all here hoping for a repeat'. In fact I did not see the launch, because one of the hundreds of computers involved malfunctioned with 30 seconds to go. But, on 4 May 1989 Magellan was finally launched, pushed out of the Shuttle and sent on its way, with some grumbling from the astronauts about the dullness of this mission.

Thereafter, everything happened at JPL. Once the spacecraft was in orbit I had to get used to going there for 4 or 5 days every 2 months or so, which lasted for about 2 years. At the beginning I tried to change my body clock, but soon found that the best way was to go to bed at 5 in the afternoon, and go into JPL for the day at about 1 a.m. when the freeways are empty and you can park. In some ways I was very glad when the project finally ended. I could go back to living quietly in a terrace house in Cambridge, and my major means of travel once again became my bicycle. But the troubles of 'The Project' were not yet over. Magellan arrived at Venus on 10 August 1990 and was successfully inserted into orbit when it was behind the planet. We all waited in great excitement at JPL for Magellan to reappear, going more slowly. When it duly did and at exactly the right speed, and a cheer went up. I did not understand why everyone was *so* happy, until Mars Observer failed, doing exactly the same manoeuvre. Everyone's contracts with NASA were conditional on successful orbit insertion. When Mars Observer failed, perhaps a hundred planetary scientists were left with 3 weeks to find another job, and the entire subject almost disappeared! After Mars Observer, I decided that people who worked with me on planetary geology would never be funded in this way, but would be poor but safe!

After orbit insertion was successful there was a huge party at a Country Club in the hills behind La Canada, which is where JPL *really* is: it only has a Pasadena address because of gerrymandering. Everyone involved in 'The Project' came, about 450 people altogether. The total cost of 'The Project' was about $1000 million, by *far* the most expensive experiment in which I had ever been involved. At the party I remember asking someone I was introduced to whether everything was now likely to be OK. 'No' she replied, 'I don't think the software has been properly tested: "The Project" is short of money and has cut corners'. I thought nothing more about this remark until the following week, when I was back in Cambridge and Magellan made the lunch-time news. All contact had been lost and the spacecraft was out of control, after having collected only a few, beautiful, test images. When I next returned to JPL, and after communications and control had been restored with Magellan, one of the people involved in the rescue explained how this type of disaster had been planned for, and how the spacecraft had been programmed to search for the Earth if it went out of control. This particular problem happened several times, but was not so alarming after the first event. It was finally traced to a software bug, using the exact copy of the spacecraft that had been built when it was manufactured, which had been kept on Earth.

To my astonishment Magellan worked perfectly for the first Venusian day. The only problem was that one of its tape recorders failed. Then, slowly, different parts of the electronics started to fail. For me this was the worst time. The spacecraft was returning wonderful images of the surface, almost everyone in 'The Project' was happy, NASA was giving press conferences and making noises about closing down 'The Project' to save money that it would spend on the space station, and the spacecraft was slowly dying. But I could not understand how most of the astonishing features on the planet had been produced, and knew that I needed the gravity field to have any chance of doing so. When the spacecraft was collecting images, it could not also be used to measure the gravity field. So I needed the imaging to stop. The big dish could then be pointed towards the Earth, not at Venus, when the spacecraft was nearest Venus. The line of sight component of the gravity field can then be obtained from the Doppler shift of the carrier frequency, by measuring its rate of change.

I was saved by the failure of part of the imaging system, which made it impossible to use Magellan to produce pictures of the surface. A different problem now arose, which was that so few people were interested in the gravity that NASA wanted to shut down 'The Project'. But a lot of letters and phone calls kept it alive for another 3 Venusian days measuring gravity, before NASA Headquarters finally ran out of patience and destroyed the spacecraft by burning it up in the atmosphere of Venus. They did so to prevent any further efforts by the gravity wallahs to keep it going 'for just *one* more day: please! It *only* costs $10 million a year'. And so the mission ended, having returned more data than all the previous deep space missions combined. Magellan must rank with Voyager as one of NASA's most successful deep space experiments ever. Those on 'The Project' felt that Magellan never got the attention it deserved, from either the press or from NASA Headquarters. Perhaps the problem was that it slowly built up the spectacular images, rather than returning a photograph in the glare of the television cameras.

But, scientifically, Magellan was an outstanding success. One small spacecraft had made images, at a resolution of about 150 m, of almost the entire surface of Venus. This is better coverage than we have of the Earth. Because 70% of our planet is covered by water, the sea floor must be imaged using sound waves, and many parts of the oceans have almost no coverage.

The altimeter measured the shape of almost the entire planet with a resolution of about 20 km, which again is better coverage than we have of Earth, because of the oceans. The gravity field that came from the Doppler tracking was also better determined than was that of the Earth 2 years ago, although the geophysical measurements have since improved. These are remarkable achievements, and the two people who I think are most responsible for this success are Steve Saunders, who was the project scientist at JPL, and Gordon Pettengill, a radar scientist at MIT who had total grasp of every aspect of the spacecraft. Steve has now moved to NASA Headquarters, and Gordon took early retirement in a hurry when Mars Observer failed. They are two of the most impressive scientists with whom I have worked. Much of the success of Magellan was due to their clear understanding of the scientific issues, and an ability to keep these in focus through the extraordinary logistic complications of managing a mission of this size. One of the great strengths of modern science is its independence of the individual. All of the major advances in which I have been involved would have been proposed by someone else a year or two later, and it is only with some effort that the community remembers to attach my name to these ideas for a few years, until they are either discarded or become so much part of the framework of the subject that no one bothers to say where they come from. This tendency is especially strong where, like Gordon and Steve, scientists are leaders of large teams. Yet, the success of projects, such as Magellan, depends entirely on such people devoting a large chunk of their lives to keeping the overall objectives clearly in mind throughout the whole process of the design and operation of the spacecraft, and its interaction with the DSN and JPL itself, whose activities for a year or two were dominated by the problem of dealing with the huge amounts of data that Magellan generated. I was entirely an onlooker throughout this operation, although I was included in all parts of it in exactly the same way as were my American colleagues who were US nationals working at US universities with NASA support. This lack of interest in where people work and originate is an aspect of US science that has always impressed me. Although I had expected that it would operate for Magellan also, I thought the stress on peoples' relationships as the key data became available might lead to some hint of nationalism, but it never did. This spirit has yet to arrive in Europe!

Synthetic aperture radar

The key idea underlying the Magellan Mission is that of synthetic aperture radar, or SAR. Ordinary radars on ships and aeroplanes work by pointing a radar dish in a known direction and measuring the time taken for radar waves to travel from the dish to the target and back. For good resolution they require big dishes and very large amounts of power, which is in short supply on a spacecraft. A SAR uses a different principle illustrated in Fig. 2, which depends on the Doppler effect rather than the size of the dish to obtain good resolution. On a SAR the dish is only used to focus the radar on a patch of the planet to one side of the ground track of the spacecraft. The frequency of echoes that come from parts of this patch that are exactly perpendicular to the orbit of the spacecraft are the same as that of the outgoing signal. However, that of echoes from behind the spacecraft are shifted to lower frequencies. Conversely, those from targets ahead of the spacecraft are shifted to higher frequencies. The distance of a target at right angles to the track can be found from the time taken for the echo to return. Because the SAR uses all the echoes that return to the spacecraft from a patch on the surface, it requires less power to map a planet than does a conventional radar, and can obtain good resolution with a small dish by using the Doppler effect. Although the radar images from the SAR look like photographs, they contain certain artefacts not present in ordinary images. One of the most trouble-

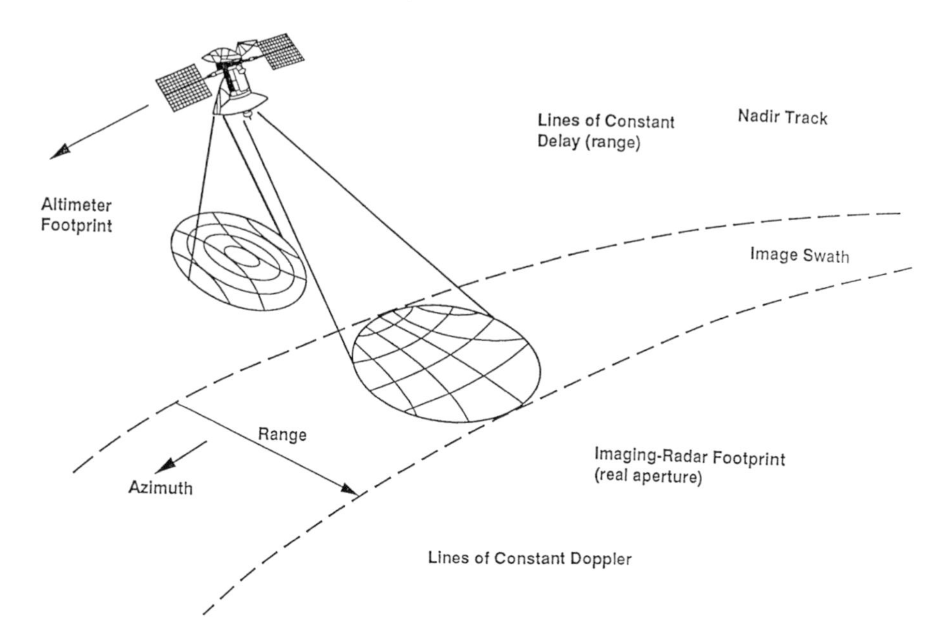

Fig. 2 How the SAR works. The nadir track is the path of the spacecraft on the surface of Venus. The SAR illuminates a patch of the surface (the footprint) to one side of the ground track, and uses the two way travel time of the echo (the delay) to measure the distance from the spacecraft, and its Doppler shift to measure its position along the strip that is being mapped (the image swath) (Ford 1993, *in* Ford *et al.* 1993).

some is called 'overlay', and arises because the radar return from the top of a steep hill arrives before that from its bottom. Figure 3(a) illustrates the problem, which causes the top of the hill to appear to be closer to the spacecraft than its base. Figure 3(b) shows a SAR image from a steep-sided glacial valley in Alaska. The glacier in the bottom of the valley is visible as a series of bright lines parallel to the sides of the valley. It is obscured in places by the returns from the top of the valley side towards the bottom of the image. In some areas of Venus this effect makes the images very difficult to interpret.

Figure 4 shows the sequence of operations that had to be carried out on every mapping orbit. Each such orbit mapped a strip about 25 km wide and 18 000 km long. Images like those reproduced here are built up by assembling many individual strips, and the dark lines on some show where there are data gaps. As this assembly continued, the extraordinary features on the surface of Venus were slowly revealed.

(a)

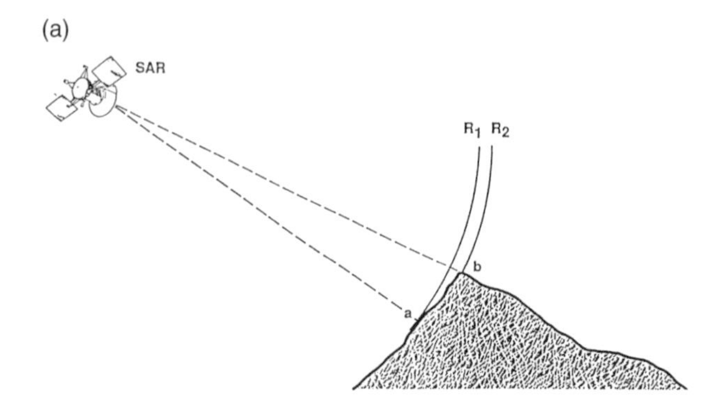

(b)

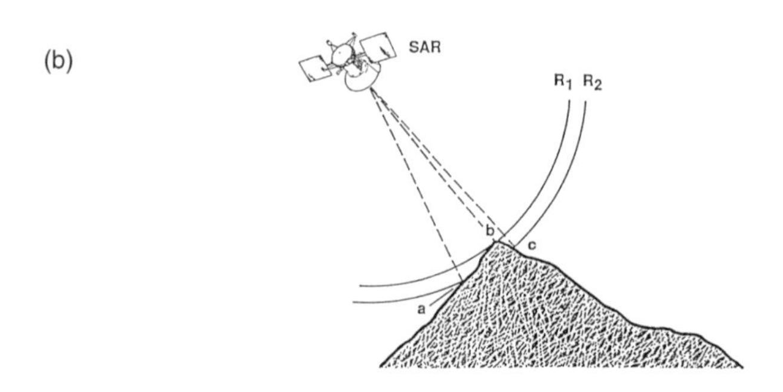

(c)

Fig. 3 (a) When a steep hill is illuminated at a shallow angle, the echo from the top, b, arrives after that from lower down, a. However when the illumination angle is steeper, (b), the reverse is true. The echo from the top of the hill then arrives first and overlies that from the base (Ford *et al.* 1989). (c) A SAR image from Alaska, showing a glacier at the bottom of a steep valley. In places the glacier is overlain by echoes from the valley side. Illumination is from the top (Farr 1993, *in* Ford *et al.* 1993).

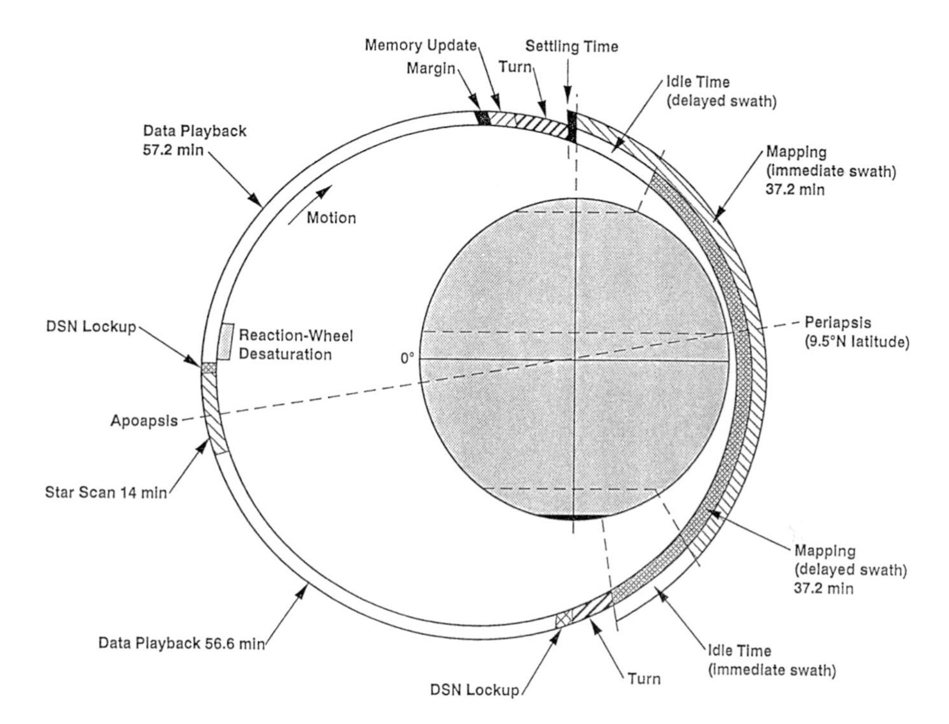

Fig. 4 The sequence of events that occurred during every 3-hour orbit of Magellan for 2 years (Ford 1993, *in* Ford *et al.* 1993).

Craters

All planets in the Solar System have been bombarded by meteorites which have produced craters, and Venus is no exception. Figure 5 shows an image of a medium sized crater, with a central peak, a flat floor with a circular crater wall, surrounded by lobate ejecta. The dark regions show areas that do not return radar energy, while the bright regions are strongly reflective. Whether or not a patch of the surface returns radar energy depends on its roughness at the radar wavelength of 12.6 cm, and is not related to its optical properties. The ejecta are radar-bright because they are rough, and contain boulders of all sizes. Some of these are sufficiently large to be visible as individual features on the SAR image shown in Fig. 5. Figure 6 is an image of the largest crater on Venus, known as Mead, with a diameter of 280 km. Like some large craters on the Moon, it has more than one ring. The craters on Venus differ in two ways from those on the Moon, Mars, and Mercury. There are few small (< 35 km diameter) craters, because small meteorites break up as they pass through the dense atmosphere: the pressure at the surface of Venus is 90 times greater than that on Earth. When meteorites break up in this way they deposit thick layers of dust like those visible as dark splotches in Fig. 7. The other difference is the ejecta, which flows out as lobes because it is fluidized by gas when the dense atmosphere is compressed by the impacting meteorite. These lobes are clear in Fig. 5.

Meteorite craters become invisible if their shape is changed by extensive deformation, or if they are buried by lava. The second process seems to be more important on Venus, perhaps because deformed craters are hard to recognize. Figure 8 shows an example of a crater that has almost disappeared beneath lava flows, whereas only the deepest part of the crater is Fig. 5 is flooded.

Impact craters are important because at present they provide the only method of estimating the mean age of the surface of Venus. The principle of the method used to do so is straightforward. Because of continual bombardment, the density of craters larger than some radius increases with the age of the surface. If the impact rate is known, the surface age can be estimated. Various such studies have been carried out, and show that the average surface age of Venus is about 500 million years (Ma). Surprisingly, all regions that have been dated in this way appear to be of similar age. In this respect Venus is quite unlike the Earth, whose ocean floor has a mean age of about 60 Ma, with a systematic variation from an age of 0 Ma on the spreading ridges, to about 180 Ma for the oldest ocean floor. In contrast the mean age of the continents is about 1800 Ma. If similar age variations were present on Venus they would have been discovered by counting craters. So it seems as if the entire planet was resurfaced 500 Ma ago, and that the process involved then stopped. No one yet understands how this took place, and why the history of Venus has been so different from that of the Earth.

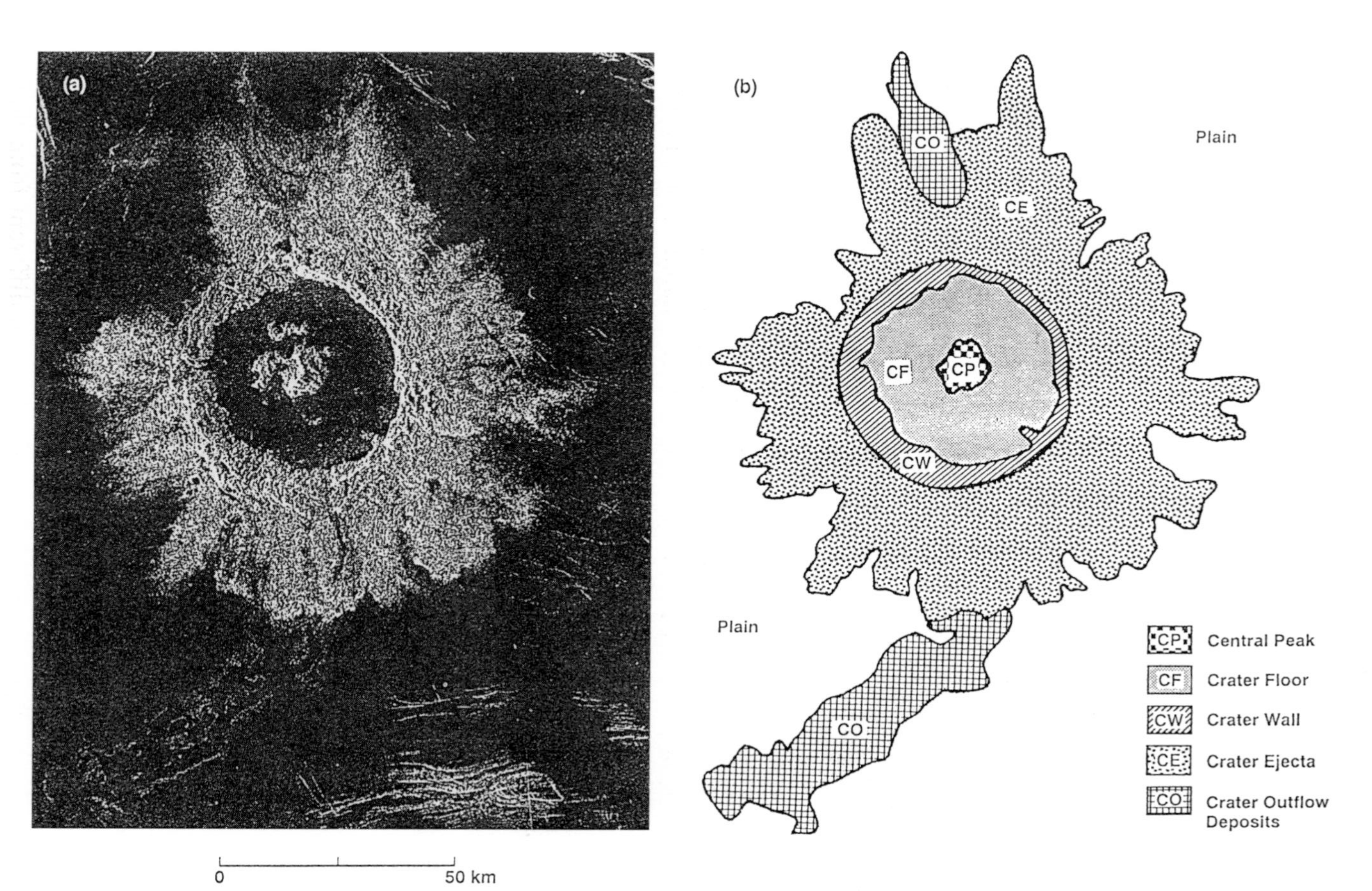

Fig. 5 Image and sketch map of the crater Danilova. Illumination is from the left with a 35° incidence angle (Weitz 1993, *in* Ford *et al.* 1993).

Fig. 6 Mead crater, the largest on Venus. The stripes are artefacts of the processing. Illumination is from the left with a 35° incidence angle (Weitz 1993, *in* Ford *et al.* 1993).

Volcanism

Many parts of Venus are covered with huge lava flows. Figures 9 and 10 show images of two such features. The bright and dark regions correspond to variations in surface roughness, which varies with the speed with which the lava flowed. So the boundaries of the bright regions do

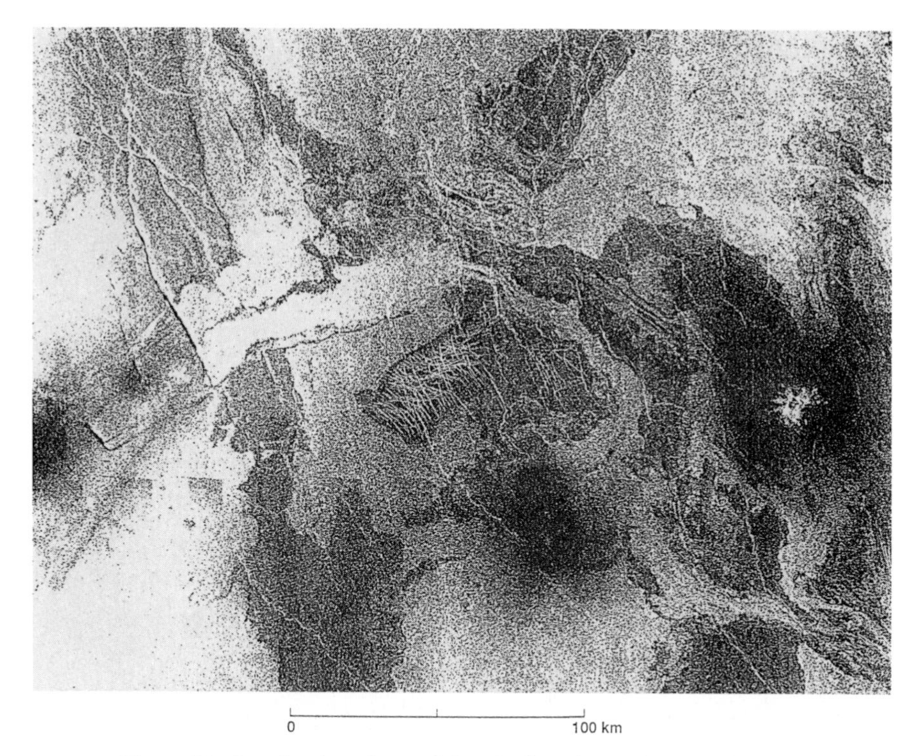

Fig. 7 Dark splotches formed by the breakup of meteorites in the atmosphere (Weitz 1993, *in* Ford *et al.* 1993).

not always correspond to the edges of individual lava flows. Small features such as channels, levees, and ponds of lavas caused by pre-existing features, are clear in Figs 9 and 10, because of the high resolution of the SAR images. Similar features are found on Earth. Although the flows in Figs 9 and 10 are larger than any recent flows on Earth, in the past many flows of this size or larger have been erupted. Because they have now been eroded or covered with sediment, they are less obvious on Earth than they are on Venus, where erosion and sedimentation are almost absent.

Although Magellan had no means of measuring the composition of the surface of Venus, several Soviet landers had done so earlier. They found that it consists of basalt, which is also by far the most common material that makes magma flows on Earth. It is therefore likely that the flows in Figs 9 and 10 are basaltic. The surface details of the flows are preserved so well because the surface temperature of Venus is about 430°C, so no liquid water is present to erode the surface or deposit sediment. The only sediment present is wind blown dust and dust from the breakup of

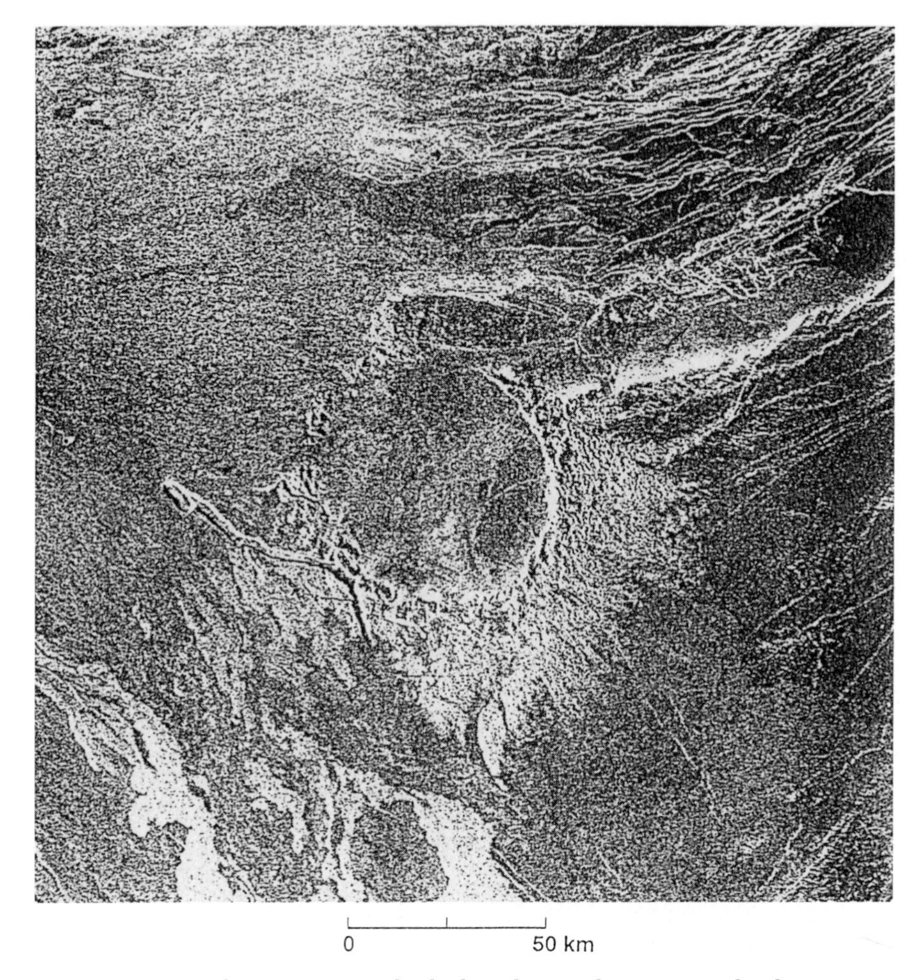

Fig. 8 Alcott crater, which has been almost completely flooded with lava (Weitz 1993, *in* Ford *et al.* 1993).

meteorites. A dark patch of dust from a meteorite is visible in Fig. 9, just outside the bottom left corner of the rectangle labelled 5.

The gradient at which a lava flow comes to rest depends on its size. Very large flows can flow down a gradient of 1° or less, and so form huge shallow domes. But some smaller volcanoes on Venus have produced smaller flows that form large volcanoes with steeper sides. Figure 11 shows one of the largest, called Sapas Mons. The individual flows are clearly visible in the SAR image. The highest parts of the volcano rise more than 3 km above the average planetary radius. One reason why the higher parts of the volcano in Fig. 11 are so bright is

Fig. 9 Mylitta Fluctus, a lava delta formed from many overlapping lava flows (Ford *et al.* 1993).

that they are rough. But all high regions of Venus are good reflectors of radar. Above a certain elevation the electrical properties of the surface change, although no one yet knows why. Its effect is to make all high mountains in the SAR images bright white, as if they were covered

Fig. 10 A large lava flow with a lava channel (Ford *et al.* 1993).

in snow (which of course they are not: the surface temperature is about 430°C).

Most of the lava flows on Venus are likely to be made from basalt. However, some must have had a much higher viscosity than most basalt flows on Earth, because the edges of the flows are cliffs 100 m or more high. On Earth the viscosity of lavas is largely controlled by the SiO_2 concentration. A lava with a SiO_2 concentration of 65–75% can produce features like those in Fig. 12.

Figure 13(a) shows examples of a different type of highly viscous lava flow that is not found on Earth, called a pancake dome. These domes are formed by single eruptions, unlike the volcano in Fig. 12. The shape of these domes is that expected for a drop of very viscous material spreading over a rigid surface. Figure 13(b) compares the profile across a dome like those in Fig. 13(a), measured by the altimeter on Magellan, with the shape expected for a spreading viscous drop. The agreement between

Fig. 11 Sapas Mons, a large shield volcano (Ford *et al.* 1993).

the two is good, and allows the viscosity of the lava to be estimated to be about 10^{16} Pa s.

Basaltic melts sometimes intrude into planar cracks that form by brittle failure in solid rock. If the sheets are vertical they are called dykes, whereas if they are horizontal they are known as sills. On Earth basaltic dykes are 30–50 m thick, and extend for as much as 2000 km horizontally. Similar features are visible in Fig. 14(a), because a shallow linear depression forms at the surface where the two sides of the dyke separate. Figure 14(b) shows a larger-scale image of one of these depression, just south of the domes at the centre left (which are the same as those in Fig. 13a). The melt in these dykes moves at about 5 m/s when

Fig. 12 A large volcano constructed from very viscous lava (Ford *et al.* 1993).

it is flowing, and takes a few days to travel 1000 km. In this time the temperature of the melt falls by about 50°C.

Another volcanic landform that is common on Venus is formed by flowing lava. Figure 15 shows a number of short channels that were formed when melt flowed out of calderas. Much longer channels also occur. Figure 16 shows a segment of the longest, whose length is almost 8000 km. Although its meanders look like those of a large river, it was formed by flowing basalt, not water. Surprisingly, the gradient along this channel does not slope consistently in either direction, and it is not known which way the melt flowed.

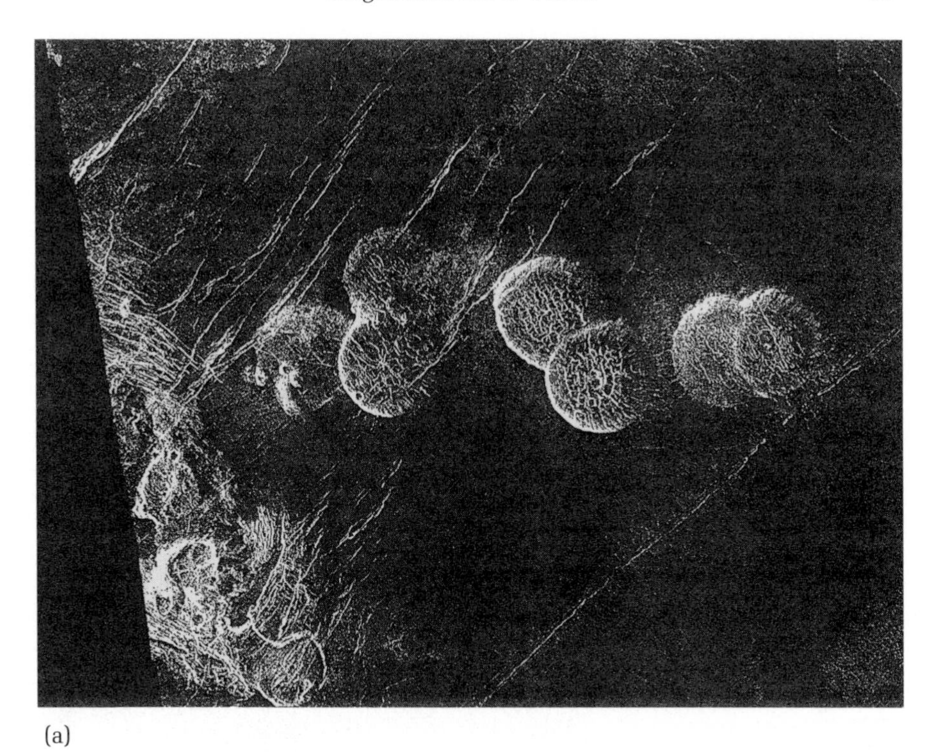

(a)

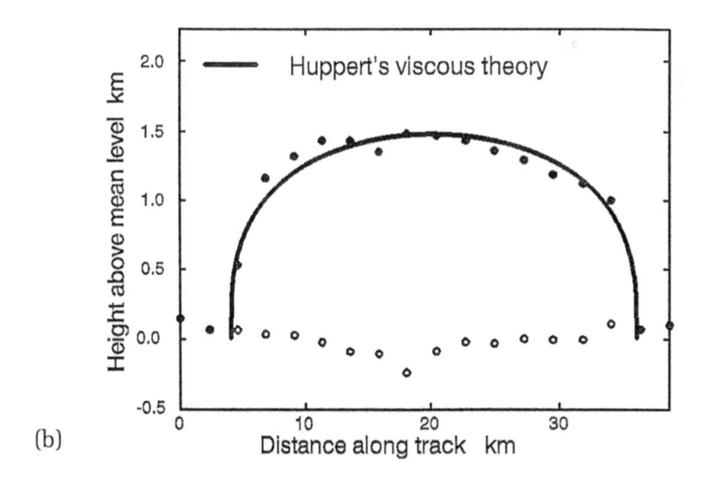

(b)

Fig. 13 (a) Overlapping pancake domes, constructed from very viscous lava. (b) The solid dots show a profile across a dome like one of those in (a), and the solid line shows the expected shape of a viscous spreading drop, calculated from Huppert's (1982) expression.

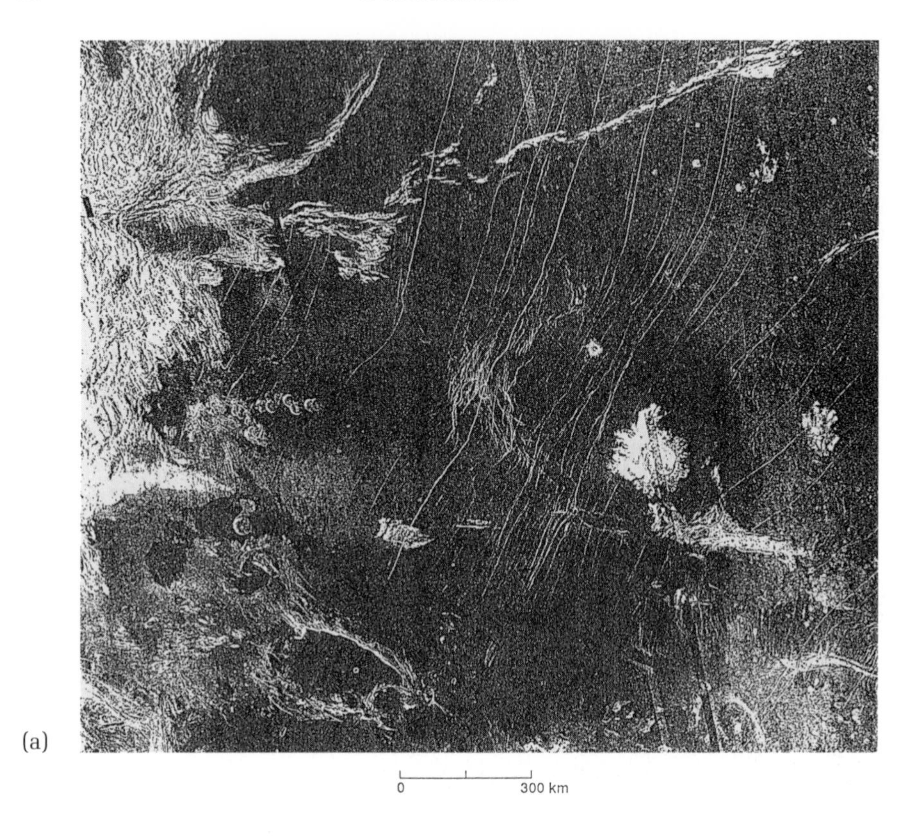

(a)

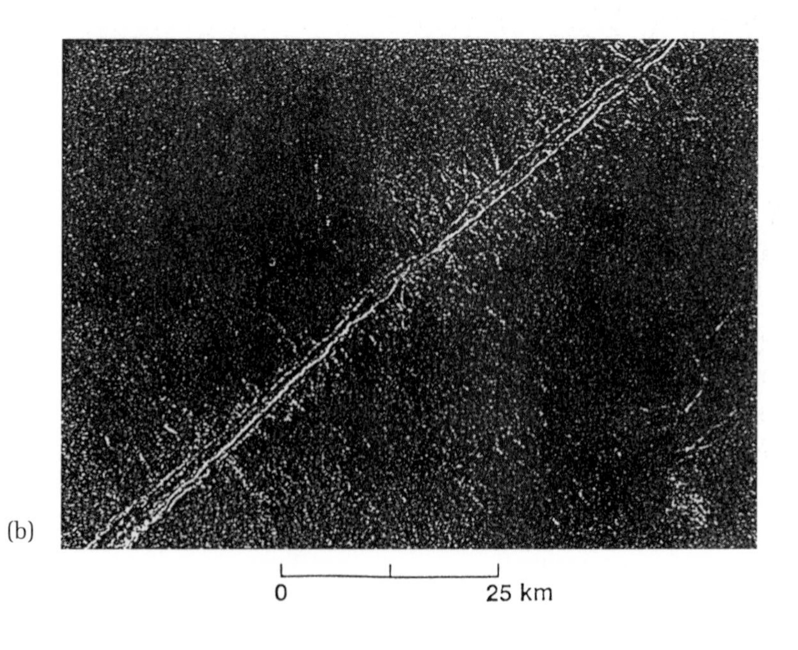

(b)

Fig. 14 (a) The bright linear features are the surface expression of dykes that extend for more than 1000 km across the dark plains. The domes in Fig. 13(a) are visible at the centre-left (Weitz 1993, *in* Ford *et al.* 1993). (b) a detail of one of the small grabens formed above a dyke (Stofan *et al.* 1993, *in* Ford *et al.* 1993).

Fig. 15 Lava channels flowing out of calderas (Ford *et al.* 1993).

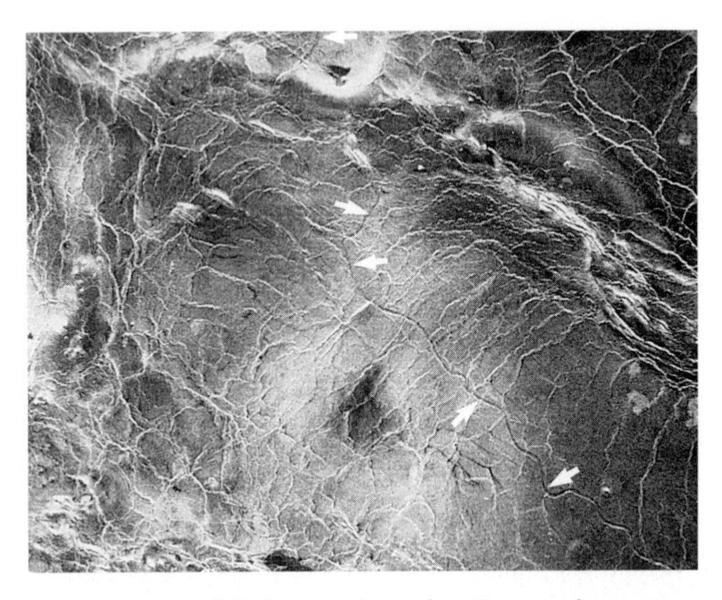

Fig. 16 A detail of the longest channel on Venus, whose course is marked by white arrows.

Tectonics

The evolution of oceans and continents on Earth largely results from plate motions. Before Magellan, various authors had attempted to identify features on Venus corresponding to the three types of plate boundary known on Earth. The Magellan SAR images showed that none of these suggestions were correct, and that features such as terrestrial plate boundaries are not common, although they do exist. The most obvious tectonic features on Venus are some large rifts that are formed by extension of the outer layers of the planet. Figure 17 shows one example. The long dark and bright linear features are faults. Where the movement produces a fault that slopes to the left, the SAR illumination produces a bright line, and the image of the fault scarp is often overlain (see Fig. 3). Fault scarps that slope to the right are not illuminated by the SAR, and show up as dark lines. These patterns can be used to estimate the amount of extension, which is only about 45 km. The rift in Fig. 17 is

---

Fig. 17 A large graben system produced by extension. Left facing faults show as bright lines because of overlay, whereas right facing faults are not illuminated, and are therefore dark (Stofan *et al.* 1993, *in* Ford *et al.* 1993).

(c)
0
200 km

therefore more like the East African Rift than the oceanic rifts that produce major ocean basins, where the amount of extension is often thousands of kilometres.

On Earth there are three types of plate boundaries: where plates separate from, slide past, or move towards, each other. On Venus sliding boundaries have been identified in only one small part of the planet. Structures like those formed on Earth where plates move towards each other are much commoner. It is not known whether any of these boundaries are still active, or how much movement occurred between the plates on either side. However, the distribution of craters suggests that the surface of Venus is not now being replaced by spreading ridges on a large scale. This conclusion is consistent with the rarity of features like spreading ridges in the SAR images.

Mantle convection

On Earth the plate motions themselves transport about 80% of the heat that is lost from the mantle. New plate is formed on ridges, where hot mantle upwells to fill the space between the separating plates. As the plates move sideways they cool and contract. It is this process which causes the places where plates are separating to form ridges, and the depth of the ocean to increase with age. Where plates are destroyed, by one plate overiding the other, one sinks into the mantle. Because it is colder than its surroundings, it is denser and hence sinks under its own weight. This process of cooling at the surface, followed by the creation of cold sinking regions by plate destruction, is called thermal convection, and is familiar to all cooks! It can occur in the mantle, even though the mantle is made from solid rock, because solids at temperatures close to their melting point can flow. Such flow is called creep. The best known example of such flow is a glacier, which moves downhill even though it consists of solid ice, because the ice is close to its melting temperature and can therefore creep. The rock of the mantle also flows because its temperature is close to its melting temperature.

On Earth the mantle circulation is dominated by plate creation and destruction. There are, however, places like Hawaii where a different type of flow occurs that is largely unaffected by the plate motions, and where a hot plume rises beneath the interior of a plate. Where such plumes approach the surface, their temperature exceeds the melting temperature and large volumes of melt are produced. It is this melt that has constructed the islands of Hawaii. Figure 18 shows a section through a convective model of Hawaii that matches the observations. As

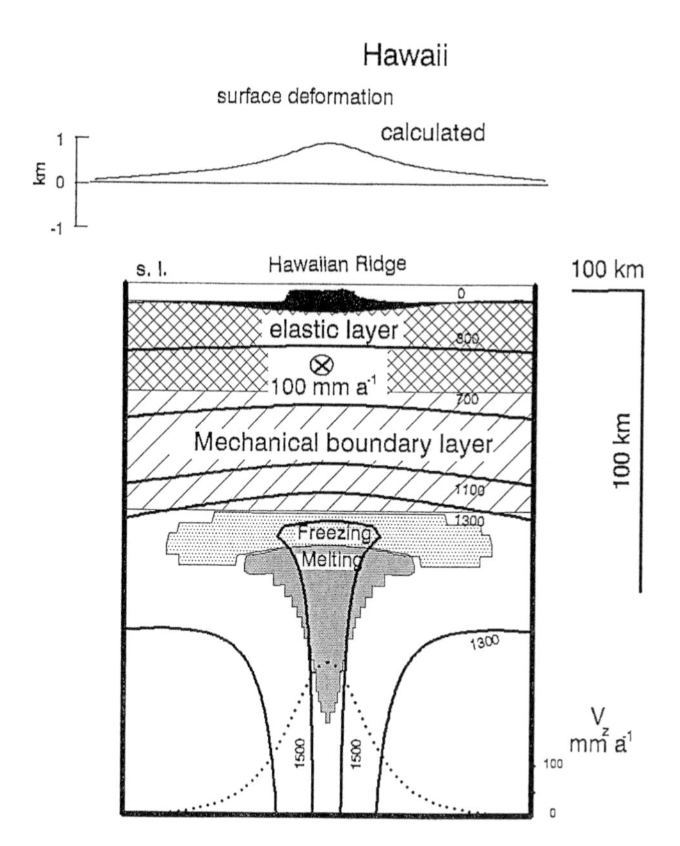

Fig. 18 A model of the axisymmetric plume beneath Hawaii. The numbers show the temperature of the isotherms in °C. The vertical velocity V_z at the base of the figure is shown by the dotted line.

well as generating the melt that forms the Hawaiian Ridge, the hot rising plume pushes the plate upwards to form a long wavelength bulge in the sea floor. Although features like Hawaii are impressive, the associated plumes only transport about 10% of the heat lost from the mantle, or much less than do plate motions. None the less such features are important, because they show how the mantle circulates when it is not being driven by plate motions. An important means of mapping such flow uses the gravity field, because hot rising plumes produce small long wavelength positive gravity anomalies. Figure 19 shows how they do so. The plume itself is hot, so its density is less than that of the surrounding mantle because of thermal expansion. It therefore produces a negative gravity anomaly. But it pushes up the plate above it, and the extra material produces a positive gravity anomaly, which is slightly larger than the

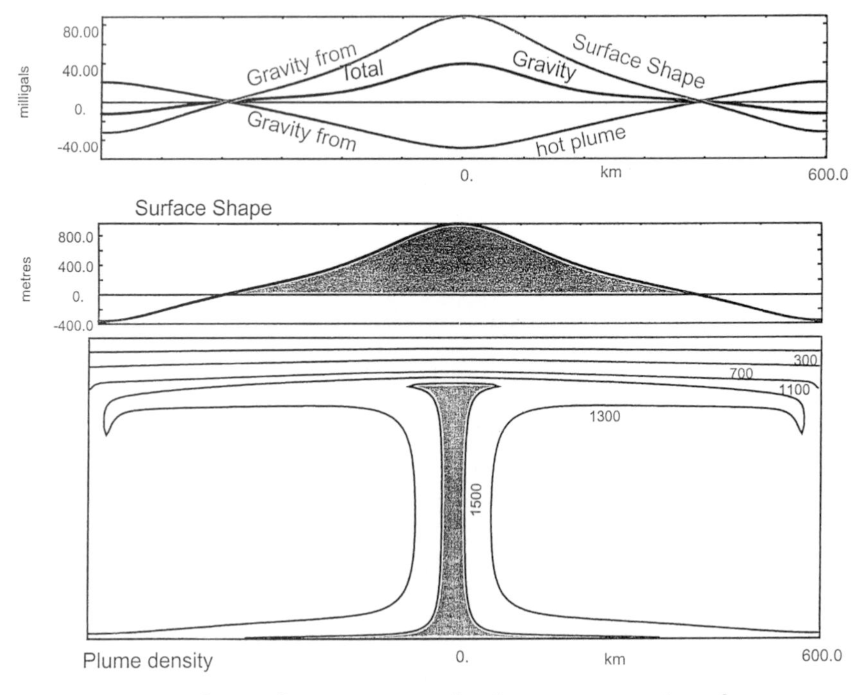

Fig. 19 The total gravity anomaly above a convecting plume results from a positive anomaly from the surface deformation, which is larger than the negative anomaly from the hot plume.

negative anomaly from the plume. The total gravity anomaly is therefore positive, and about a factor of three smaller than would be expected from the topography alone. On Earth variations in crustal thickness dominate the surface topography, which therefore cannot be used directly to map the convective circulation. However, crustal thickness variations are compensated. Continents have a crustal thickness of about 33 km, compared with 7 km for oceans and up to 70 km for high mountains. When the topography is compensated by variations in the thickness of the low density crust, the resulting gravity anomalies are small; much smaller than those due to the plumes. So the gravity field can be used to map the geometry of mantle convection and to distinguish those topographic features that are maintained by convection from those that are supported by crustal thickness variations.

The same approach can be used on Venus if the gravity field can be mapped. Fortunately the gravity field affects the orbit of any satellite, like Magellan, that is in orbit round Venus. Because Venus is so far from the Earth, the orbit of Magellan cannot be obtained from the pointing direction of the big dishes of the DSN. Instead the Doppler shift of the carrier frequency is used to measure the line of sight velocity of the spacecraft, to an accuracy of about 0.1 mm/s. This velocity is then used to model both the Magellan's orbit and the gravity field of Venus. Figure 20 shows maps of the gravity field and of the shape of the planet. The third map shows what remains when the convective topography corresponding to the gravity field is subtracted from the observed topography. The difference is called the residual topography. Regions with positive residual topography are like continents on Earth: because they lack gravity anomalies, the surface elevation must be maintained by an increased crustal thickness, rather than by mantle convection. The large positive gravity anomalies in Fig. 20(a) are all associated with elevated topography, and show the locations of rising plumes. They are surrounded by linear negative anomalies, which mark regions where colder mantle is sinking. Therefore, the gravity field from Magellan directly maps the mantle circulation within Venus. Because of the absence of plate motions, and because the viscosity of the convecting region is about 10 times greater than it is on Earth, the map in Fig. 20(a) is easier to interpret than are corresponding maps for the Earth.

The large size of many of the convectively supported topographic features on Venus compared with those on the Earth is surprising. It suggests a possible explanation for the puzzling gradients in the long channel in Fig. 16. Convection in the mantles of both Earth and Venus is likely to be time dependent, partly because new plumes start from the bottom of the convecting region, and partly because the whole convection pattern slowly changes. the surface topography supported by convection must therefore also change with time, and the gradients along the channel may also change after the channel itself has formed. This idea is easily tested, by generating a profile of the residual topography along the channel. Figure 21 illustrates profiles of the measured and residual topography along the channel, and shows that the long wavelength reversals of gradient are indeed the result of convection. The shorter wavelength irregularities correspond to features on the SAR images that are produced by thrusting.

The gravity field can also be used to measure the thickness of the elastic layer that supports short wavelength topography on both Earth and Venus. Although the method used is complicated, its principle is simple. If a volcano is constructed from basalt, whose density is less than that of the mantle, on top of a thick rigid plate, it produces a large positive gravity anomaly, because of the extra mass of the volcano. If, however, the elastic part of the plate is thin and bends under the weight of the volcano, the topography is compensated by the low density root of basalt, and the gravity anomaly is much smaller. Therefore in principle

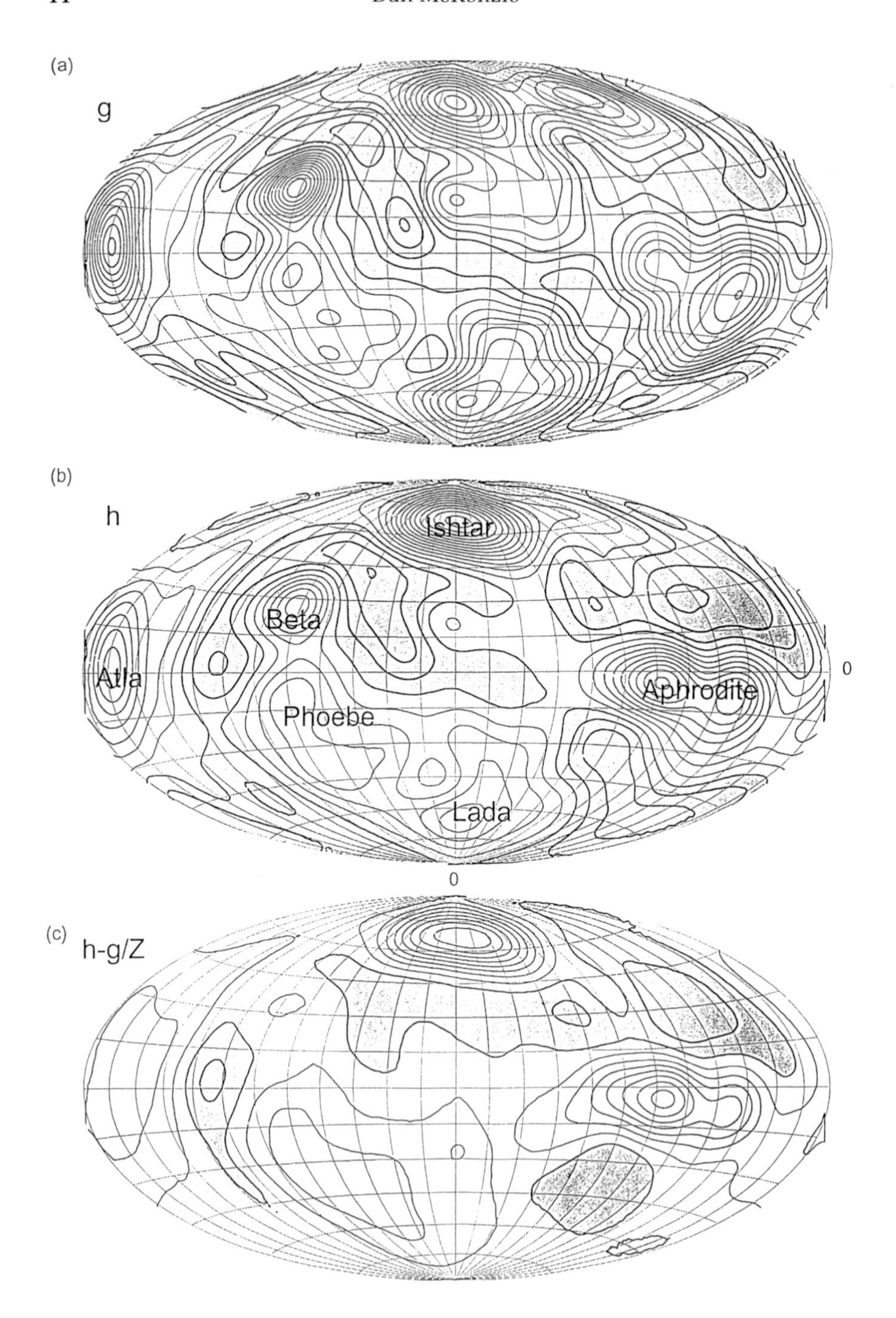

Fig. 20 (a) Gravity, (b) topography and (c) that part of the topography which is not convectively supported (known as the residual topography).

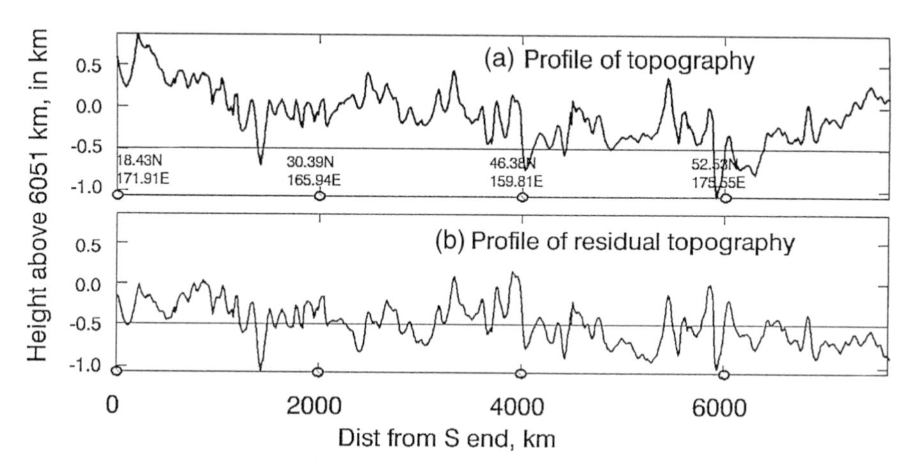

Fig. 21 (a) Observed profile of the topography, and (b) residual topography, along the channel shown in Fig. 16.

the relationship between gravity and topography can be used to estimate the thickness of the outer elastic layer on both Earth and Venus. It is not straightforward to use this method on Venus, because the short wavelength gravity anomalies of interest are not measured accurately by Magellan. Complicated processing of the data is required to extract the signal from the noise. When this is done the elastic thickness is found to be about 30 km, or very similar to that for Hawaii, even though the surface layers are hotter than they are on Earth.

Figure 22 shows a section through a plume on Venus for comparison with that shown in Fig. 18 for Hawaii. There are a number of important differences between plumes on Earth and Venus. On Venus the surface layer that forms the lid on top of the convecting region is thicker than it is on Earth, even though the surface temperature is 430°C instead of °C. Also, the Venusian plume is larger than that beneath Hawaii, because of its higher viscosity.

Venus and Earth

Magellan has given us our first look at a planet that is similar to the Earth, and whose mantle is still actively convecting. As we expected before the spacecraft was launched, the gravity field has given us most information about what is going on now, and we have (just!) enough information to construct a convective model of the mantle circulation of Venus at the present day. But as yet we understand almost nothing

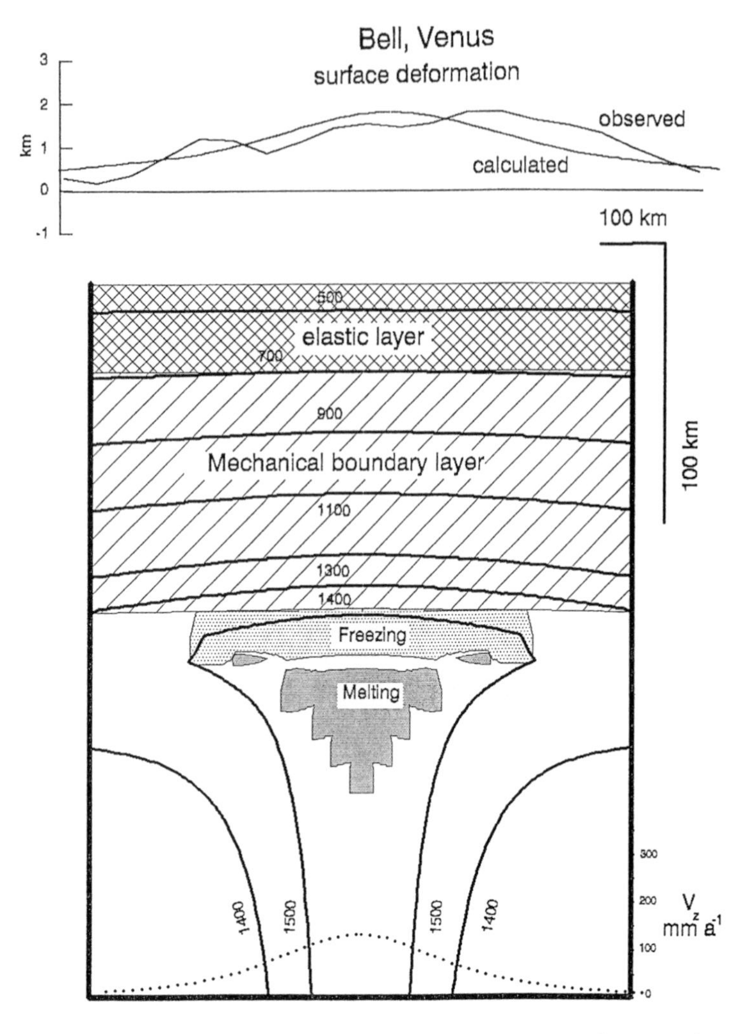

Fig. 22 A model of a venusian plume (compare with Fig. 18). The numbers show the temperature of the isotherms in °C. The vertical velocity V_z at the base of the figure is shown by the dotted line.

about its geological history. The major difference between the Earth and Venus is that the mantle of Venus is more resistant to deformation than is that of the Earth at the same temperature. This difference allows Venus to support elastic stresses at temperatures below about 650°C, compared with an upper limit of about 500°C for the Earth, and causes the viscosity of the mantle to be about a factor of ten greater. The same effect can also account for the absence of large scale plate tectonics on

Venus, by increasing the strength of faults. The most probable cause of this difference is that Venus contains much less water than does the Earth, both in its atmosphere and mantle. Even though the water content of the Earth's mantle is only about 50–100 parts per million, its presence has a major influence on the strength and viscosity of the mantle. It is not yet clear whether the water in the Earth's mantle has remained there since the Earth formed, or whether it has been transported downwards from the oceans by sinking plates. Nor is it clear why Venus is now so much drier than the Earth. None the less the fact that Venus *is* dry can account for the principal differences between it and the Earth.

When I explained all this to a friend of mine who is a geochemist he said 'so all you have discovered by spending $1000 million is that Venus is dry.' In one sense his implied criticism is fair. However, until Magellan no-one had any idea that the small amount of water in the Earth's mantle controlled its tectonics. We have only understood this because of the Magellan mission. Whether the mission was 'worth $1000 million' is a question that I am often asked. It is one which I can see no sensible way to answer. I am glad that such decisions are made by politicians, and not by me. But what is clearly silly is to spend so much money on the mission and so little on analysing the data. I have been able to obtain about £10 000 a year to do so with Francis Nimmo, a graduate student at Cambridge with whom most of this work was done. I am grateful to the Royal Society and NERC for this support. Even such modest amounts of money have not been available to most of the US planetary geologists working on Venus.

Notes and bibliography

Most of the data from the Magellan project is discussed in two special issues of the *Journal of Geophysical Research (Planets)*, **97** (1992), 13063–13689 and 15921–16380. Many of the papers in these issues are rather technical.

Ford J.P., Blom R.G., Crisp J.A., Elachi C., Farr T.G., Saunders R.S., Theilig E.E., Wall S.D., and Yewell, S.B. (1989). *Spaceborne radar observations: a guide for Magellan radar-image analysis*, JPL Publication 89–41 (12/89), Pasadena, CA explains how to interpret SAR images, and how various artefacts arise. It was written before any Magellan images were available.

Most of the images in this chapter are taken from Ford J.P., Plaut J.J, Weitz C.M., Farr T.G., Senske D.A., Stofan E.R., Michaels G., and Parker T.J. (1993). *Guide to Magellan image interpretation*, JPL Publication 93–24, Pasadena, CA. The images themselves are available on CD-ROMs from NSSDC (see the web page http://nssdc.gsfc.nasa.gov/cd-rom/cd-rom.html NSSDC CD-ROM CATALOG Dec 1996).

Roth E.L. and Wall S.D. (ed.) (1995). *The face of Venus*, NASA SP-520 Washington DC, is a photo-atlas with many spectacular images of Venus.

More technical accounts can be found in Venus II, shortly to be published the University of Arizona Press, which reviews much of the work that has been carried out on the Magellan data. A shorter review, by Nimmo F. and McKenzie D. (1998) Volcanism and tectonics on venus, *Annual Reviews of Earth and Planetary Science*, **26**, 23–51, is especially concerned with the gravity observations.

DAN MCKENZIE

Born 1942, graduated in Physics from the University of Cambridge, and completed a PhD at Cambridge under Sir Edward Bullard. He then worked at a number of oceanographic laboratories in the US before returning to Cambridge. He is now a Royal Society Research Professor and a Fellow of King's College. His present interests are planetary geology and melt generation, but is best known as one of the originators of the theory of plate tectonics. He was elected to the Royal Society in 1976, and also belongs to the US National Academy of Sciences.

Meteorites: messengers from the past

MONICA M. GRADY

Abstract

Meteorites are an important part of our Solar System. They have a direct effect on the Earth. Our planet is bombarded constantly from space by the tiniest of dust grains, and less frequently by enormous impact crater-forming bodies. Meteorites are ancient, older than the oldest rocks now present on Earth, and are credited with bringing both life and death to our planet. It is possible that organic molecules and water were introduced to the early Earth by meteorites and comets, sowing the seeds which eventually led to the evolution of life in its myriad of forms. But meteorites also have an awesome potential for destruction: a huge body fell at the end of the Cretaceous (65 million years ago), resulting in a large crater, and inadvertently led to the demise of the dinosaurs and many other groups. Meteorites fall all over the Earth, and have been collected from the icy wastes of Antarctica, the hot sands of the Sahara, and even the back garden of a senior citizen in middle England. The location and timing of meteorite falls are unpredictable: we do not know where or when the next one will land.

This chapter explores the nature of meteorites, from their formation at the birth of the Solar System to their final resting place on Earth. It highlights some of the different types of meteorites: the most primitive stony meteorites, which are related to comets and contain 'stardust' produced in the out-flowing wind of ancient stars; the dense iron meteorites, some of which are the nearest accessible analogues to the Earth's core, and the group of igneous meteorites thought to come from our neighbour, Mars, and have helped us to unlock the secrets of the Martian surface.

Introduction

Meteorites are fragments of ancient material, natural objects that survive their fall to Earth from space, and are recovered. They are not radioactive, and are almost always cold when they land. Meteorites are distinguished from *meteors*, or 'shooting stars', which are pieces of dust that burn up high in the atmosphere. No material is recovered on Earth from a shooting star. A meteoroid is a small body travelling through space, that may, or may not, land on Earth as a meteorite. All these words derive from the Greek 'μετεωρος', meaning 'things on high', the same root as meteorology, the study of the weather (a truly atmospheric phenomenon). The study of meteorites is known as *meteoritics*.

Meteorites were formed at the birth of the Solar System, ~ 4560 million years ago. Although the Earth, along with the other planets, was also formed at this time, none of the original material remains: it has been removed by bombardment or otherwise eroded, or recycled through geological activity (plate tectonics, volcanism, etc.). It is only by studying meteorites that we can learn about the processes and materials that shaped the Solar System and our own planet.

Meteorites are divided into three types on the basis of their composition: stones (composed of minerals often found in rocks on Earth); irons (made up of iron metal alloyed with nickel); and stony-irons (as their name suggests, a mixture of stone minerals and iron metal). A distinction is made between meteorites which are observed to fall (mostly stony meteorites, reflecting their abundance on meteorite parent bodies) and those that are found. The latter tend to be mainly iron meteorites, because they readily stand out from surrounding terrestrial rocks. Meteorites are generally named after a place near to which they fall or are found (usually the nearest Post Office). Exceptions occur in desert regions, where locality names are few; in these cases, meteorites are given the name of the geographical area in which they were found, followed by a number.

How big is a meteorite and how often do they fall?

Meteorite falls are not rare events: in fact, they are a lot more common than people realize. Approximately 40 000 tonnes of extraterrestrial material fall on the Earth each year—this is about four particles per hour per square km of the Earth's surface. Fortunately, almost all of this material arrives as dust, (known as micrometeorites or cosmic dust), small fragments (< 1 mm across) that are captured by the Earth and come from asteroids and comets. A small proportion of micrometeorites might also be from outside the Solar System, from interstellar space. The very smallest of micrometeorites (< 50 μm) do not melt as they pass through the Earth's atmosphere, but remain as fluffy aggregates of silicate minerals. This cosmic dust has been collected by aircraft equipped with specialized collector plates, flying high in the stratosphere. Slightly larger particles, like the one in Fig. 1, melt as they fall and form tiny rounded droplets. Micrometeorites such as these 'cosmic spherules' were first separated from deep sea sediments collected from the ocean floor by the *HMS Challenger* expedition of 1872–76. Subsequently, cosmic spherules and unmelted micrometeorites have been collected by filtering the water produced from melting large volumes of Antarctic ice, and also from melt-water streams in Greenland.

Generally speaking, larger meteorites fall less frequently than small ones. Meteor Crater in Arizona (Fig. 2) was produced by the impact of

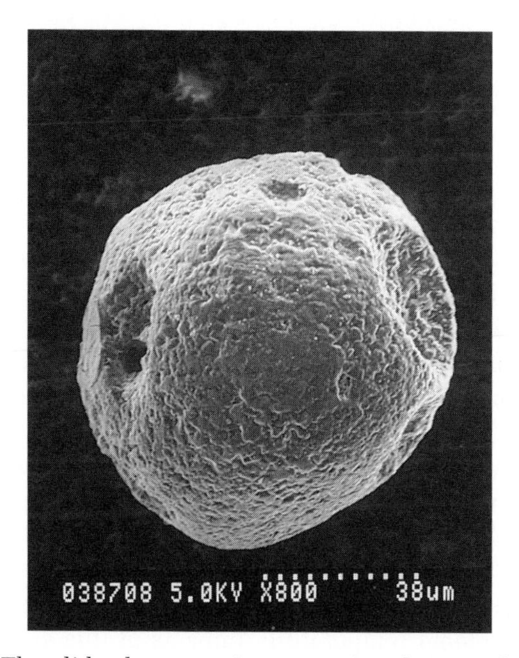

Fig. 1 The slide shows a micrometeorite photographed in a scanning electron microscope. The grain is ~ 100 μm in diameter, and has been magnified 800 times. The micrometeorite was one of several hundred recovered by melting a large volume of Antarctic ice, and filtering the resulting water. (Photograph Courtesy of Dr M.J. Genge, Natural History Museum.)

Fig. 2 Meteor crater in Arizona, ~ 1.2 km across and ~ 200 m deep. It was formed by the impact of an iron meteorite, ~ 5–40 m across, approximately 50 000 years ago. (Photograph courtesy of USGS.)

an iron meteorite ~ 50 000 years ago. The crater is ~ 1.2 km across; the original meteorite was estimated to weigh between 15 000 and 25 000 tonnes, and would have been ~ 35–40 m in diameter, but most of it was vaporized by the impact. Over 30 tonnes of iron meteorite, named *Cañon Diablo*, have been recovered from the vicinity of Meteor Crater. Any environmental effects associated with an impact this size would be mostly localized or short-term. For instance, living creatures in the vicinity of the impact site would have been killed instantly, but the impact would not have led to the world-wide extinction of any species. A more extensive phenomenon associated with an impact of this magnitude might have been reddening of sunsets due to dust in the atmosphere, carried very quickly around the globe, but this would rain out on a time-scale of a few weeks or months, before it could have a significant effect on global temperature. Impacts of this size occur approximately every 10 000 years. Larger impacts occur even less frequently: every 50 million years or so. A huge bolide fell at Chicxulub, in the Yucatan Peninsula in the Gulf of Mexico at the end of the Cretaceous period (65 million years ago). The crater is now buried, but geophysical surveys estimate its diameter to be between 180 and 320 km. Environmental effects caused by an impact of these dimensions include a darkening of the sky, due to ejected rock dust, followed by a rapid, global drop in temperature. In the case of Chicxulub, the impact was into a sedimentary rock formation, including evaporite deposits (i.e. sulphate-bearing rocks), resulting in tonnes of sulphur oxides ejected into the atmosphere. The energy of the impact fused nitrogen and oxygen from the atmosphere into nitrogen oxides. As the temperature dropped, sulphur and nitrogen oxides washed out of the atmosphere as acid rain, leading to a change in the pH (acidity) of the oceans. These consequences affected the entire globe, not just the local region, and for an extended period of time. It is entirely possible that the end result of these global environmental changes was the extinction of many species, including the dinosaurs, although this is by no means accepted by many palaeontologists.

Between the micrometeorites and the crater-forming meteorites lies the range of 'recoverable' meteorites. Several thousand meteorites, about the size of a football, fall on the Earth every year. Many break up in the atmosphere, and three-quarters fall in the oceans and are lost. Usually, only five or six are seen to fall and end up in collections. No one is known to have been killed by a falling meteorite, apart from the alleged death of a dog by a stone from the *Nakhla* Martian meteorite that fell in Egypt in 1911, although several people have been injured. For example, a young boy was hit on the arm by a meteorite while playing football in the town of Mbale in Uganda in 1992. He was fortunate to suffer only a bruise from the 30 g stone that caught him after it ricocheted off the leaf of a banana tree: the *Mbale* shower consisted of several tens of stones, the largest of which weighed 27 kg. If the boy had been hit by any of the larger stones, he would probably have been killed. Meteorites can, however, do much damage to property: several have hit houses and broken roofs and windows, and cars are also regular targets, for example, the *Peekskill* meteorite fell in New York State in October 1992, landing in the boot of a parked car.

The last meteorite to fall on England landed in 1991, in Glatton, near Peterborough, in the garden of a Mr Pettifor. It is a stone meteorite, and weighs almost 0.7 kg. It fell through a hedge of conifers, and landed below a wooden fence. It was recovered immediately after it landed, but only because Mr Pettifor was in the garden at the time. If he had been inside his house, he would not have heard the whining noise made by the meteorite as it fell through the air. This startled him, and caused him to investigate the commotion made by the meteorite as it subsequently crashed through the hedge. The meteorite is now part of the UK national meteorite collection, curated at the Natural History Museum in London.

Where are meteorites found?

Meteorites fall almost randomly over the Earth's surface, but often are lost—by falling into the ocean, or among other rocks. It is not possible to predict where and when the next one will land. Many meteorites are recovered from deserts (both cold and hot), as the dry environment ensures their preservation, and the lack of vegetation and other rocks enhances their chance of being found.

More meteorites have been found in Antarctica than anywhere else in the world. This is not because more meteorites fall in Antarctica: there is, in fact, a slight decrease in the flux of meteorites at the poles compared with the equator, but because those that do fall are preserved in the ice, for as long as a million years. As the ice moves towards the edges of the continent, the meteorites are carried along. When the ice meets a barrier that it cannot cross (such as a mountain chain), the ice is forced upwards. Subsequent ablation (or stripping) of the ice is caused by the winds that constantly blow downwards from the South Pole, scouring the ice surface. An equilibrium is reached, between flow of ice outwards and removal of ice by ablation. The continual removal of ice results in a gradual build-up of the solid materials it carries with it (mostly meteorites, but also including fragments of scoured bedrock), leading to concentrations of meteorites in these so-called 'blue ice' areas (Fig. 3). The first concentration of meteorites in Antarctica was recorded in 1969. Since then, meteorite collecting expeditions by US, Japanese and European scientists have returned ~ 15 000 specimens, representing approximately 8000 individual meteorites. Included in this total are some of the rarest of all types. The total compares with ~ 2000 'regular' meteorites from more temperate latitudes amassed in the 200 years since meteorites were recognized as natural phenomena. The success of the Antarctic collection has necessitated a change in the procedure for naming meteorites: Antarctic meteorites are named by giving them a number detailing the year of the field expedition that recovered them, followed by a curatorial analysis number. The whole is prefixed by an abbreviation of the field area, e.g. ALH 84001 was recovered in 1984 from the Allan Hills region of Antarctica, and was the first specimen examined on its return from the field.

Meteorites have also been found in several hot deserts, for example the Sahara in Africa and the Nullarbor Plain in Australia, in all cases on hard, rocky surfaces, rather than sand. Although they are not concentrated by a specific mechanism, such as the ice movement described in Antarctica, many meteorites have been recovered from hot deserts, as these deserts are relatively more accessible than the Antarctic. Meteorites

Fig. 3 A large stony meteorite found on the ice in the Lewis Cliffs area of Antarctica (~ 200 km north of the South Pole, near the Beardmore Glacier in the Queen Alexandra mountain chain). The dark surface of the meteorite is readily visible against the blue background of the ice. The meteorite weighs about 11 kg.

that have fallen over the past 50 000 years are preserved on these old land surfaces and are readily distinguished from the surrounding terrain. Again, they are given a combined geographic and numeric name.

Where do meteorites come from?

Our local star, the Sun was formed out of a nebula, a rotating cloud of gas and dust. As the cloud rotated faster, it collapsed and flattened into a disk with the Sun at its centre. Within the disk, dust grains joined together to form bigger and bigger bodies, eventually producing the planets. The Sun and planets formed approximately 4560 million years ago. There are many active regions throughout the universe where star formation is still occurring, e.g. the Orion Nebula (Fig. 4).

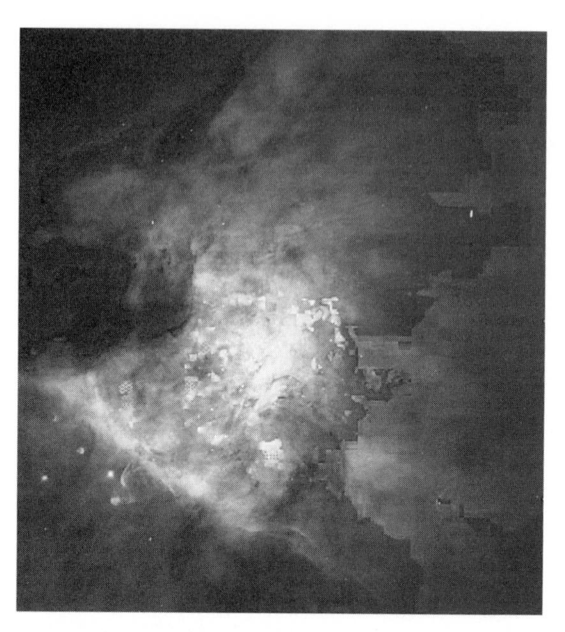

Fig. 4 The Orion Nebula (750 000 million km or 1500 light years away) is visible (through binoculars) as a glowing cloud just below the 'belt' of the constellation of Orion. The area imaged is 6500 million km across, and shows an active region of newly forming stars. It is likely that our Sun, like many stars, formed in an environment similar to that of the Orion Nebula. (Photograph courtesy of Space Telescope Science Institute.)

The Sun is the star at the centre of the Solar System, and all the planets orbit around it. The four inner planets (Mercury, Venus, Earth, and Mars) are mainly made from rock. Then follow the giant planets Jupiter and Saturn (composed mainly of gas) and then the outer planets Uranus and Neptune (gas and ice). The outermost planet, Pluto, and its satellite Charon, are small compact icy bodies that have an affinity with a disk-like array of similar objects known as the Kuiper Belt. There are thought to be approximately 35 000 Kuiper Belt objects orbiting the Sun at the distance of Pluto (5913 million km or ~ 40 AU; 1 AU is the mean Earth–Sun distance, approximately 150 million km) and beyond. When objects from this belt are dislodged by gravitational perturbations, they might enter the inner Solar System as comets. Between Mars and Jupiter, at the hiatus between rock and gas in the Solar System, lies the Asteroid Belt, the place from which most meteorites come.

The asteroids lie in the region between Mars and Jupiter and orbit the Sun at a distance of approximately 450 million km, which is three times that of the Earth from the Sun (3 AU). There are several thousand asteroids, the largest of which, named Ceres, is ~ 914 km across (for comparison, Earth has a diameter ~ 13 000 km, and Moon ~ 3500 km). Asteroids are rocky, metallic or carbonaceous bodies. They are material remaining after the planets formed: Jupiter's gravitational pull prevented the bodies from joining together to form a single planet. Occasionally, influenced by Jupiter, the orbit of an asteroid is altered such that it might collide with another, and break up. Images of asteroids obtained recently by the Galileo probe show that asteroids themselves have cratered surfaces indicating that collisions are frequent within the Asteroid Belt. Fragments of disrupted asteroids fall to Earth as meteorites.

How is it known that most meteorites originate from the Asteroid Belt? One way is to photograph incoming meteoroids and measure their speed and entry path—the orbit can then be calculated. There have been three separate investigations of this type, each involving a network of ground-based cameras. For example, the Canadian Meteorite Observation and Recovery Program operated from 1971 to 1985, photographing the night sky using an automatic camera and taking photographs at 0.25 second intervals. As a result of this programme, the fireball of the *Innisfree* meteoroid was photographed in 1977 (Fig. 5), and 3.79 kg of meteorite recovered as six specimens and several fragments. The main piece was found 12 days after the fireball observation, on snow.

One of the most recent meteorite falls to be recorded was that of *Peekskill*, to which reference was made in the second section, as it landed in the boot of a car. The track of the extremely bright fireball was recorded on several video cameras, mostly by members of the public attending outdoor football games on the Friday afternoon of the event. The video footage has been edited together to produce a film of the fireball travelling over the north-eastern USA. It is clear that the meteoroid broke into several pieces during its flight, each of which became a fireball. Only one meteorite was recovered, and that was the 12.4 kg object that hit the car. Orbits of the *Innisfree* and *Peekskill* meteoroids have been calculated, as have those of two other observed fireballs associated with the *Pribram* (Czech Republic) and *Lost City* (USA) meteorites, and all four orbits are elliptical and extend into the Asteroid Belt, confirming that this is from where the meteorites originated.

In addition to meteorites from the Asteroid Belt, there are currently 18 meteorites from the Moon in the world's collections. Lunar meteorites can be compared directly with samples brought to Earth by the Apollo and Luna missions between 1969 and 1976. The surface of the Moon is covered in craters caused by impacting bodies. If the impactor arrives with the requisite velocity and on a favourable trajectory, then the force

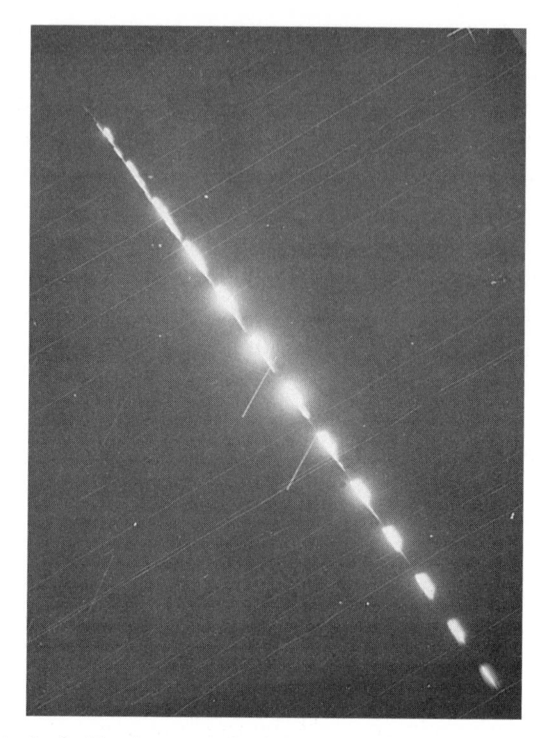

Fig. 5 A fireball photographed by the Canadian Meteorite Observation and Recovery Program, using an automatic camera. The photographs were taken at 0.25 second intervals. The *Innisfree* (Alberta) meteorite was recovered as a result of this observation. (Photograph courtesy of CMORP.)

of the impacts will be sufficient to eject material from the surface with a velocity great enough to overcome the Moon's gravity and be launched into space. Subsequently, the material goes into orbit in interplanetary space, and some of it eventually lands on the Earth as a meteorite. In the same way, rocks have come to us from Mars: we have 12 meteorites that have been ejected by impact from the surface of our neighbouring planet (see later—What can we learn from meteorites?).

Comets are also part of our Solar System. They formed at the cold outer reaches of the Solar System, where ices could condense. The icy bodies that produced comets are now thought to inhabit a spherical region of space extending to ~ 50 000 AU, called the Oort Cloud. Other icy bodies formed in the Kuiper Belt, at or beyond the orbit of Pluto. Like the planets, comets orbit the Sun: for example, Halley's comet takes 76 years to complete one orbit. Comets have been described as 'dirty snowballs', a mixture of ice and dust that has never completely melted.

Each time a comet approaches the Sun, part of its ice melts, and streams away from the central portion, or nucleus of the comet, carrying with it some of the dust. This expanding cloud of gas and dust gives rise to a comet's characteristic appearance of a head and a tail. Each time the comet draws near to the Sun, more of the ice and dust is lost. The dust becomes spread out along the entire orbit of the comet. For several comets, we can predict when the Earth will pass through this dust, giving us a brilliant display of shooting stars, or a meteor shower, e.g. the Perseid meteor shower that can be observed on clear nights between 9 and 12 August. It is also possible that comets are the parent objects of a group of five very special meteorites (see later—What can we learn from meteorites?)

The bright fireball often associated with an incoming meteoroid is the result of frictional heating as the body travels through the atmosphere. Only the outermost surface melts; the resulting droplets of molten meteorite are carried away by the speed of passage. Finally, as the meteoroid is slowed down by the atmosphere, the molten surface cools rapidly to a glassy coating, or fusion crust (Fig. 6). The presence of a fusion crust is often characteristic of meteorites. It is very important to stress that it is only the outermost surface of the meteorite that melts: the interior remains cool and unchanged. Meteorites are cold when they land.

Fig. 6 The very glossy black fusion crust on the outer surface of a broken stone of the *Stannern* achondrite. The fusion crust is less than 1 mm thick, and the unchanged, pale grey, basaltic meteorite is readily distinguishable from the glassy crust.

What are meteorites made from?

Meteorites are divided into three main types (stone, iron and stony-iron), reflecting their composition. Most meteorites are stony (96% of all falls), made up of the same minerals (olivine, pyroxene, plagioclase) as many terrestrial rocks, minerals which contain silicon, oxygen, magnesium, iron, calcium, and aluminium. The stony meteorites can be subdivided into chondrites, those which have remained unmelted since formation (or aggregation) of the parents. These almost all contain small rounded droplets of once-molten material, or chondrules (from the Greek '$\chi o \nu \delta \rho o \varsigma$', meaning 'granule' Fig. 7). Chondrites retain a chemical signature close to that of the original material from which they aggregated. The other division of stony meteorites is the achondrites. These are igneous rocks, like basalts, that formed from melts on their parent bodies. Achondrites do not contain chondrules, and, as a consequence of the melting process, are chemically differentiated, i.e. no longer exhibit a primordial signature. Figure 6 is of the Stannern achondrite, a typical basalt from the Asteroid Belt.

The other large division of meteorites, the irons are made dominantly from iron metal typically with 5–15 wt. % nickel. These meteorites have all been formed during extensive melting processes on the parent bodies

Fig. 7 A slice of rock about 30 μm thick, viewed in plane polarized light through a petrologic microscope, showing the appearance of chondrules. Field of view ~ 2 mm.

from which the meteorites originated. The heat source for melting was, in some cases, the result of impacts, but for many iron meteorites the heat source was most probably from the decay of short-lived radioactive isotopes, such as ^{26}Al. The iron meteorite parent asteroids were sufficiently large that this heat built up and was retained, allowing reduction reactions (similar to smelting in a blast furnace) to occur within the parents. Iron–nickel metal, produced from the reduction of silicate minerals, migrated under gravity to the centre of the parents, forming a core, while the less dense remaining silicates rose to the surface, forming a crust. Iron meteorites are the closest physical analogy we have to the material which forms the Earth's core. (In contrast, either short-lived radioactive elements were absent, or heat from radioactive decay of these elements was dissipated in the smaller chondritic parents, thus melting did not occur.)

Figure 8 shows a slice of the *Gibeon* iron meteorite. The slice has been polished, then etched with acid to reveal a pattern, the Widmanstätten pattern, named after the scientist who first described it. This pattern is

Fig. 8 A polished and etched slice of the *Gibeon* meteorite (from Namibia), showing its Widmanstatten pattern.

characteristic of most iron meteorites, and is produced from the intergrowth of two alloys of iron metal with nickel (kamacite and taenite), each alloy containing different nickel concentrations. It is a result of the very slow rate at which the metal cooled: between 1 and 100°C per million years, allowing nickel atoms to diffuse through the iron lattice. The final pattern is 'frozen' in when the nickel no longer has sufficient energy to move. The slice also shows dark areas, which are patches of iron sulphides and elemental carbon as graphite.

The final main subdivision of meteorites is the stony-irons: a mix, as the name suggests, of stone and metal. The pallasite subgroup of these very rare meteorites, composed of almost equal volumes of stone and iron, have one of the most beautiful of appearances, produced from the intergrowth of iron–nickel metal with olivine (a magnesium–iron silicate mineral, common on Earth as the semiprecious gemstone, peridot). Pallasites were also formed by melting in their parent, and represent an intermediate stage between iron meteorites and differentiated silicates, a snapshot of material from the core/mantle boundary of the body. Figure 9 is a slice of the *Brenham* pallasite.

Fig. 9 A polished slice of the *Brenham* pallasite (from Kansas, USA), composed of orange-brown olivine grains set in a network of iron–nickel metal.

What can we learn from meteorites?

Different types of meteorite provide evidence about events that have occurred as the Solar System formed and evolved.

One important class of stony meteorites is that of the carbonaceous chondrites. These have chemical compositions (apart from hydrogen and helium) that are close to that of the Sun. The parent bodies of the carbonaceous chondrites probably formed towards the outer edges of the Asteroid Belt, where it was cooler, allowing some ices to condense. The carbonaceous chondrites can be subdivided into several groups, the most primitive of which is that of the CI chondrites, very rich in water, sulphur, and organic compounds. There are only five of these rare meteorites, and it is possible that this small group might be from the remains of burnt-out comets, rather than from the Asteroid Belt. Each time a comet approaches the Sun, it loses ice. It is possible to envisage an episode when the last vestiges of ice are evaporated from a cometary nucleus, leaving behind just the stony material. This, then, having no 'fuel' to drive it, falls towards the Sun, and may be captured by the Earth as an unusual meteorite.

There are several other groups of chondritic meteorites, also with compositions unfractionated since their aggregation, but with lower volatile contents. Although these meteorites have not melted since their formation, they do contain materials that were once molten (Fig. 7). These spherical silicate assemblages (chondrules) were produced by rapid cooling of droplets of molten stone. The droplets came from collisions between clumps of dust grains in the early stages of the formation of the Solar System, so meteorites such as these represent the materials from which the Solar System grew.

In addition to chondrules, chondrites contain organic compounds in varying quantities. Some groups contain amino and carboxylic acids and complex hydrocarbons, while others contain elemental carbon. It is meteorites like these, together with the ice- and volatile-rich comets, which probably brought volatile materials to the newly formed Earth, and helped establish our planet's atmosphere and oceans. Without them, there would be no life on Earth.

Also buried within chondrites are tiny grains of dust that came from stars other than our own Sun. These grains are diamonds (invisible to the naked eye, only 3 nm across) and silicon carbide (or carborundum). The diamond grains in Fig. 10 are too small to be seen individually. They were separated from the meteorite by dissolving 10 g of sample in strong acids, to remove all the stony material, leaving behind clumps of an acid-resistant residue containing diamonds, weighing only ~ 40 μm

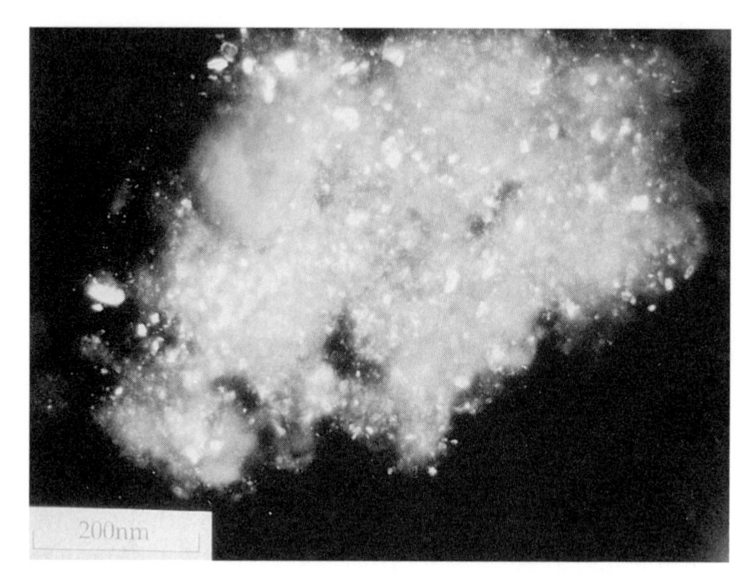

Fig. 10 A clump of interstellar diamonds isolated from a meteorite by dissolving away all of the stony and metallic components. The individual diamond crystallites are too small to be seen in this image taken using a transmission electron microscope. The material has been magnified approximately 150 000 times (Photograph courtesy of Dr M. Lee, University of Edinburgh.)

and thus representing about 4 p.p.m. of the starting material. The interstellar origins of the diamonds have been inferred from the isotopic compositions of nitrogen and the noble gas xenon trapped within the diamond lattice, and released when the diamond is burnt in the laboratory. The diamonds were blown from the surfaces of neighbouring stars, and carried on the stellar wind into the collapsing dust cloud that formed our Solar System. From these grains, we learn that our Sun did not grow in isolation, but had neighbours.

In an earlier section ('Where do meteorites come from?'), it was reported that almost all meteorites come from the Asteroid Belt, and that a few come from the Moon. There are also 12 meteorites from Mars. How is it known that they come from Mars? We know much about the composition of Mars' atmosphere and surface from data from two NASA spacecraft that visited the planet in 1976 (the *Viking* probes). Experiments on board the craft measured the atmosphere as they fell towards the planet. The probes also scratched the surface and analysed the composition of the soil. In 1979, the 8 kg stone meteorite EET A79001 (Fig. 11) was found in the Elephant Moraine region of Victoria Land in Antarctica.

Fig. 11 The EET A79001 martian meteorite (8 kg in weight), in which pockets of shock-produced glass were found to contain trapped Martian atmosphere (Photograph courtesy of NASA-JSC, Houston.)

EET A79001 contains numerous dark patches of glass, distributed throughout the meteorite in pockets. This glass was made by localized melting during the event when the rock was thrown off its parent's surface by an impact. When the glass is analysed in the laboratory, trapped gases are released. These gases have the same composition as Mars' atmosphere (as measured by Viking), which demonstrate the meteorite's Martian origin

More recently, another Antarctic Martian meteorite (ALH 84001) has been studied. It contains patches of orange carbonates (Plate 2) throughout its entire 1.9 kg mass. The carbon and oxygen in the carbonates indicate that the grains were produced at low temperatures from Martian atmospheric carbon dioxide dissolved in water, probably just below the surface of Mars, when water circulated through it. Scientists from NASA think they might have found evidence for fossilized Martian bacteria inside these carbonate patches, showing that life might have existed on another planet. By studying meteorites from Mars, we can learn about events that have taken place in the past on our neighbouring planet, when it had a thicker atmosphere and could support running water, even though the surface of the planet now seems to be dry.

Conclusions

Meteorites are a diverse set of extraterrestrial materials, representing planetary and Solar System material in its many forms. By studying meteorites, we can study processes that have taken place as our Solar System evolved. We can also learn about the evolution of other stars that contributed to our solar neighbourhood. Without meteorites and comets, it is likely that life would not have evolved on Earth; studying Martian meteorites might possibly allow us to study the primitive beginnings of life on another planet. But meteorites are not only associated with the seeds of life: they are also instrumental in influencing evolutionary pathways, as a consequence of catastrophic impact and associated environmental changes.

The study of meteorites is a constant reminder of our mortality, and can best be summed up by the sober phrase that warns us: 'dust to dust....'

Acknowledgements

I would like to thank Matthew Genge, Robert Hutchison and Ian Wright for reading the manuscript in its several manifestations, and improving the text with their comments.

MONICA M. GRADY

Born 1958, received an honours degree in chemistry and geology from the University of Durham in 1979, then went on to complete a PhD on carbon in stony meteorites at the University of Cambridge in 1982. She has continued to specialise in the study of meteorites, carrying out research at Cambridge, then the Open University, prior to joining the Natural History Museum in 1991, where as Senior Scientist in the Department of Mineralogy, she carries out research on and curates the national collection of meteorites. She was fortunate to be part of the 1988/89 US expedition to Antarctica to collect meteorites. Her particular research interests are in the fields of carbon and nitrogen stable isotope geochemistry of Martian meteorites, interstellar components in meteorites and micrometeorites.

Television beyond the millenium

WILL WYATT

Introduction

In dutifully observing the rules relayed to me for my appearance before you tonight, I don't know whether to be encouraged or intimidated by the knowledge that they—the discourses and indeed the rules—have an unbroken history going back to 1826. I may perhaps, as a non-scientist, take comfort in the knowledge that previous audiences have sat through presentations on Dr Marshall Hall's reflex function of the spinal marrow, the manufacture of pens from quills and steel illustrated by modern machinery, and the condition and ventilation of the coal mine goat.

But then, they didn't have television in those days ...

Television is what I propose to discuss, and it's possible, without exaggeration, to say that many of the essential elements of the technology of TV were first demonstrated at these discourses. The first electrical flash photography in 1851, the announcement of the discovery of the electron in 1897, and on 13 March 1882, Eadweard Muybridge demonstrating 'the employment of automatic apparatus for the purpose of obtaining a regulated succession of photographic exposures.'

These were the first moving pictures that most there present had ever seen, and in the dignified interest with which they were received there is little to suggest a future for the medium beyond scientific research and exploration. A hint, however, of something more is contained in the guest list for that night; the Prince and Princess of Wales, Gladstone, Huxley, and Tennyson. If Muybridge is the stepfather of cinematography, does that make them the forefathers of the couch potato?

Television craft and its technology are inextricably linked and I hope to show where they have brought each other and to speculate a little on where they may be heading—from an uncompromising single channel presided over by the autocratic paternalism of one man, to a future where control may be not only impossible and perhaps unwelcome—but where responsibility, excellence and creative vision will always be required.

There are some ideas that are so great that, once somebody has thought of them, an unstoppable snowball of research and experiment begins to roll to will them into reality. So it was with TV. The word and idea of TV entered the world long before the technology brought it into being.

The word 'television' first appeared in 1900 but the theory was around from 1875. Scanning imagery—still the basis of TV was invented before World War I on the theory that light can generate electricity, so a lit image can be converted into electrical signals and sent via radio to be reconstructed elsewhere. The early experiments suffered because they couldn't generate signals strong enough, but then valves came along, and by the early 1920s there were several competing potential systems.

John Logie Baird, the name always associated with TV, started his experiments with TV in 1923. In 1925 he gave his first demonstration of 'the transmission and reception of moving images' using his spinning disc system. This was the original Baird televisor: essentially, the system shined a light through a disc perforated with a spiral of holes which stored the lit fragments of the image in front of it and reconstructed them as a series of vertical strips.

At the time the BBC was a private company owned by a consortium of radio manufacturers. Baird used their facilities informally to make his test but couldn't make any public announcement of it.

Things hardly improved for Baird in 1927 when the BBC was turned into a corporation. Lord Reith, its first Director General and a fellow Scot, was more than sceptical of this new fangled technology. In October 1928 Reith's assistant, Gladstone Murray, attended a demonstration and made the following report:

> *The Baird system ... provides an interesting laboratory experiment ... The demonstration considered in terms of service might well be considered an insult to the intelligence of those invited to be present ... We have a primary duty to the listening public to do what we can to promulgate the truth and to prevent the excitement of false hopes.*

Baird, however, was undaunted and was already advertising his receivers for sale, even though the only programming he offered were tests sent from his lab to his receiving station in North London. The cost of a set was £150—more than the price of a fancy car—but nevertheless some 'televisors' were sold. In September 1929, the BBC was reluctantly

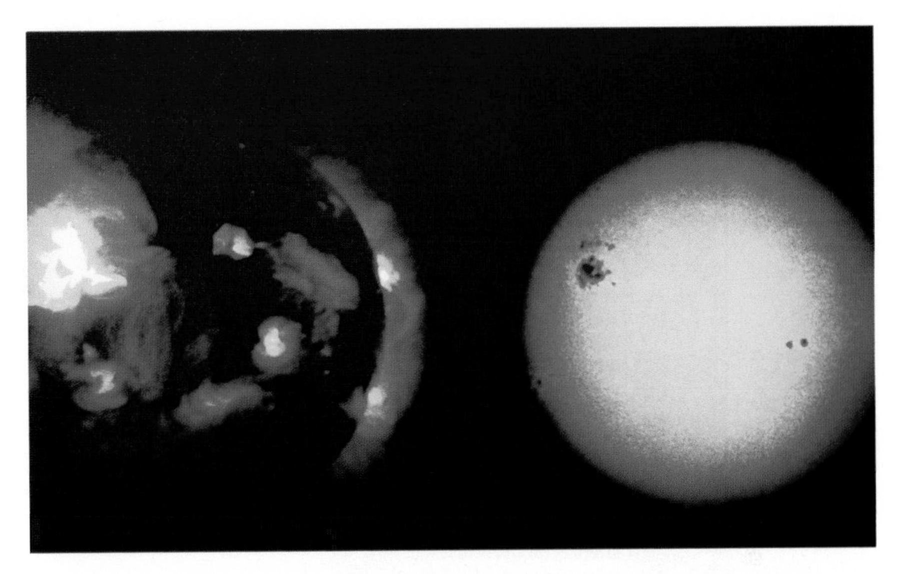

1. The sun in both visible light (*right*) and in X-rays (*left*). The dark areas visible (sunspots) are seen to be the seat of energy which manifests itself as X-rays in the solar corona. (Yokoh satellite.) (See p. 11.)

2. Patches of orange carbonates on the surface of martian meteorite ALH 84001. The field of view is about 0.5 mm across. The grains were produced below the surface of Mars when water circulated through it. (See p. 65.)

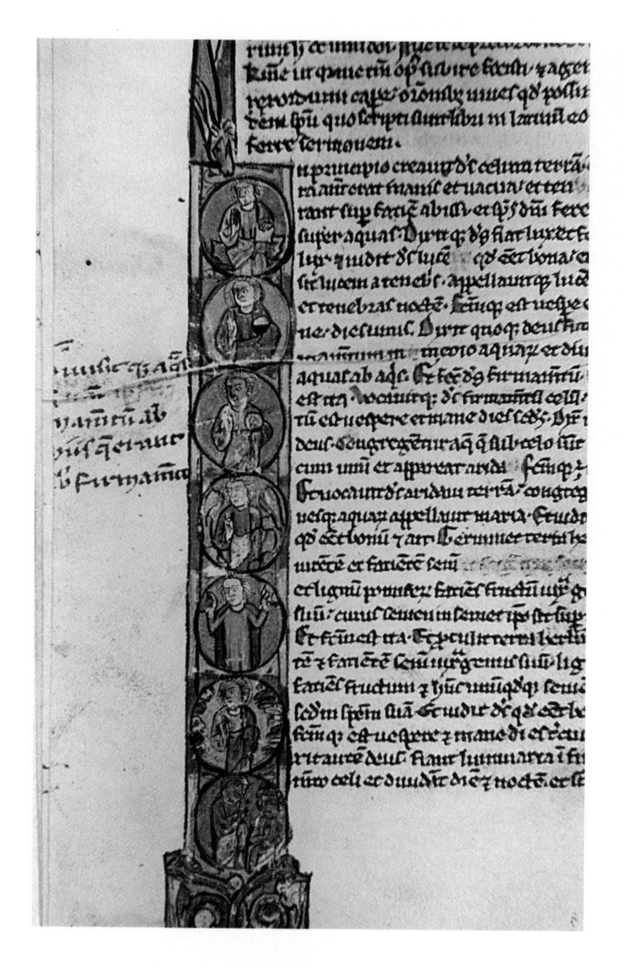

3. The historiated letter 'I' from *Genesis* in the Lucka bible. (See p. 159.)

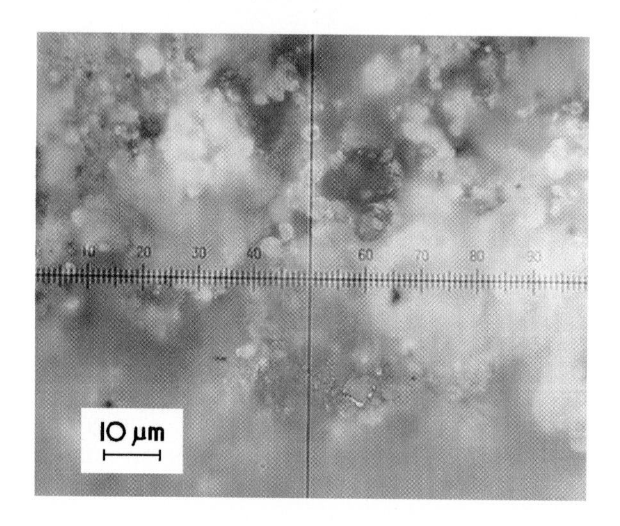

4. Magnified (× 1000) portion of the dark grey column depicted in a sixteenth century German manuscript showing the presence of at least seven pigments. (See p. 160.)

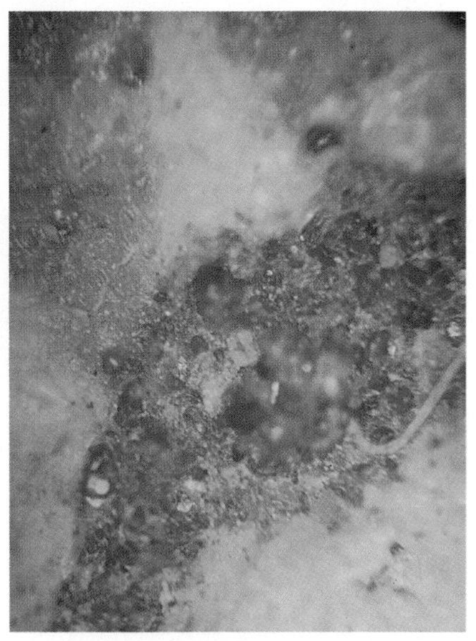

5. Photographs and microphotographs of elaborately decorated initials on a fifteenth century German manuscript MS Ger 4 (f 2r and f 28v) from the DMS Watson library. (See p. 161.)

6. Faces on the Byzantine/Syriac lectionary (upper f 188v, lower f 67v) blackened by degradation of white lead to black lead sulphide. (See p. 161.)

7. Partially reconstructed glazed bowl painted in black on a blue background (CF 1412). The scale section on the left is 2.5 cm long. (See p. 164.)

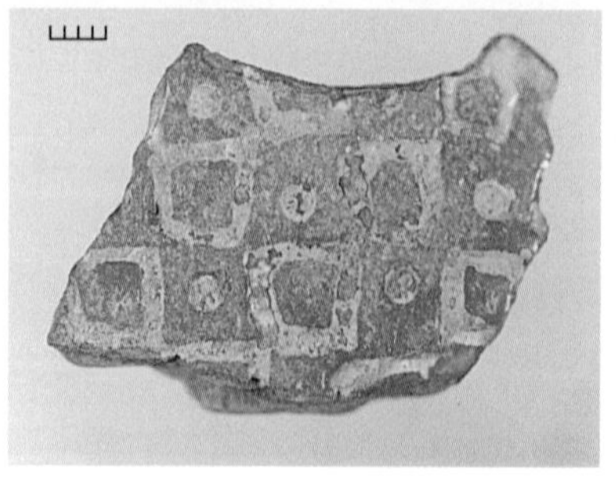

(a)

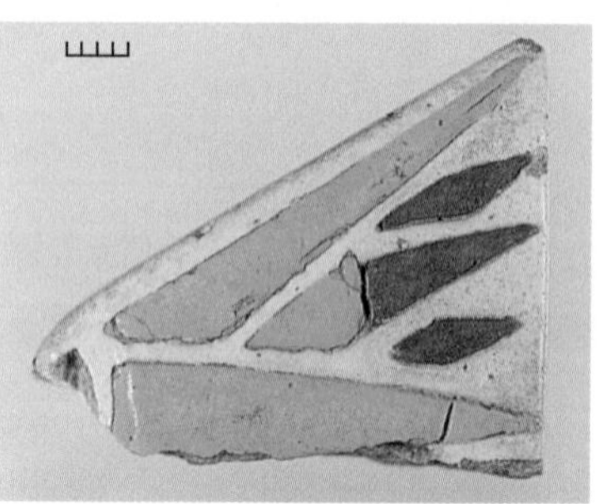

(b)

8. Polychromatic Egyptian faience samples from the Petrie Museum, (a) UC. 686 red pigment identified to be red ochre and (b) UC. 888 (lotus) yellow pigment identified to be lead(II) antimonate, $Pb_2Sb_2O_7$ (scale in mm). (See p. 164.)

persuaded to give him access to its one medium wave transmitter for experimental public broadcasts from his workshop, now in Long Acre. They were still keen to distance themselves:

> *In granting facilities for the experiment demonstration in which the public can, if they so desire take part, neither the Postmaster General nor the BBC accepts any responsibility for the quality of the transmission or for the results obtained.*

Broadcasts only happened on weekdays from 1100 to 1130 in the morning. Because of the limitations of the technology and his budget, most of them were simple solo performers of various sorts, but in 1930 Baird televised a play, *The man with a flower in his mouth*

1930 saw another first—the earliest recorded example of a published quote about TV's supposed pernicious effect on public values. And, what's more, it was in the BBC Yearbook:

> *Television might provide the most wonderful entertainment the world has ever seen, but might alternatively, if the control fell into the wrong hands, see all entertainment debased to the level of international millions or used for vilest propaganda.*

The reluctant entry of the BBC

In August 1932 the BBC took matters into its own hands by taking over production in a small studio in the basement of Broadcasting House.

This was still mechanical, or 'low definition', television using the spinning disc system, producing a flickering image with only 30 vertical lines, repeating at 12 and a half frames per second, tinted orange and, when viewed at the receiving end, measuring only 3 inches by one and a half. This experience may be familiar to those of you who have tried to download video clips from the Internet. The parallel is suggestive. Even though the pictures were extremely crude by modern standards, people at the time were surprised to find that the images were recognizable.

Once the superior editorial capacities of the BBC were brought to bear on it, content improved considerably. Though adaptations of great works of literature were still in the future, the high editorial purpose achieved a landmark when, late in 1932, they featured a seal which played the saxophone. (A triumph only excelled in 1979 by *Nationwide*'s skateboarding duck.) Naturally, the seal was considered something of a celebrity and arrived at the studio in style in the back seat of the engineers open tourer. The engineer's mistake was to escort this artiste in through the front door of the newly-built Broadcasting House. Memo writing was evidently already a high art in the Corporation, and on this occasion one was unleashed to enshrine a new internal regulation: 'Performing animals must in future use the goods entrance'.

Other problems were caused by the fact that the producers had been put in a studio also used by Henry Hall and his band for rehearsing. For a while they tried to fit round it, but eventually Hall was sent upstairs and the TV people were allowed to install a hard floor over the carpeting and a screen that hid the clock and the lavish, but inappropriate, decor of the room.

Things were not easy in front of the camera either. Let me introduce, Dee Barron. Dee is made up with the heavy blue-black, white, and yellow ochre make-up which emphasized the features of the early performers. The costumes and decor were also determined by technology; any large areas of single colour would weaken the signal and lead to fuzzy, smeary images. So the production staff had to design very graphical, decorated backdrops, costumes—and even masks. In case performers arrived inadequately patterned, a dressmaker was on call to sew on trim in emergencies.

The first night's broadcasts on 22 August 1932 were reviewed in *The Times* of the following day with decorous restraint: '*The broadcast was quite successful and one could easily recognise the artiste ... it should encourage a wider interest*'.

So the make up worked.

In November 1933 a 'caption machine' was devised to generate the first graphics: a mini-spot system for fixed titles and then a system which rotated a drum to position a selected card in front of the spot.

The audience at the time were called 'lookers in' or 'lookers' and a lot of them, presumably those who didn't have an extra £100 to spare, made their own sets from kits, including one sold by the *Daily Express*. Cross media ownership goes back a long way. And some of these do-it-yourself lookers in were very far afield: depending on the weather and time of year, people were getting pictures in Madeira and North Africa.

High definition and cathode ray technology

But all the enthusiasm could not overcome the primitive technology. Baird's scanning beam did not give off enough light, the amplifiers were not powerful enough to transmit a strong signal, and the device in the set for reproducing these images, was inadequate. The solution again derived from an idea dating back to 1908, which proposed the use of a cathode ray tube, invented 11 years earlier.

The first demonstration of cathode ray reception was given again by Baird in 1933. It was still small—4 by 3 inches—greenish tinted and only visible in a darkened room. A simple cathode ray tube was a vacuum tube with a phosphorescent screen at the front. A cathode at the back fired a stream of electrons across the tube to the phosphor coating which could be seen to glow green when the lights were turned down.

The early tubes didn't have 'charge storage' to keep the dots on the screen long enough to make the illusion of a whole continuous image, but the British government had begun to develop radar and they needed cathode ray tubes for that.

In 1931 money had been given to the EMI/Marconi Company to start a laboratory in order to develop better tubes, and in January 1935 the government recommended the initiation of a 'high definition' TV service. But Baird, not knowing about this deal, had continued to improve his system. By this time he had got his definition up from 30 to 240 lines and his image size up to $14\frac{1}{2} \times 10$ inches—but EMI's was 405 lines and scanned more rapidly, producing less flicker.

This EMI camera was also lighter and more mobile and flexible. Essentially, a mechanical system was always going to lose out to electronic technology, but Baird's persistence and his status as a pioneer, earned him the right to compete. So the service was launched with two competing technologies—used on alternate days.

The site chosen for these was the roof of Alexandra Palace, a nineteenth century exhibition centre some 350 feet above sea level in North London. The staff who moved in there in 1936 to start up the new system initially thought they had 4 months to familiarize themselves before they went on the air. However, the move coincided with the annual Radio Olympia show and a panic message arrived that nobody was buying TVs. This was not unreasonable because there was nothing to watch on them. So, the 4 months became 9 days, and they got on the air in time to broadcast to Radio Olympia twice a day for 10 days from 26 August.

Early days at Alexandra Palace

The service was officially inaugurated by the postmaster general at 3 p.m. on 2 November 1936. The short time span of TV's history was brought home to me when I sat next to Dallas Bower, the producer of the main programme of that first day, a half hour variety show, at a commemorative lunch a few weeks ago.

Reith, nothing if not consistent, was still grumbling and still contemptuous: his diary for 2 November reads: '*To Alexandra Palace for the television opening. I had declined to be televised or to take any part. It was a ridiculous affair and I was infuriated by the nigger stuff they put out. Left early ...*' But later he began watching despite himself—and his worst fears were realized. He wrote: '*Television is an awful snare...*'

The advantages of the Marconi system soon became very obvious and on 30 January 1937 the Baird transmissions ended. At that time broadcasting still only lasted 2 hours each day and not at all on Sundays. From 1937 it jumped to 16 hours a day. This sudden jump in hours led many people to wonder how all that air-time could possibly be filled. Among them was Jean Bartlett, an assistant producer from the early Baird days, who wrote: '*There still remain vast gulfs of time to be filled by programmes. Technical developments to the lay reader seem nothing short of miraculous. Can the imagination of the programme builders keep pace with them?*'

Technical constraints were still hefty enough to put quite a brake on that imagination. The cameras were still big, and without zoom lenses the only way to change the size of the shot was to change the lens on a rotating turret each time. So programmes still tended towards those where the action, if there was any, happened directly in front of the camera. So into the studio were brought variety acts straight from the London hotels and theatres where they were currently playing. Among them in the early days was a very young Margot Fonteyn, an unknown dancer at the VicWells company whose impresario Lillan Bailiss, had offered the services of as many of her singers and dancers as the BBC required.

One of the most popular shows was *Picture Page*, which in the way of new technologies modelled itself on a familiar one—the telephone exchange 'switching' viewers through to the next act.

Mickey Mouse Cartoons were also popular. The relationship between TV and the cinema was for long a delicate one, with the film business wary of the rival medium which threatened to cannibalize its audiences which of course it did, yet slowly realizing that this was a new and separate market for their productions. This suspicion started early and the only company that released to the BBC its films to show was Walt Disney, who let them have as many cartoons as they liked for nothing. The Disney Company today takes a rather different view.

Outside broadcasts and first sports coverage

From the beginning, teams of production and technical staff had been eager to get out of the confines of the studio, and had experimented with the idea of an 'outside broadcast' (OB). In the early days this simply consisted of ever longer lengths of cable, which could extend from the studio out of the building into Alexandra Park. Within a year, however,

TV engineers were building the first lorry equipped with mobile recording and transmitting equipment. It was called an OB unit and it was a major step in the development of modern TV. When it was only 4 days out of the factory, the unit was used for the first ever OB—the coronation in 1937 of George VI. There was no producer, the cameras and lorry load of equipment, was manned by the R&D team who had put the unit together. It was the chief engineer who stood looking down on the procession and suggested to the people in the lorry when they should cut to the pictures of a different camera

Many people bought their first set for the live coverage of the Coronation; viewing figures were estimated at 10 000—a huge audience at the time, but still far, far fewer than the people actually present on the ground.

It was the birth of the OB units that brought into being what is still one of the great benefits of TV—bringing an event of national importance to the whole nation, allowing viewers to share experiences and build a common stock of shared memories, as part of the national culture.

It wasn't long before the BBC OB lorries became a familiar sight at major sporting events. By 1937 the OB unit, with a radio link mounted on a lorry, was first at Wimbledon. In 1938, it was at Wembley for the Cup Final, by the Thames for the Boat Race and at Epsom for the Derby. Events which previously had only been recorded on film were now broadcast as they happened. When Chamberlain returned from Munich in 1938 the OB unit was on the tarmac at the airport to cover the event; the first example of electronic news gathering. War was looming and the BBC TV was about to be put on ice.

The war and TV shuts down

On 1 September 1939, broadcasts ceased with the immortal words 'I tink I go home' from one, Mickey Mouse. War was imminent and TV could not continue because the broadcast frequencies were needed for that other new development—radar.

It was nearly 7 years, on 8 June 1946, before TV came back with a bang for the OB of the Victory Parade—this time seen by 100 000 people.

After the war—birth of TV news

The 8th of June 1946 also saw the first broadcast of TV news. The BBC's radio news department was extremely suspicious of its frivolous sibling and thought TV lacked the gravitas for its weighty subject matters.

The radio people reluctantly agreed to a deal whereby they'd supply the content—including the news readers—and TV would supply the pictures. However, the cameras weren't allowed to show the news readers for fear their facial expressions might betray some bias, so the public was treated to an endless series of caption cards and maps.

Things improved, but not much, when the policy was changed. Now the people reading the news were visible, but they were deliberately chosen to lack personality and as they were also reading from scripts without Telepromters it remained a fairly arid viewing experience. Worse, it made the news readers look shifty rather than solemn, suspicious rather than serious, and the tabloids promptly dubbed them 'the Guilty Men'.

The 1950s and arrival of ITV

Technology continued to drive programming: 1950 saw the first OB from abroad—a link (by microwave) from Calais. I remember watching it. TV also continued to reach out to new audiences. The Coronation of our present Queen in 1953 was the first broadcast event which had a bigger TV audience than a radio one. It was a critical event in British TV history, because from this time, TV ceased to be a novelty and became a central part of Britain's creative and informational landscape.

It became literally part of the landscape, for the prices of sets were falling and by 1955 aerials were sprouting all over the rooftops of Britain. And there were now 13 transmitters with the potential to reach 92% of the population.

The BBC no longer had a monopoly; there was a new kid on the block—ITV, a rival commercially funded channel. So began the competition for audiences. ITV brought new quiz shows and filmed series, and its news arm, ITN, was a personality driven affair from the start with Chris Chataway and Robin Day, whose complete lack of deference on screen knocked even world leaders off their pedestals.

The competition led to a number of new kinds of popular BBC programmes: *Panorama, Tonight, Your life in their hands*, and John Freeman's *Face to face*. These programmes made use of studios and OBs as before, but in addition there was a new lighter 16 mm film camera easier to carry, easier to move altogether a more flexible tool than the bulkier 35 mm cameras.

Producers were soon making short documentary films for TV, and a new kind of TV reporter was born. People like Fyfe Robertson and Alan Whicker reported on the everyday events from up and down the country, bringing the experiences of British people to a TV audience. They presented stories which were sometimes serious, sometimes funny, but almost always reflecting a unique angle on some part of British life.

Later, as film makers wanted to capture on film yet more intimate portraits of life in Britain, reporter-less, cinema verite films and the concept of the fly on the wall documentary film was realized. I say films, because until 1958 there had been no videotape. TV was either live broadcast or film which had to be shot, processed, physically cut together with razor blades and sellotape, and finally taken to the transmitting station to be run from a telecine machine.

The reason we have a record of the early transmissions and some of the other historic events is that they were also covered on film, or film cameras were turned on to the TV screen to make a record of the transmission. When plays were repeated, the cast had to come back to the studio a few days later and do it all over again.

But all that changed in 1958 when the Ampex company introduced a system whereby an electronic signal from a camera could be stored on magnetic tape. The reels were heavy, the tape itself 2″ wide, so the recording equipment was also big and cumbersome but at least it meant that TV programmes could be stored for later transmission and/or for repeats—and for sales abroad. From then on the BBC vaults became a prized resource.

Miniaturization and simplification of technology

It has become a familiar constant with electronic technology that things always get smaller and usually cheaper. It's true of everything from calculators to mobile phones, and it's certainly true of TV cameras. Smaller and especially lighter cameras are obviously conducive to mobility and flexibility, both of which are good for most purposes. But the need for a cathode ray tube put a physical limit on how small they could be.

The breakthrough arrived with the charge coupled devices (CCD). Instead of a tube, the CCD cameras encode the information electronically, on three silicon chips. So now, the size of the camera is limited only by the size of the lens and the size of the tape it's recorded on to.

Analogue video tape degrades fairly sharply as it gets smaller and the signal has to be squeezed on to a smaller area. Even the best Hi8 camera with the best lenses produced a pretty rough image—certainly not the equivalent of professional modern Beta cameras, still less film. But these small cameras could be mounted in unlikely places and capture events which would otherwise be impossible.

The medical profession took advantage of miniature cameras, and this enabled them, their patients, and indeed TV viewers to explore inner space.

Later, when more traditional TV cameras became smaller, cheaper and more user-friendly new programme ideas emerged, notably *Video diaries* which became a regular series where people captured their own experiences, without any need for a technical crew.

But recently videotape has gone digital and the results of that are pretty spectacular. This small digital video camera produces an image about the same quality as a Beta camera which is much bigger and very much heavier.

What makes this revolutionary in the history of TV is that for the first time, we have a high-quality camera that can be used not only by specialist camera teams but by anyone who can afford just £2500. This camera was made for the domestic market. But BBC producers are already embracing this new technology enabling them to make some programmes at a fraction of the cost of traditional methods.

These small cameras also provide programme makers with all kinds of new possibilities. On the one hand we have the capability of getting quite extraordinary views of events but on the other hand the possibility of filming secretly becomes easy and more tempting. As with genetic engineering and other new technologies we have to be wise to the implications and aware of the uses and potential misuses of it.

A recent unlikely location is a cricket stump with a camera, which consists only of a lens and a circuit board that encodes the information. It is linked by microwave to a recorder, and the camera itself can go almost anywhere. The Australians came up with the idea of putting a camera in the stump for cricket coverage, but the BBC improved on the technique by replacing hollow wooden stumps (liable to splinter disastrously) with more or less shatterproof fibreglass. Naturally, this iconoclastic step was taken in consultation with the cricket authorities and the skills of our highly trained technicians were able to produce a stump indistinguishable from its fellows—even to the England wicket keeper. The one outstanding problem remains the tendency of umpires to replace the stump facing the wrong way. Work, however, is underway on a remote steering device that should solve that too.

This year a BBC TV camera travelled to an even more unlikely location—inside the pith helmet of a Marine during the Beating Retreat in June. The helmet happened to come with a convenient badge hole and a judicious dab of gold felt tip completed the disguise. The BBC is again covering the bob sleigh event for the Winter Olympics in early 1998, so you can expect some vertiginous shots by then.

Editing and the action replay

But to go back to the first video recording on tape: once it had become possible, all kinds of production opportunities presented themselves. Richard Dimbleby had demonstrated the notion of immediate playback for the first time on air in *Panorama*.

But of equal significance, the editing of studio and OB programmes was now possible. This in turn led to new production techniques: we saw the introduction of pre-recorded material, added as inserts to otherwise live programmes. Plays could not just be recorded but performed and recorded out of sequence, and the birth on videotape of the 'retake' or 'shall we do that just one more time?'

At first editing was a clumsy technique. The 2″ tapes had to be physically cut with a razor blade, and stuck back together at the appropriate junction often with a visible bump on screen. But the invention of electronic editing transformed the process from being a primitive safety device—a way of correcting mistakes—to a tool of production, which changed the way that programmes could be made.

If you look at news and current affairs programmes today, they are built entirely using this technology: some reports are recorded hours before, some minutes before, others are edited as the programme is being broadcast, to be slipped in later in the bulletin. And the running order, the sequence in which the items are shown, can, and indeed frequently does, change as the programme progresses. Modern digital editing allows us this flexibility.

In sports production, video recording provided the means to offer edited highlights as soon as an event was over. It provided the replay. They have developed to become part of the narrative of televized sports and in so doing, challenged the authority of the referees or, as in cricket now, provided them with a new aid. The action replay is, of course, under the control of a computer. And it was computer technology which was used to create dramatic advances in graphic design. Today, graphic designers have all but abandoned their pens, pencils, paper, and ink. Long gone are the days when someone was employed to change stencilled caption cards by hand. The computer can capture still frames, combine pictures, alter existing images, add text—with a software programme ironically named 'Paint Box'. The designers used this new found chip power to invent a new art form—the opening title sequence.

The front titles of the Royal Institution Christmas Lectures, use the latest computer technology, the Hal, a successor of Harry. The explosion of the cracker was created by photographing 30 still frames of cotton wool shaped progressively by the designer to create the effect of an explosion, when played back in real time.

There is no doubt that computer technology has given rise to a new industry in television: post-production. This is the packaging, adding new dimensions and a new sophistication, to recorded programmes. It's a sophistication which is still expanding as we move from complex animation to the realm of virtual studios.

But now, let me now move on to look at the changes outside of the production experience, which have also dramatically altered TV viewing for the audience.

Standards

When TV was half as old as it is today—On 21 April 1964 to be precise, the technical quality of broadcast TV went up a notch with the launch of BBC2, which broadcast on UHF frequency with 625 horizontal lines on the screen instead of 405. To receive it, the public had to be persuaded to buy a new dual standard set. The software—a greater choice of programmes—was the bait to get the public to buy the hardware—the receiver and a new aerial which became another accessory adorning Britain's skyline.

After buying and fitting this new apparatus the public waited expectantly in front of their new sets on the appointed evening of the launch. They were disappointed. A failure of the existing technology—a huge power cut in London obliterated the evening's viewing, so BBC 2 began with the first ever *Play School* the following morning. However, after the initial shock, the robust opinion of the television manufacturers was that the spectacular calamity of the opening night was the best publicity they could possibly have had.

In 1967 the lure of TV was strengthened by the introduction of colour, initially only on BBC2, but then in 1969 also on BBC1 and ITV. Colour had its first great showcase with the 1969 investiture of the Prince of Wales, which used just about every colour camera in the country. It was watched by 16½ million people.

TV has an unparalleled role in uniting the population through mass experience and the internationalization of that experience was furthered by satellite transmission. It meant that events around the world could be broadcast live from even the most remote locations.

Sputnik 1 launched by the USSR in 1957 heralded a new era in communications. And in April 1965 Intelsat 1 or Early Bird, as it was known, became the first satellite to transmit television traffic on a commercial basis. Before that experimental systems had been launched and the most successful of these was Syncom 3 over the Pacific which was used for the Tokyo Olympics in 1964.

These could be seen live by satellite as far as California but had to be relayed by line and finally by aeroplane to people watching in Britain. By 1966 satellites were in place to link all five continents. The first big event to celebrate this, a monster programme, was *Our world* in 1967. 24 countries saw it live, hosted in Britain by Cliff Michelmore.

By 1968, the Mexico City Olympics were live, in colour, around the globe and in 1969 the world witnessed the furthest OB yet—from the moon.

Satellite technology has of course been the enabler, in this century, of a burgeoning of new TV channels. The existing distribution from terrestrially based transmitters was using most of the available frequencies. Distributing by satellite direct to peoples' homes via a special dish bypassed this frequency traffic jam. Distributing via cable does the same but the capital investment required to lay the cable is huge and here it has spread itself more slowly than in mainland Europe.

Both distribution routes have, by providing more choices, helped move the balance of power from broadcasters and producers to viewers. Easy access to these choices would be impossible without another piece of hardware—one of the most significant to date—hardware which has been helping to shift that balance of power.

This suitcase is the latest in satellite technology as used by BBC News. Within the past 10 years, news crews have been able to travel with a suitcase carrying the latest in satellite technology. The pictures shot and edited on location can be transmitted back to base, from wherever in the world they happen to be. And, of course, the physical material has no need to pass through national borders, so the control of the reports lies with the broadcaster. The Gulf War correspondents had two kits similar to this one, and for the first time some of the coverage of that war was instantaneous and no longer censored by the military.

The remote control

Transforming the way programmes are watched is this—the remote control.

No longer do you leave your seat, move to the set, select a channel, return to your seat, settle down to watch. And if you wish to change channel you have to do it all over again. Now, you can sample at will. Switch the moment you lose interest if you so choose. Watch two—children manage three—programmes at once. Graze the services available; nibble at the passing images. Within each household there are tales of 'the battle for the remote control.' Many of the losers buy new sets which they take to bedrooms, or kitchens, for viewing alone, a phenomenon hitherto common with radio listeners.

Viewing habits changed with the advent of the remote, but I can assure you that with the explosive changes in the pipeline, we ain't seen nothing yet.

So what are these changes and how will they affect television in the future? In this last section I want to outline very briefly some of the new technology which is now under development, and speculate about the future.

Digital—what is possible

Currently, whether TV is recorded on film, on analogue or digital tape, it is transmitted to your home as an analogue signal. Channel 5, is the last analogue terrestrial channel. In future, new channels will be broadcast digitally, whether by satellite, by transmitters, or by cable, SKY will launch digital channels probably by the end of the next year. At the same time the BBC plans to start broadcasting digital services—BBC1, BBC2 in Widescreen, and a 24-hour news channel. This is not just one huge change. It is the beginning of a series of huge changes.

We have been experimenting for some time. Widescreen formats will enable viewers to see movies in the ratio they were filmed and will bring a new dimension to sports and major events. I was there, when for the very first time, we transmitted analogue and digital widescreen at the same time: The first digital widescreen simulcast—it was of the Trooping of the Colour—in 1997.

Digital allows so much more: not only is there a dramatic improvement in sound and picture quality, but many, many more digital channels can occupy the space of one analogue channel. The zoochannel world will soon be with us. For the BBC this means that extra programming can be offered alongside the digital BBC1 and BBC2, but also extra information. This will allow viewers to engage with the programmes in new and satisfying ways, following the paths of their own curiosity and interest in a way that has never previously been possible.

Just as the programme at the theatre or opera brings to life a wide range of information about the performance, so we can use the digital capacity to bring to life new levels of information and interaction. Participation is up to the individual viewer. People who just want to watch the show can still do so but people who want more can guess the value of that piece on the *Antiques roadshow* or read the score of *Young musician* in parallel with the performances.

Parallel soundtracks can allow you to choose whether to receive the original version of a foreign film or a dubbed one. You've missed the beginning of a drama or a film. At the press of a button you can have the story so far. We can offer more chances to catch up with the most talked about programmes or your favourite dramas.

And with partners it is hoped to launch some pay channels using the BBC's ever growing archive, theming and presenting the material to appeal to particular groups of viewers. Other broadcasters will be planning their offerings and we shall certainly see 'near video on demand'—the same film starting at 10-minute intervals on a dozen or more parallel channels.

Before long we will have moved from a world in which all viewers receive their TV by analogue signal over the air from terrestrial masts to a world in which TV will be received from many different exclusive routes—digital satellite, digital terrestrial, digital cable, and later down the telephone wires.

It is not just the distinction between broadcaster and audiences that is blurring. Digital technology is also blurring the distinctions between different broadcast and communications media themselves. Digital transmission technology makes no distinction between video, audio text, graphics, and pictures. The nature of any digital recording is that every piece of information is simply converted into a string of ones and zeros for transmission to a receiver which reassembles only those parts of it that its user wants to see and hear.

These receivers of digital information will continue to be known as telephone, radios, TVs, and personal computers (PCs). But at their heart all will be computers. They will differ in shape, size, and processing power, and they might offer a variety of screens, speakers, and printers, but ultimately at the heart of every receiving device in the digital world will be computing power.

What the digital world means is that in time the lakes and rivers of analogue will become one vast digital ocean. TV programmes, once they become part of the on-line digital offer, will no longer be part of the TV offer but part of the universe of all digital shopping, banking, interactive entertainment, and information services.

This means that digital media will tend to shift the balance of power from broadcasters to the owners of rights—sports, films, entertainment, and drama. In the old environment, TV was the richest analogue medium and viewers were restricted to a few terrestrial channels.

Because the supply of programmes far exceeded the channels' limited capacity to show them, channel owners could pick and choose content that they broadcast and reward it on a cost-plus basis.

Digital technology is removing those channel capacity limits and soon high-quality content will be much rarer than the capacity to deliver it. Therefore, the rights' owners will be able to pick and choose the transmission media rather than vice versa.

TV over the last 50 years has shaped the way that we think. We have come to expect and accept neatly packaged programmes in linear form. This has given enormous influence to media owners. Digital technology will partly diminish such influence. Linear TV will remain the most powerful form of communication but in the next 3–5 years the digital age will begin to have more meaning in our lives.

The TV set itself is being given a run for its money. Although worldwide there are about three times as many TV sets in use as PCs, with annual sales approaching 70 million, sales of PCs will overtake those of TV sets in the next year, and the PC is fast reaching the point where it begins to rival the quality of a TV set.

This sounds like a paradise of choice for the viewer. But, in order to get all these services, viewers will have to invest in a set-top box; both for them to receive the digital signals and for suppliers to charge for their programmes and services. That set-top box is the gateway to the digital future. The standards in that box are proprietary. So the keeper of the gateway is in a powerful position.

In Britain it is a uniquely powerful position. That is because the owner of those standards—Rupert Murdoch—also owns the rights to the most attractive programming. Sport and movies are in his own words the 'battering ram' to drive take up of pay television.

All credit to him for having had the vision and the guts to get BSkyB where it is. That has begun to change the market for good. But what happens next—and I mean over the next few weeks—will shape the future of digital broadcasting in Britain. If the government gets it wrong one player may end up dominating the market through digital satellite.

Let me give you two examples of what this might mean. First, it will determine how far interactive services will develop. The power of the PC through the TV. Potentially a tremendous force to change our lives—whether in education, information, home shopping, or banking. But without access to the proprietary standards in the gateway, broadcasters and others cannot plan and develop these services. They will happen at the pace and in the manner determined by the dominant player.

Second, with an enormously greater range of channels and ancillary services, context becomes as important as content. Thus the role of the electronic programme guide will be crucial—the browsing mechanism that helps you find what you are looking for. Currently, we have the *Radio Times*, the *TV Times*, *TV Quick*, and others. The electronic programme guide could be a Sky programme guide or it could be a *Radio Times*/BBC programme guide, just as it could be a *Hello!* magazine programme guide. It will be a powerful tool to direct viewers towards some services and away from others. For the BBC and ITV to be simply a few channels 'brought to you by BSkyB' on their guide is a grim but real possibility. That is what might happen in a dominated digital market. What is more, it will be a smaller market. Several million people who want it and can afford it will take digital satellite. But that is only a fraction of the 22 million homes in Britain who could enter the digital age.

If the government gets it right—as we are arguing for—then there will be an open market. One in which there is genuine interoperability between delivery platforms and interconnection with different service providers. The consumer will get a wider choice of new services and programmes and a choice of delivery platforms—satellite, cable, and terrestrial television. We intend to be on all these platforms. We hope that regulation helps ensure those platforms are there for us to be on.

What we don't know about the future is the impact that regulation will have on the way that these markets develop. But we are arguing for it to help the market develop not hinder it. We also don't know how the public may react. They are increasingly resentful at being forced to make technology choices to consume entertainment, and they may wait longer than anyone imagines in upgrading to the digital age. Different countries will upgrade to digital at different times, and will leap-frog each other in terms of their systems and platforms.

Broadcasters will only be meaningful if, like newspapers, they manage to bring trust, attitude, quality, and other appealing characteristics to help the viewer and listener find their way through a far more confusing—but exciting—media world.

Although there will be generational changes eventually, the one certainty is that people's tastes and sensibilities will change far less than technology or regulation. As we have seen in every other industry from food retailing to home insurance, people put a real premium on brands that they can trust because of the promise of the quality, reliability, and decency that comes with them.

In an increasingly crowded, noisy room, a friendly trusted face will become more and more valuable. The BBC will be enabled by technology not only to be a producer of much of the best material in the world, but to be a link, a broker if you like, between creators and audiences from all over the world who share our sensibilities and our values.

Finale

In this august environment, constrained by the rigours of scientific method in the search for truth, I dare not stray further into the murky waters of crystal gazing and prognostication. I've talked for long enough and as you'll have gathered my colleagues and I have quite a lot of work to do.

WILL WYATT

Born 1942, won a History Scholarship to Emmanuel College, Cambridge. He trained as a reporter on the Sheffield Morning Telegraph before joining BBC Radio News as a sub-editor in 1965. He moved to television production in 1968, working on programmes such as *Late Night Line-Up*, *The Book Programme*, and *B. Traven: A Mystery Solved*. He became Head of Presentation Programmes in 1977 and later Head of Documentary Features, responsible for such programmes as *40 Minutes*, *Around the World in 80 Days*, *Children in Need*, and *Rough Justice*. Appointed Managing Director, Network Television, in 1991, he has this year been appointed to the new post of Chief Executive, BBC Broadcast, part of a major BBC reorganization designed to meet the challenges of the digital age. He is a Governor of the London Institute and ex-Govener of the National Film and Television School. His book *The Man Who Was B. Traven* was published in 1980. He is a Fellow of the Royal Television Society.

Molecular information processing: Will it happen?

PETER DAY

Preamble: brains and foresight

Molecular information processing is going on all around us. Molecular-based information processing has in fact been going on for millions of years in its natural form, through the brains, not just of human beings and their predecessors, but in all other higher (and perhaps even lower) living organisms. For all brains, however rudimentary in evolutionary terms, operate at the molecular (or more precisely supramolecular) level. Thus nature has given us an existence theorem and the question becomes a different one: not 'Can information be stored and manipulated at the molecular level' but rather 'Can we ourselves design and manufacture artificial structures using molecules that will carry out these functions?' Sadly, at the present time the short answer to the second question is 'we don't know', but if that was all there was to be said about the matter there would be no point in continuing this exposition.

So I find myself in the rather unusual position, in tackling a subject that does not yet exist. Indeed, in some ways it is even a bit embarrassing because I am not known as a great fan of crystal ball gazing. Over the last two years the Department of Trade and Industry (now the civil service resting place for the Office of Science and Technology) has conducted a Technology Foresight exercise, and it is significant that no mention of molecular-based information processing appears in it. Perhaps this is one result of the very narrow subject base for the advice it sought: for example no chemist was invited to sit on the panel dealing with information technology (IT) and electronics, despite the fact that even the present generation of integrated circuit hardware is constructed by chemical means, using photolithography. Conversely, neither did any representative of the IT industry contribute to the deliberations of the Chemicals Panel. In fact, when the House of Commons Select Committee on Science and Technology took evidence about the procedure and effectiveness of the Foresight endeavour I have to confess to submitting (and defending under interrogation) a highly sceptical view on the subject. It is therefore starting from this sceptical and cautious base that I want to consider where the subject that has been called 'molecular electronics' has come from, what its present status is, and what a future information processing regime based on molecules might just possibly look like.

Let me begin by nailing my own colours firmly to the mast: I am a chemist, and a distinguishing feature of chemical science is to build new structures, that is, new arrangements of atoms into molecules and molecules into aggregates (which might be, but do not necessarily have to be, crystalline) in such a way as to create new properties and functions. As noted already, molecular chemistry is already contributing very significantly to the electronics industry as it exists today. Examples are the fabrication of silicon wafers, etching of fine structures on their surfaces using photoresists, deposition of thin films by chemical vapour deposition and, above all, displays made from liquid crystals, which we find in digital watches and laptop computers. However, that is not the focus of my present concern. I want to consider whether molecular assemblies themselves might be used as a means to store and process information. Put baldly like that it sounds like science fiction, but let us remind ourselves what has been happening in electronic data processing over the last 50 years, or indeed in other forms of processing for longer than that.

Decreasing size: from the abacus to (very large-scale integrated) VLSI circuits

The earliest mechanical form of data processing (if you leave aside counting on our fingers, which gives rise to the word 'digital') was the abacus. The simple form of this device, still widely used in bazaars throughout the Far East, still serves to illustrate a number of important points about computational functions that are shared by advanced electronic systems. First of all the information is stored and processed in 'bits', units symbolized by the balls that move on the wires of the frame. Second, the bits are manipulated serially, i.e. one after the other, rather than all at once. Third (a point that will come up again later), to move the bits takes a certain amount of energy, in this case muscle power, thus reinforcing the fact that each computational step consumes power.

The same considerations apply to more advance mechanical computers, which reached their apotheosis in Babbage's arithmetical engine, a device of gear wheels so elaborate that the handle could scarcely be cranked by a strong man. In more recent times mechanical calculators remained in use till some 30 years ago, overlapping chronologically with the first ones to use electrons.

The earliest devices to process information using electrons, such as the Colossus at Bletchley Park during the Second World War, were no smaller than Babbage's engine: indeed, the Colossus (which was well named) filled a whole room. Similarly, the individual components (valves) from which they were constructed were of similar dimensions to Babbage's gear wheels, i.e. a few centimetres. Furthermore, the programming and recording mechanisms remained mechanical, in the form of holes punched in cards or paper tape. Over the succeeding 50 years, however, the components of information processors have grown inexorably smaller, at the rate of one order of magnitude every 10 years. Valves were replaced by transistors, so that the electrons no longer circulated in a vacuum but through the conduction band formed from the overlapping atomic orbitals in a solid. Single transistors, linked by wires, gave way to integrated and then to VLSI circuits etched by elaborate sequences of chemical operations on to the surface of a silicon single crystal. Figure 1 gives a pictorial indication of the orders of magnitude that have been traversed.

Nowadays, the individual features found in commercially available personal computers are less than 1 μm (one-millionth of a metre) across and compare with the size of a virus. Enormously complicated patterns are achieved routinely by photolithography (Fig. 2) or etching with finely focused electron beams. Astonishingly, this evolution in the size of electronic computing elements, or features, follows an exact linear relationship with time, over quite a long period (Fig. 3). It has been called Moore's law, after the founder of the integrated circuit manufacturing company, Intel. In fact the relation shown in Fig. 3 turns out not to be the result of some hidden law of nature, but actually results from economics: the cost of building a new plant to produce microprocessors scales with their feature size, and currently it stands in the region of $1 billion, which explains why chip manufacturing throughout the world is concentrated in so few hands.

Given the vast size of the investment in time, effort, ingenuity (and hence money) to make microprocessor features ever smaller and smaller, it is legitimate to ask why all this trouble is being taken. Clearly, small size is an advantage in itself: we can carry around computing power in a laptop equivalent to the mainframe machines of 20 years ago. Individual

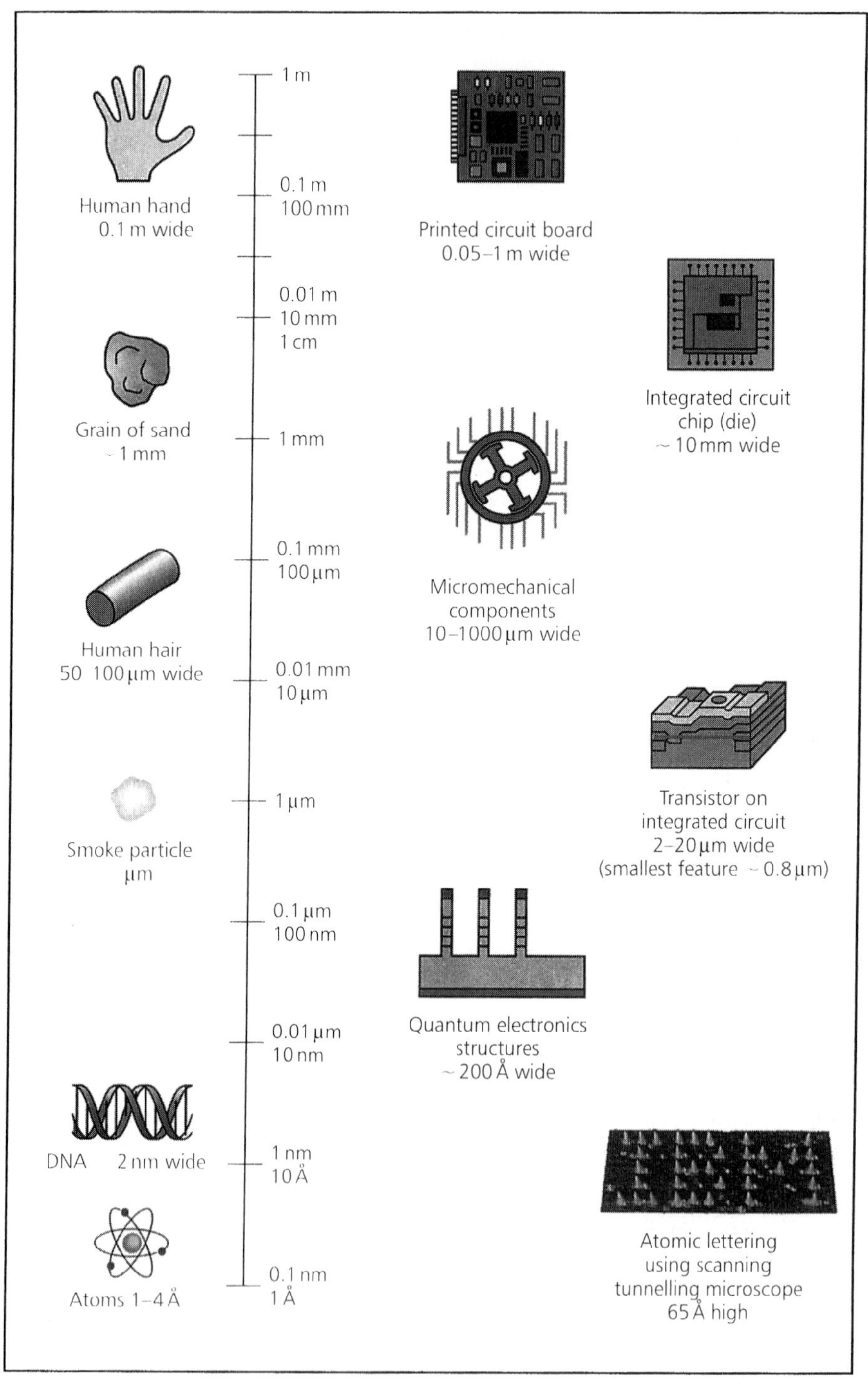

Fig. 1 A pictorial representation of the sizes of information storage media. Reproduced with permission from *Chem. Brit.*, **32**, 29–31 (1996).

Fig. 2 A modern microprocessor. Reproduced by courtesy of Sharp Electronics.

switching processes also take less power. Valve-based computers had huge power consumption but the laptop only needs a rechargeable battery. (Incidentally the latter is itself a marvel of solid state chemistry,

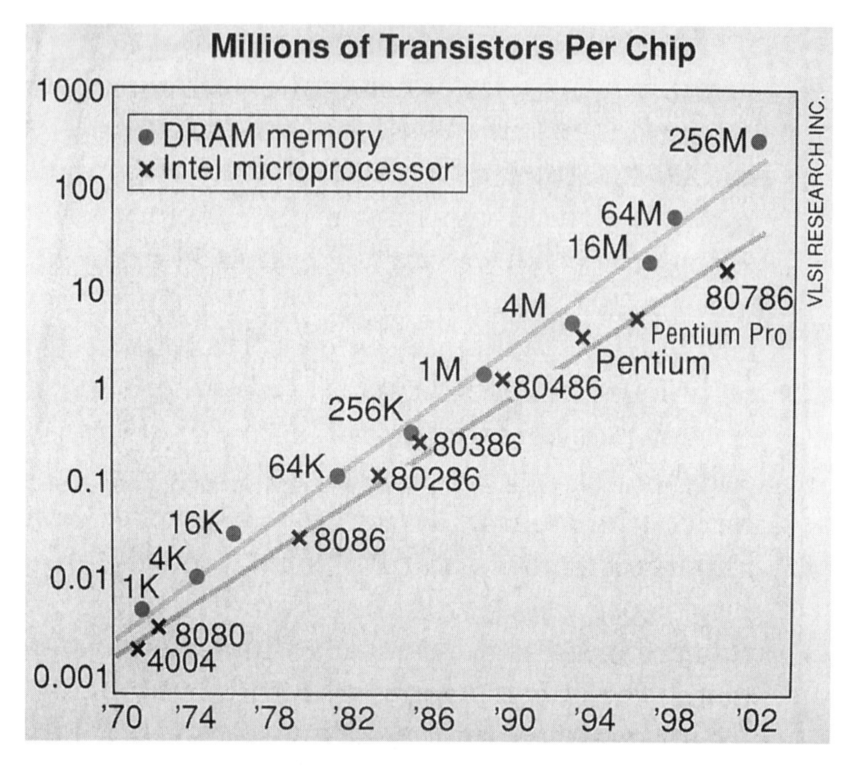

Fig. 3 Decreasing feature sizes of microprocessors with time. Reproduced by permission from *Science*, **274**, 1834 (1996).

but that is a story for another time.) In view of these advantages to miniaturization, can we imagine any limit to the process, apart from human ingenuity and cost? Well, unfortunately, yes. There are three so-called 'fundamental' limits, set by the laws of nature. The first is set by the statistics of the switching process in a binary system (i.e. on or off), which translates into thermodynamics via entropy. The second arises from the dictates of quantum mechanics, embodied in Heisenberg's uncertainty principle. The third, which is the most important one in practice is determined by the heat dissipated by each switching event. As we saw with the abacus and Babbage's engine, it takes energy to throw a switch, and it is no accident that the largest volume inside the central processor unit of a modern supercomputer consists of pipes filled with refrigerating fluid! Altogether the three limits can be displayed on a single plot of power dissipation against switching speed (Fig. 4).

The final consideration brings me at last to the central point. Single chemical events, such as the reaction of a molecule catalysed by an enzyme, consume far less energy than the present generation of semiconductor-based electronic switches (roughly 10^2 kT compared with

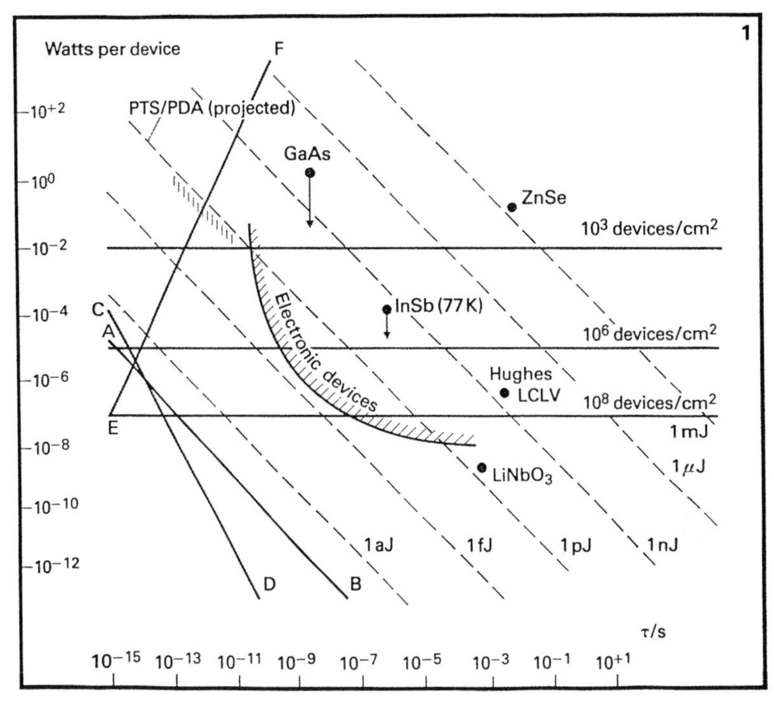

Fig. 4 The fundamental limits of computing speed and power dissipation. Reproduced by permission from *Chem. Brit.*, **26**, 52–4 (1990).

10^{10} kT). In principle, therefore, substantial advantages might follow from exploiting molecular chemical events as elementary information processing. If that is the case, several questions need to be addressed before we could contemplate any attempt to realize the potential of the new approach. First, is it going to be possible to make complex ordered structures from molecular components analogous to those of microelectronic circuits such as the ones in Fig. 3? Second, what are the kinds of physical or chemical processes that could be envisaged to store and process the information? And third (perhaps the hardest question) how could we get the signal in and out? Or, put another way, can we interrogate and detect the states of single molecules? The remainder of this account will be devoted to providing some answers to these questions.

Ordered molecular arrays: chemistry at work

Building structures out of atomic or molecular building blocks that are ordered over a long range in one, two, or three dimensions are at the very heart of the science of solid-state chemistry. For example, arrays of molecules forming domains with different orientations are shown in Fig. 5, where the components in question are long chains of carbon atoms (n-alkanes). Note the great, although superficial, similarity between this structure and that of Fig. 3. One of the most powerful methods of engineering ordered molecular arrays exploits the affinity of hydrophobic molecules, such as the alkanes above, for other hydrophobic ones, and of hydrophilic (literally 'water loving') for hydrophilic. It is termed the Langmuir–Blodgett method, and as it has been expounded in detail in Friday Evening Discourses I will not give more than the briefest of mentions here. This approach to making thin films one molecule thick on the surface of water is said to have had its origin in an experiment carried out by Benjamin Franklin, who poured a small quantity of oil on to the Serpentine pond in Hyde Park and, watching it spread out across the entire surface, was able to make a simple calculation of its thickness from the volume of the oil that he had poured out.

Apart from such rough and ready arithmetic, there are nowadays many ways of verifying that such films really are only one molecule thick. For example, one can build up multilayers by dipping a glass slide repeatedly in and out of a trough of water, which has such a

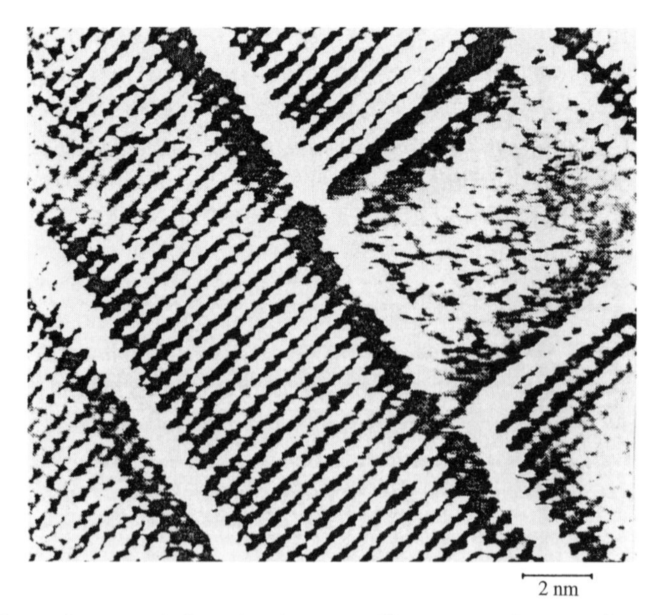

Fig. 5 Structural domains in an n-alkane crystal. Reproduced by permission from *Atomic and nanoscale modification of materials*, P. Avonis (ed.), Kluwer Academic Publishers, Dordrecht, 1993, p. 263.

monomolecular film on its surface. If the molecular chains are terminated by hydrophilic groups, they stick in the water, and are transferred to the similarly hydrophilic silicate surface of the glass. Thereafter, successive dippings produce layers on the glass with successive hydrophilic–hydrophilic and hydrophobic–hydrophobic orientations (Fig. 6). How do we know that single layers are being added at each stage? If the molecules each contain a light absorbing centre we can measure the absorption increase after each dipping: the increment turns out to be constant. Even better, atomic scale resolution in electron microscopy permits a direct view of the aligned molecules, as seen in Fig. 7.

Apart from these somewhat artificial methods of producing thin films of molecules, three-dimensional architectural assemblies of molecules are produced in wondrous variety by chemical synthesis. Figure 8 brings together a few examples culled almost at random from the literature of the last few years. Such elaborate constructions are not put together by what has (somewhat disparagingly) been called 'engineering methods', i.e. piece by piece assembly, but by the much more powerful method of 'self assembly'. The latter reaches its apotheosis in biology, where the three-dimensional structures of enzymes are completely defined by the sequence of amino acids forming the protein chain (the so-called primary structure). Hydrogen bonding coils the primary structure into the α-helix (secondary structure), which in turn coils into the final shape through the medium of charge, and both hydrophobic and hydrophilic interactions.

An example of a low-dimensional magnetic lattice, which self-assembles merely by mixing the ingredients in aqueous solution, is the phosphonate salt synthesized by Simon Carling at the Royal Institution (Fig. 8c). The organic and inorganic components in the structure segregate spontaneously into alternating layers. Further examples from our own work in the Davy Faraday Research Laboratory are molecular layer compounds showing the highly unusual property of superconductivity.

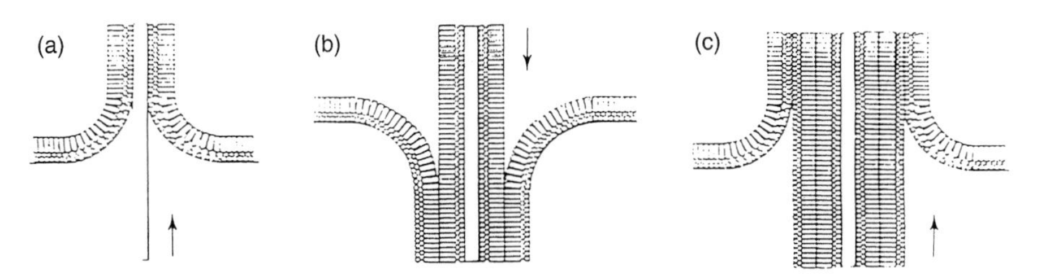

Fig. 6 Forming Langmuir-Blodgett films. Reproduced by permission from *Introduction of molecular electronics*, M.C. Petty, M.R. Bryce, and D. Bloor (ed.), London, Edward Arnold, 1995, p. 225.

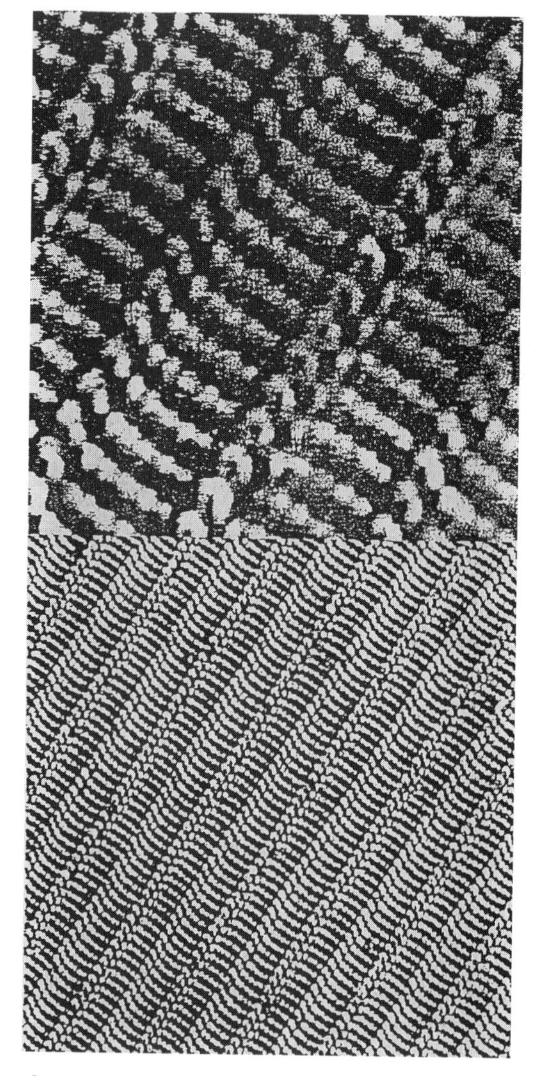

Fig. 7 Aligned molecules observed by electron microscopy. Reproduced by permission from *Adv. Mater.*, **8**, 903 (1996).

The compounds in question are so-called 'charge transfer salts', formed by oxidizing a heterocyclic aromatic molecule called bis(ethylenedithio)tetrathiafulvalene (or ET for short) and combining the resulting cations with inorganic anions. The oxidation is carried out most conveniently by electrochemical means and indeed, it has often occurred to me that the small glass cells used for growing crystals of these materials would be recognized quite easily by Sir Humphry Davy, should his ghost return to our laboratory!

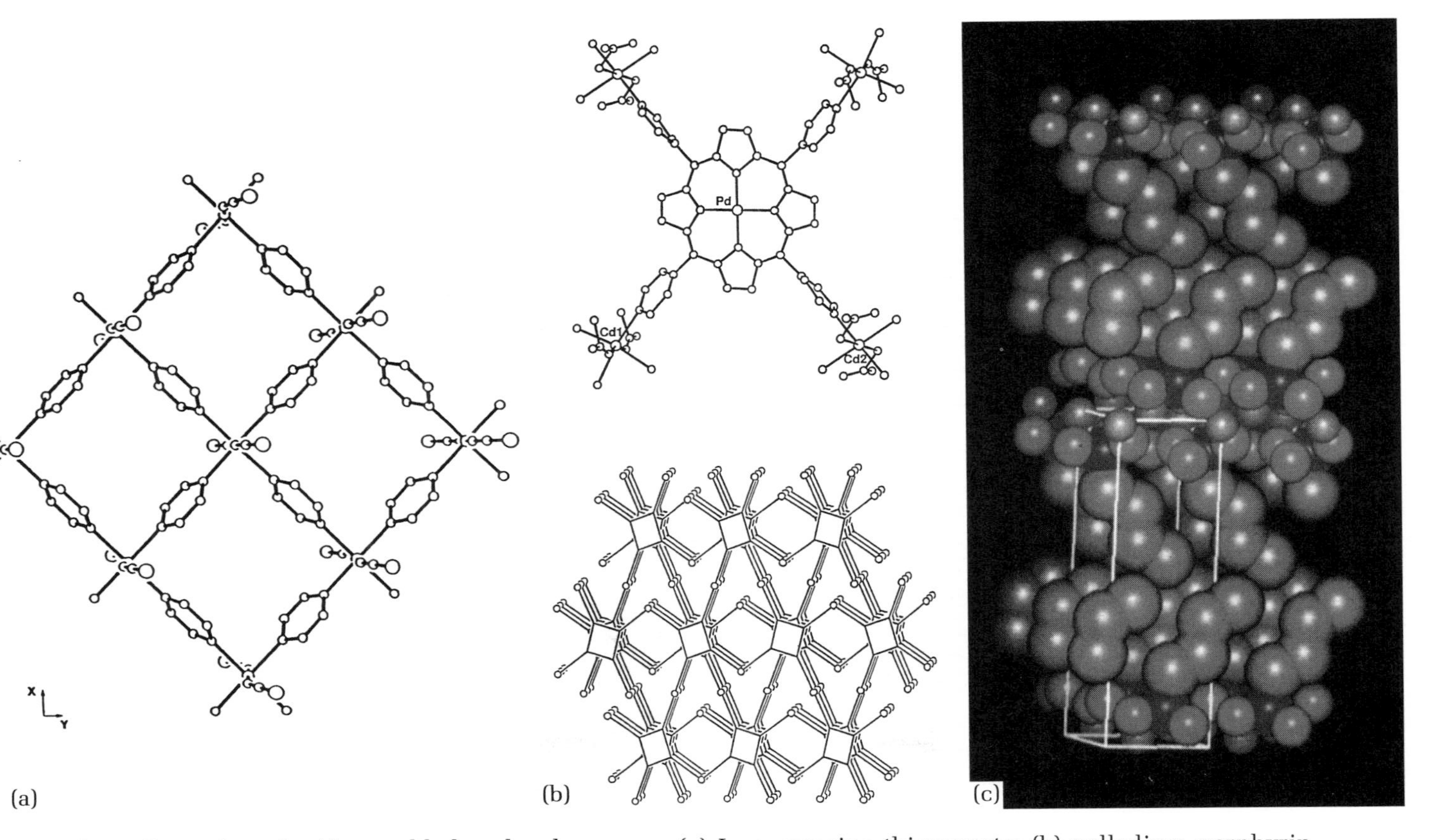

Fig. 8 Examples of self-assembled molecular arrays. (a) Iron pyrazine thiocyanate; (b) palladium porphyrin cadmium adduct; (c) manganese *n*-butylphosphonate.

The structures of one of the compounds prepared in this way is shown in Fig. 9. Again we see the organic part of the structure (ET) segregated from the inorganic part, $(H_2O)Fe(C_2O_4)_3.C_6H_5CN$. The compound is metallic, and becomes superconducting at low temperature. It is worth noticing in passing that we can now make superconductors containing water inside the crystal lattice, something that would have been thought quite extraordinary a few years ago. The compound in Fig. 9 also contains a high concentration of paramagnetic iron atoms, again something that would have been thought quite incompatible with superconductivity before, because magnetic moments are supposed to disrupt the Cooper pairs carrying the superconducting current.

Apart from self-assembly of molecular units into infinite arrays and networks, another development of recent years that has given a big impetus to the new field of nanotechnology and molecular electronics is the controlled preparation of clusters of atoms with defined sizes, and hence properties. So far as small metal particles are concerned, the origins of this topic lie far back in history, and actually have their origins in the Royal Institution. Michael Faraday made a colloidal solu-

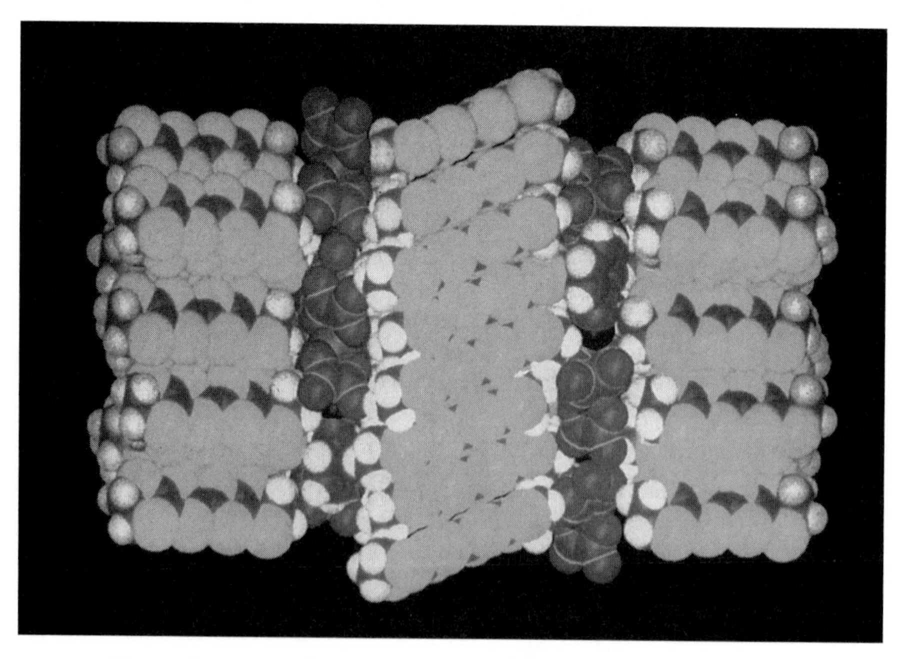

Fig. 9 Organic and inorganic layers in a superconducting molecular charge transfer salt synthesized in the Davy Faraday Research Laboratory of the Royal Institution.

tion of gold whose beautiful purple colour remains perfectly clear and uncoagulated to this day: it rests in a cupboard in the Director's study. Much more recently it has become possible to synthesize particles containing quite distinct numbers of atoms; for example, the one shown schematically in Fig. 10 contains 586 gold atoms. That seems a strange number, but you have to bear in mind that the packing of the atoms is the same as in bulk gold, that is, a hexagonal close-packed lattice of equal spheres. Starting from a single central atom, we can imagine it surrounded by 12 neighbours, then a further layer is added, and so on, generating a series of 'magic numbers' corresponding to the number of shells added. Finally, the particle is sheathed in an organic coating of alkanethiol molecules, whose sulphur atoms attach themselves to the outermost gold atoms. Similar particles can also be made from semiconductors such as CdS. and in that case the luminescence that takes place when electrons are excited across the band gap varies very markedly in

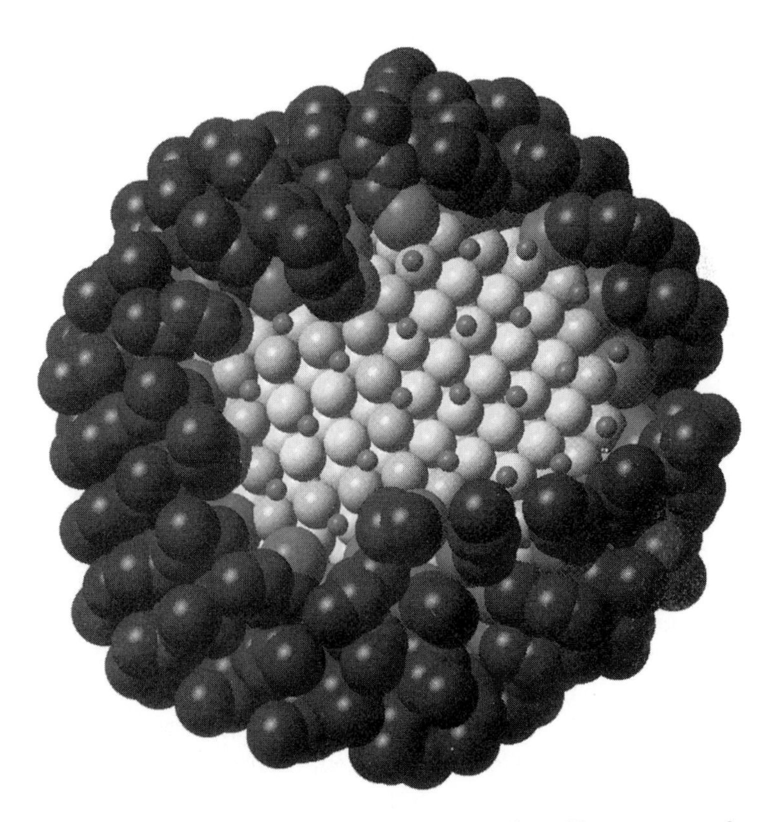

Fig. 10 A cluster of 586 gold atoms. Reproduced by courtesy of Professor D. Whetten, Georgia Institute of Technology.

frequency with the size of the particle. Particles about 20 Å in diameter emit blue light while those 40 Å emit red. Sizes in between emit every other colour in the visible spectrum. These are examples of what physicists now call 'quantum dots': electrons being confined in such small volumes that their energies are quantized.

Even more remarkable than the ability to manipulate the sizes of such tiny objects is the way in which it is now possible to move them around on a surface. Even atomic scale patterns have been made using the atomic force microscope: a famous early instance was writing the name of the company IBM in xenon atoms on a silicon surface. A circle of iron atoms on such a surface also acts as a 'corral' for electrons trapped inside it, whose wave-like energy states are reminiscent of the waves seen on the surface of the tea in a cup when it is shaken. Unfortunately, in many cases the atoms do not form very strong bonds with the surface, so they have to be placed and maintained at low temperatures. A significant recent exception, however, is the icosohedral carbon cluster molecule buckminsterfullerene (C_{60}): it turns out that they stick hard enough that they can be manipulated at room temperature, as, for example, into the S-shaped array illustrated in Fig. 11. From here it is just a short step back to the abacus mentioned earlier, and C_{60} molecules have actually been moved back and forth in a groove on a silicon surface to make a nano-scale model of this ancient counting device.

Mechanisms

If the answer to the question 'Can we make ordered atomic and molecular arrays with interesting electronic behaviour by chemical methods?' is a resounding 'yes', then what about the second question about the mechanisms that could be envisaged to store and process information? It was in this area that much of the early hype about molecular electronics gave the subject a bad name and led to justifiable scepticism. About 15 years ago an American theoretical physicist. Forrest Carter, began to speculate about the construction of gates and switches (necessary building blocks for information processing) out of molecular components. Many possible switching mechanisms were taken into account such as jumping of protons from one side to the other of unsymmetrical hydrogen bonds, electron transfer between donor and acceptor groups, and so on. Carter's ideas created a big stir and several conferences were held to discuss them. However, being a physicist (and a theorist at that) Carter had not really given much close thought to how such elaborate molecular struc-

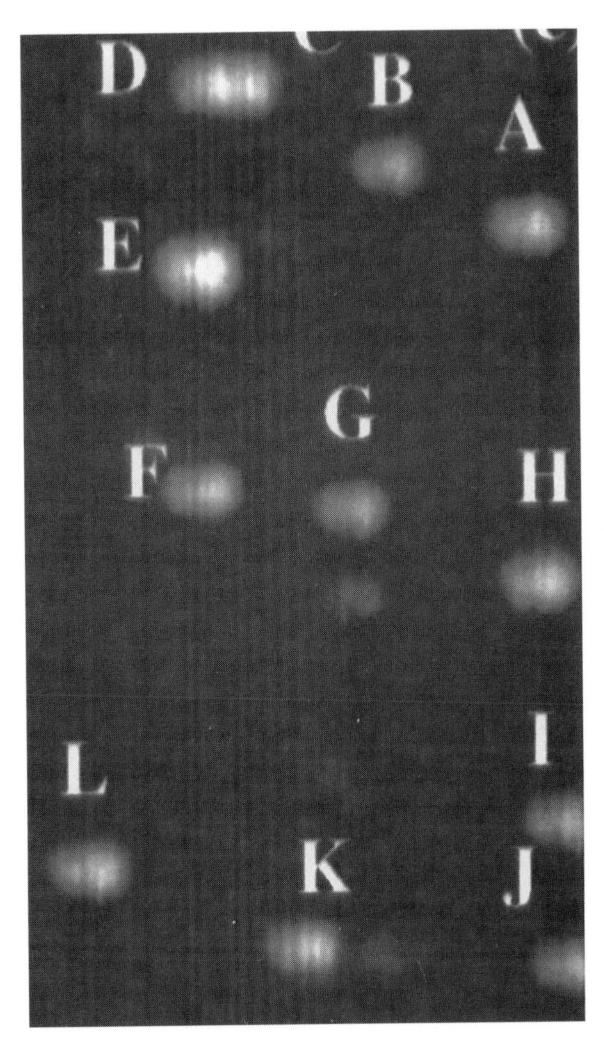

Fig. 11 Fullerene molecules placed in an array. Reproduced by permission from EPSRC Newsline, November 1996, p. 7.

tures as he conceived might be synthesized. His ideas therefore met with much scepticism, especially among chemists. An example of one of his proposals is shown in Fig. 12, which shows a hypothetical molecular wire with a gate attached to it.

Now in fact there is some practical basis behind the idea of a molecular wire, as such wires exist in nature in the form of conjugated polymers, of which the simplest prototype is polyacetylene, a chain of carbon atoms connected by alternating single and double bonds. If we

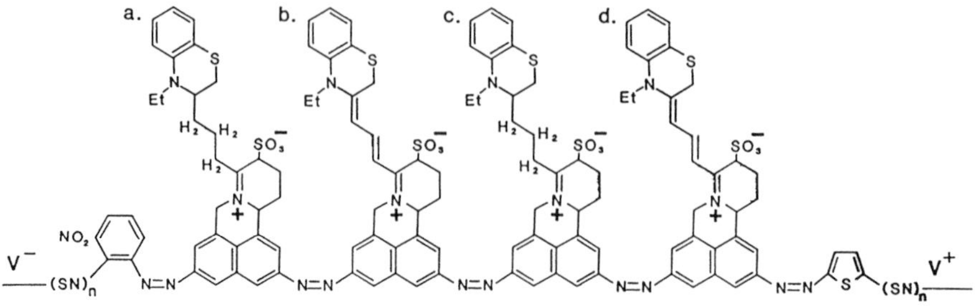

Fig. 12 A hypothetical molecular circuit gate (after Forrest Carter, 1984).

represent it in the *trans* configuration, it is clear that we can write two equivalent structures, in a fashion analogous to benzene:

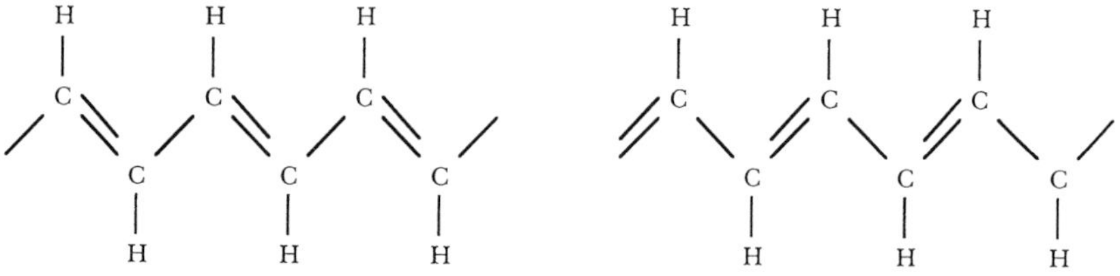

In an infinitely long chain the two would be indistinguishable but consider what happens if we reverse the conjugation in the middle of a chain:

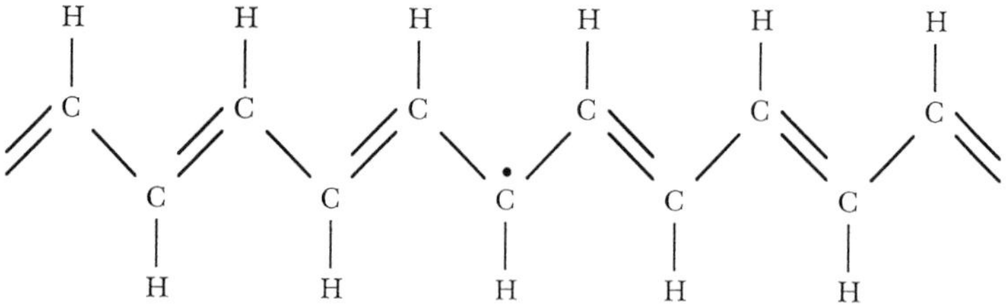

The dot signifies the presence of an unpaired electron separating the two parts of the chain where the alternation has different phases. By exchanging single and double bonds this singularity can move along the chain, taking the unpaired electron with it. It turns out the motion is analogous to that of a solitary wave, of which the best known example is the Severn Bore, the single tidal wave that sweeps up the River Severn every 24 hours. In the context of wave–particle duality, the solitary wave can be identified with a pseudo-particle called a soliton.

Apart from solitons, several other developments in molecular science in the last few years lend hope of purely molecular data processing events. The heart of a transistor is a p–n junction, a region of semiconductor separating positively and negatively doped material. Such junctions act as rectifiers in the sense that on applying a potential in one direction current can be made to flow, but not in the reverse direction. A hypothetical molecular switch was the subject of a conceptual patent by Aviram and Ratner as long ago as 1974, but recently British scientists Ashwell and Sambles have realized the effect by building molecules that combine electron donating and accepting components into Langmuir–Blodgett films. A second, and quite different, approach has been taken by Stoddart, who has synthesized what he calls 'molecular shuttles'. A cyclic molecule forms a loop around a linear one or another much bigger cyclic one, and the loop moves from unit to unit up and down the chain. A method called dynamic proton nuclear magnetic resonance shows that in an example such as that of Fig. 13 the 'train' moves on the 'Circle Line' at about 300 turns per second. Two trains can be introduced using high pressure synthesis, when it turns out that they rotate at the same speed, never colliding! Other possibilities could involve the motion of ions, which is what happens when nerve impulses travel, or the transport of chemicals, which in biology is accomplished by neurotransmitters. All of these are chemical options for transmitting impulses on a molecular dimension, but for all of them there remains a fundamental question, the third on my list.

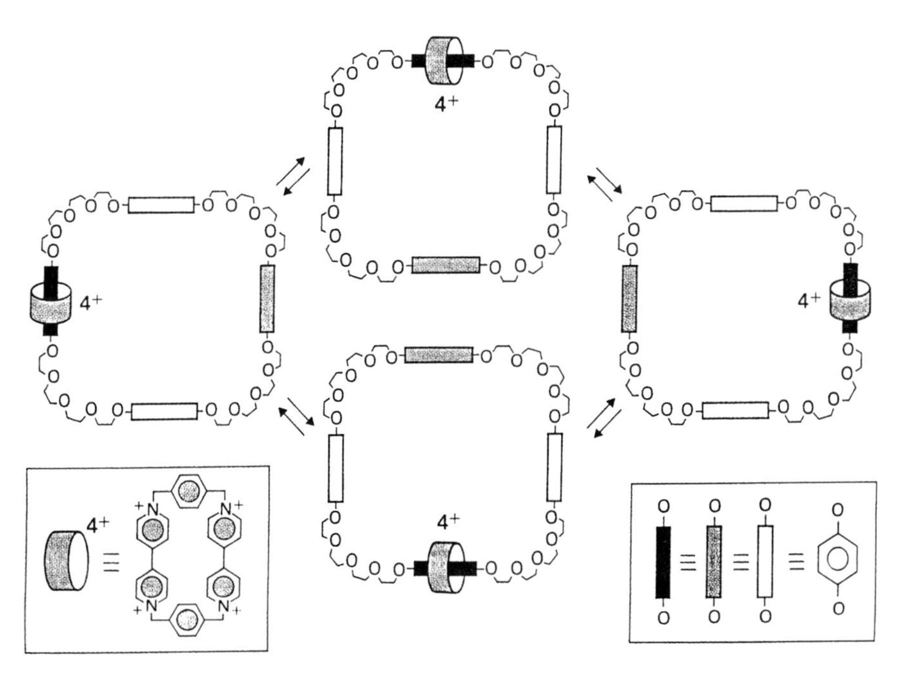

Fig. 13 A molecular shuttle. Reproduced by permission from *Chem. Brit.*

Addressing molecules

If the ultimate goal is to carry out data storage or processing at the truly molecular level, we have to face the question 'How do we get the signal in and out?' or, more starkly: 'Can we envisage ways of addressing the status of a single molecule?' At this juncture we have arrived at the point where analogies with conventional semiconductor electronics break down, despite the fact that (as we have already seen) it is already perfectly within the realm of possibility to manipulate molecules, or even single atoms by atomic force microscopy. It is also the point at which the conceptual models like Forrest Carter's fail, even if one could ever persuade a team of synthetic chemists to try and make one. The problem arises from a simple failure of imagination: Why should the architecture of a molecular information processor mimic that of one made from silicon?

Let us consider what tasks an elementary information processing unit has to perform, in an entirely abstract way. Such tasks are often set out in the form of a so-called 'truth table', summarizing the relation between the signal going in and the one coming out. Usually, the signal is assumed to be binary (i.e. either 0 or 1), but this is not essential. Two of the simplest of such gates, shown below, mimic the functions AND and OR:

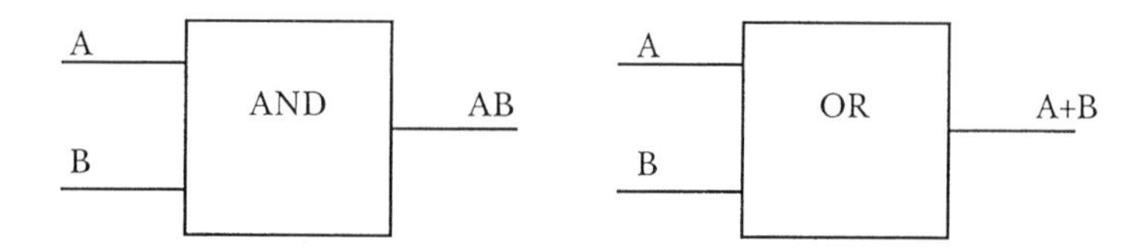

Now the rectangles are 'black boxes': what goes on inside them does not matter. From our point of view, are they single molecules (à la Forrest Carter) or something more complicated?

In conventional microelectronics the signal, consisting of a bunch of electrons, passes around the circuit from one element to another. Another way of expressing this is to say that the processing is serial. Now consider a set of individual elements that interact, not in a one-dimensional (serial) way but in two or three dimensions. In that case the state of any one element is determined by the number of its neighbours in the same state. Thus is born the notion of cooperativity, leading to bistable states when the whole set undergoes a transition from one state to another. Such phenomena have been used to store data on a much larger length scale for many years, for example in magnetic memory devices. The ultimate requirement is for the individual elements to be capable of existing in two states (charge, spin, vibrational: it doesn't matter), with a mechanism for each element to be sensitive to the state of its neighbours. One practical example of this kind, where the elements are transition metal ions bridged by organic molecules, already exists in the form of lattices in which metal ions change their spin states. Indeed, it is entering service with France Telecom as a means of reading phone cards.

Even in the realm of conventional microelectronics, structures are coming to resemble more closely the requirements of a molecular array. As the 'feature size' (jargon for the individual junctions and processing elements) becomes smaller and smaller, one arrives, willy nilly, at a situation where they all interact with one another, so that it is no longer possible to define a unique pathway carrying the electrons across the surface of the chip. Furthermore, to optimize the manufacturing yield and minimize the number of lithographic errors, designers have to make the units simpler and their arrangement more regular. Such a regular connected array, each node of which can exist in either of two states (on, off, or 0, 1) is an example of what has been called a 'two-state cellular automaton' (Fig. 14). The point here is that the entire lattice behaves as a single system. To store or process information you do not have to interrogate each element to find out its state: the information is stored in the form of a pattern. This idea forms the basis of

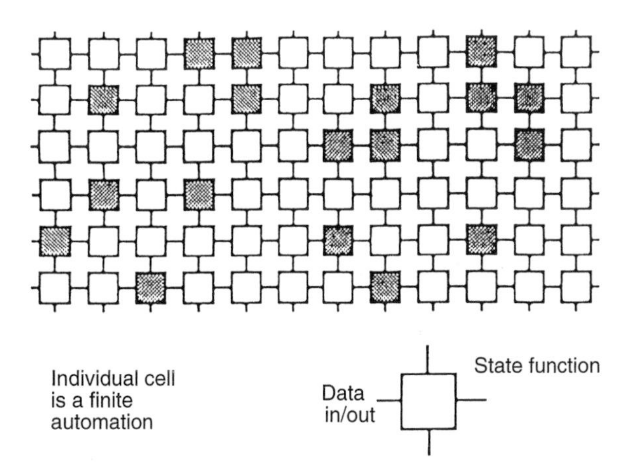

Fig. 14 A lattice behaving as a two-state automaton. Reproduced by permission from *Introduction to molecular electronics*, M. Petty, M.R. Bryce, and D. Bloor, Edward Arnold, London, 1995, p. 357.

Conway's famous *Game of life*, a program available for most personal computers.

Having set up the lattice on the screen one defines an algorithm which switches the state of a given element when a given number of neighbouring elements are in the same state. Starting with a group of elements in the 'on' state (symbolizing for us the injection of a signal, for instance at the edge of the array) the pattern evolves from generation to generation, perhaps exiting on the other side of the array. Figure 15 shows an example. This idea is not new; it has been investigated theoretically by Hopfield and Wolfram in the USA, and in the UK by Barker. What has it got to do with molecular-based information processing? Well, the smallest kind of regular array of objects is a crystalline one formed from molecules, each of which might have a different charge, spin, vibration, conformation, protonation, or any other kind of state we might dream up. The beauty of molecular arrays

Fig. 15 Evolution of patterns in a two-state automation. Reproduced by permission from *Introduction to molecular electronics*, M. Petty, M.R. Bryce, and D. Bloor, Edward Arnold, London, 1995, p. 357.

is that, as we have seen, for example in Fig. 8, given a particular shape and charge distribution among the constituents, they assemble themselves, in contrast to the need for lithographic writing to create patterns on silicon. But now we are straying towards science fiction, and it is the moment to sum up.

Facts for the future

As a platform for thinking, and perhaps also action, towards an artificial information storage and processing regime based on molecular ingredients, let us consider a number of clear facts. First, since the dawn of computation, the individual information storage and processing entities found in computing machines have been inexorably getting smaller and smaller. Second, chemistry is good at making precisely defined and structured arrays of molecules, capable of existing in different electronic, magnetic or whatever kinds of distinguishable states. Third, methods do exist whereby single molecules can be addressed, although fourth, and more subtly, perhaps it is not necessary to do so. Does all that make it likely that we shall see molecular-based information processing in practise one day? Finally, of course, we do not know but, for myself, if I could be here in 50 years time to collect, I would bet on it. After all, our brains make a pretty good job of it already!

Acknowledgements

For lending materials and examples to illustrate this presentation, my warmest thanks go to Professors Richard Catlow and Olivier Kahn.

Bibliography

The following are some general references to molecular electronics and information processing:

M.C. Petty, M.R. Bryce, and D. Bloor (ed.), *Introduction to molecular electronics*, London, Edward Arnold, 1995.

P. Day, Future molecular electronics: towards a supramolecular information processor, *Chemistry in Britain*, 1990, **26**, 52–4.

P. Day, Room at the bottom, *Chemistry in Britain*, 1996, **32**, 29–31.

R.M. Metzger, P. Day, and G. Papavassilliou (ed.), *Low dimensional structures and molecular devices*, New York, Plenum Press, 1990.

PETER DAY

Born 1938 in Kent and educated at the local village primary school and nearby grammar school at Maidstone. An undergraduate at Wadham College, Oxford, of which he is now an Honorary Fellow. His doctoral research, carried out in Oxford and Geneva, initiated the modern day study of inorganic mixed valency compounds. From 1965 to 1988 he was successively Departmental Demonstrator, University Lecturer and An Hominem Professor of Solid State Chemistry at Oxford, and a Fellow of St John's College (Honorary Fellow 1996). Elected Fellow of the Royal Society in 1986; in 1988 he became Assistant Director and in 1989 Director of the Institut Laue-Langevin, the European high flux neutron scattering centre in Grenoble. In October 1991, he was appointed Director of The Royal Institution and Resident Professor of Chemistry, and Director of the Davy Faraday Research Laboratory, and in September 1994, he became Fullerian Professor of Chemistry. His present research centres on the synthesis and characterization of (mainly molecular) inorganic and metal-organic solids in the search for unusual magnetic and electron transport (including superconducting) properties.

'There or thereabouts'

ANDREW WALLARD

One of the books which fascinated many youngsters in the 1950s was Lancelot Hogben's *Man must measure*. In it, Hogben ranged freely and his message was that the ability to measure precisely was an important aspect of society and that it helped social, cultural, and industrial progress. He pointed out how important measurement was to trade and how it underpinned or helped develop many of the major inventions and innovations on which we have now come to depend, or has enabled companies to improve their manufacturing processes so as to make better products. One my themes will be the impact measurement technology has on almost every aspect of our lives.

Measurements' impact on daily life

As far as daily life is concerned, most people have rarely, if ever, given measurement more than a passing thought. Why should they? One of the reasons is because, one way or another, rather a lot of care and rather large sums of money are spent to ensure that daily measurements are reliable, routine, and nothing goes wrong. But by the same token a lot of things on which we rely implicitly have immense potential for 'going wrong' if the system doesn't work property.

For example:

1. When paying an electricity bill or filling up a motor car consumers instinctively rely on the meter reading or the pump gauges for the price they pay. Petrol pumps are accurate to about 0.3% which translate to 50p or so a year for the average motorist; just a bit less, incidentally, than each man, woman, and child pays to run the National Physical Laboratory (NPL).
2. One-quarter of a million people have cancer therapy each year. The dose needs to be measured to about ± 5% to make sure the cancer

cells and not the live tissue is damaged. Precise measurement is important and even the best standards laboratories are actually hard pressed to calibrate machines to better than 1.5%.

3. Aircraft altimeters need regular calibration by the world's airlines to the same specification.
4. Better quality and precision engineering have ensured that motor cars now have much longer service intervals.

Measurement, then, is all around us and we automatically rely on it as one of the 'givens' of today's society. To begin with, let us look back at the evolution of measurement before we set it into today's context.

Early measurement: use of the body

Nobody really knows much about the sort of early measurement which featured in the first few pages of Hogben though the two most important factors which made it important were construction and trade. The *Bible* is littered with references to various measures of length and weight (Noah's ark was 300 cubits long, 50 wide, and 30 high—*Genesis*) and we certainly know of sophisticated measurements which go back to between 3000 and 4000 BC. The Egyptian Pyramids were built about 2500 BC and required length measurements to about one-tenth of a 0.1%. Most early length measurements were based, of course, on the body with the cubit being the distance between fingertips and elbow. This use of the body to measure things is still reflected in some of today's language, for example, the hand, the pace, the mile (*mille passum* or 1000 paces) were all dimensional measures.

Weighing was important because of its link with buying, selling, and trading. Rather like parts of the body being used for length measures the ancient civilizations turned to other 'artefacts of nature' for other quantities. In the case of weight, the most common 'standard' was a seed. The carob seed was common in the Mediterranean and is now what we call the carat used to weigh gold and silver. Other popular, and in fact quite reproducible seeds, were grains of wheat or barley. (They were also used for length—three seeds laid end to end was 1 inch.) The most widely used heavier weights were based on the shekel—256 barley or wheat ears. On a practical scale there was a link with coinage and a pennyweight in the year 1200 was based on 32 grains of wheat, although 250 years later the pennyweight was only 12 grains because the amount of silver in it had been reduced. Other systems built on this standard, and 2560 pennyweights were the weight of 1 gallon of wine—so in actual fact barley corns could act as the basis of measures for length, weight, and volume.

The hallmark, though, of early measurement was that there was no single system and even great Empires like the Greeks and Romans failed to establish consistency. Of the two, the Greek civilization (700 BC to AD 400) had the greater scientific culture and although the Athenian or Attic standards dominated domestically, the ancient Greeks, because they were major traders, compared their own standards carefully with others and also kept copies of the alternative Persian and Phoenician systems in Athens. The Romans, on the other hand, (300 BC to AD 400), although they were superb engineers, had virtually no scientific traditions so they borrowed most of their weights and measures infrastructure from the Greeks. In the case of the countries they conquered they merely took the nearest local equivalent to their own standard measure, and called it by the name of their own unit. This led to immense confusion and there was, for example, no single 'Roman pound' in the sense that there *was* an 'Athenian' one, and no attempt to reflect the Greek practice of denoting their official standards with some sort of embossed seal.

Some semblance of order

Closer to home in Britain, there was a similarly rather chaotic situation until, somewhere about the time of the Norman Conquest, William I tried to establish some consistency on to a mix of regional, Saxon and Roman-based systems and ordered that the existing weights and measures had to be 'duly certified' or compared with each other. Even then, all was a bit arbitrary and it was about this time that William of Malmsbury recorded Henry I's declaration that the yard was the distance between the Royal nose and finger. We had to wait until the *Magna Carta* (1215) for the first government policy statement on metrology ('one measure to be established throughout the land') and real attempts at some degree of consistency. Like most government statements it was applied patchily but unique 'national' standards began to be kept and were authorized by the Crown. A hundred years later in 1340, Edward III issued a Statute to authorize the 'Treasurer' (or the then Chancellor of the Exchequer) to make one set of bronze standards for the gallon, bushel, and of certain weights and to make copies for regional use. Most Monarchs then re-issued or confirmed these 'national' standards usually by stamping them. Several examples remain, notably the 1497 Winchester Yard of Henry VII, made of bronze and later endorsed by Elizabeth I.

Better standard measures

Most measurements only needed to be done to a per cent or so and bronze and iron yards, and weights were good enough for most practical purposes. But the mid-eighteenth century brought scientific and technical advances which began to put standards on to a more sound footing.

Fire destroyed Parliamentary copies

Different metals also began to be used because the length of the popular bronze or brass bars were too susceptible to temperature. In fact, the last formal bronze standard was the 1824 'Primary Standard Yard' kept in the House of Commons. This yard was destroyed in 1834 together with the pound when the Palace of Westminster was burned down.

Imperial standards

So, as a result of the fire, and in one of Parliament's brief flashes of interest in metrology, it set up a committee to derive newer, better standards and to ensure continuity with the old length and mass standards. Three Fellows of the Royal Society (Reverend R. Sheepshanks, Mr F. Baily, and Professor W. Miller) worked until 1854 to produce scientifically respectable national standards.

1. The *Imperial Standard Yard*, a solid square gunmetal bar made from what came to be called 'Mr Baily's metal' (16 proportions of copper, 2.5 of tin, and one of zinc) and which was stiff and less sensitive to temperature than bronze (eighty-five copper, fifteen tin) or brass (67 copper, 33 nickel). The critical dimension was between fine lines in two gold plugs, when the temperature was 62°F and the atmosphere pressure 30 inches of mercury;
2. The *Avoirdupois Pound* of platinum.
3. The *Imperial Gallon* which was 10 Avoirdupois pounds of pure water at 62°F and 30 inches of mercury.

The job of looking after these standards was given to the Board of Trade but prudence—the hallmark of all good committees—came into play and several copies were made and deposited with Parliament, the Royal Greenwich Observatory, the Royal Mint, and the Royal Society. To some extent, these remain our 'national' measures and copies are in the wall of Trafalgar Square.

Evolution of the metric system

To understand something of the evolution of our current system of scientifically based standards, we have to look to our Gallic neighbours. As far back as 1742, French and British scientists compare the *pied* and *livre* of Paris with the British foot and pound. Not surprisingly, given the way national systems had evolved, they differed by about 7%—far in excess of the uncertainties each was thought to have. The scientific challenge was to find a reference, preferably drawn from nature so that it was not bound to any particular country—a challenge, in fact, we still work towards today.

Length

The first standard to be tackled was length. There were two choices: one a pendulum which had a period of 2 seconds; the other a standard of length based on the length of the meridian which joined the earth's North and South Poles. The latter won the day because scientists already knew enough about pendulums to realize that the pendulum period depended on the acceleration due to gravity and so would change over the world. The French therefore pressed ahead with a decree which established the metre as one ten-millionth of the length of the portion of the earth's meridian from Dunkirk to Barcelona. Other 'metric' units followed—the mass unit, for example, was the mass of a cubic decimeter of water at 4°C.

Having stated, or defined, the standard metre, the task of actually measuring it also fell to the French. This they did by precise and careful triangulation (the technique still used by surveyors), and produced a platinum standard metre as well as a kilogram in 1799.

International standards/BIPM

These matters rested until the Paris World Exhibition in 1867 when a group of scientists met to assess the state of metrology. As a result, an improved metre with an x-shaped section (made famous in school textbooks) and a new kilogram were made from the same block of platinum–iridium metal. These formed the basis of the world-wide 'metric' system established under the *Convention du Metrè* by 20 countries in 1875. The 'international' standards were entrusted to the care of the *Bureau*

International des Poids et Mesures—or BIPM—in Paris where they reside to this day. The UK signed the *Convention du Metrè* in 1884. This is the foundation of today's international measurement system based on the 7 units of the metric system, themselves used to build up all other 'derived' units like pressure, force, electrical resistance, density, etc.

Twentieth century standards—practical, portable measurements available to all

The twentieth century saw rapid advances in the world's capability to measure. It also enabled metrologists to move away from unique 'physical artefact' standards to transportable devices, which can give access to highly accurate measurements in the scientific laboratory or in industry. As an example of this, I will concentrate on five of the seven 'base' or fundamental units.

Twentieth-century metrology

The *Convention du Metrè* signified the acceptance by most of the industrialized countries that they had to have consistent weights and measures. Most, if not all, traced their national standards to those at the BIPM and periodically took them there for re-calibration and checking.

New industry

During the early part of the century two important things happened. First, there was a considerable rise in industrial activity, and a rush of new products (transportation, aviation, communication, domestic equipment), which placed increasing demands on measurement standards as precision engineering improved and people needed new standards for a whole variety of things (radioactivity, lighting). The National Standards laboratories had to find new ways of improving the accuracy or stability of top level standards so as to satisfy the needs of the new users. In fact, this century has seen a regular and systematic improvement in the performance of national standards—so much so that we have had to improve them twofold about every 10 years. That is why we have, as some say, to keep on adding more decimal points!

Length standards—why was the metre bar not good enough?

The second thing was that quantum physics gave us the potential to replace material standards by ones based on the fundamental laws of physics. The things which limited the performance of standards like the metre were the fact that they expanded if the temperature rose, the metal itself changed length as a result of metallurgical changes, and the precision of the optical microscope techniques used to compare bars limited the precision with which intercomparisons between primary and secondary standard bars could be carried out. In addition it was extremely hard to define the 'shape' of the finely engraved line which marked the reference point.

Quantum effects—the Bohr model

To understand why quantum effects are important we need to spend a few moments looking at simple atomic theory. In an atom electrons spin around a nucleus. When the atom gains energy in some way—for example, it is accelerated, heated, or put in an electric discharge—then the electrons jump to wider orbits and to higher energy levels. They return to their natural state by losing the energy, usually in the form of light. That is what happens in a lamp or a fluorescent light tube. What quantum physics did was to teach us that the energy levels to which the electrons could jump were fixed for a particular atom and so the differences between them were precise, were 'set' by the natural laws of physics and were the same everywhere in the world. Mathematically, we describe this by the $E=hf$ relationship where E is the energy difference between the electron energy levels, h is a constant called Planck's constant and f the reciprocal of the wavelength or colour of the emitted light. The actual wavelength is about half a millionth of a metre and so we have, in principle, a very small and 'natural' standard or ruler.

Light sources

These discoveries encouraged considerable work into the development of lamps as light sources and wavelength standards. In most lamps the spectrum of the light emitted is a series of discrete 'lines' rather than the continuous spectrum which is what makes 'white' light. The scientific problem which had to be tackled was the breadth of the 'spectral lines' from early lamps. The lines were broad because the atoms in the lamp were moving in all directions—the so-called 'Doppler' effect. Much scientific effort was therefore directed towards the study of the stability of the lines from certain lamps to find out what determined their width and to try and find ways of making the lines (and hence the wavelength, or the divisions of the 'ruler') as narrow as possible.

Interferometers

It was worth persevering with these painstaking researches because the optical technique called 'interferometry' made it possible to carry out practical, precise length and distance measurements using the new narrower light sources. The principle of interferometry is simple if we think of light as a wave. The energy associated with light varies like a wave as it moves through space. If we arrange things so that two similar light waves overlap each other then the 'wave nature'—the ups and downs—are reinforced. If we shift the two waves by exactly half of the regular wave pattern (half a wavelength) then they cancel each other out. To create this effect the light waves are only shifted by about one-third of a millionth of a metre. A combination of narrow light sources and interferometry therefore offered the opportunity to measure 'standard' metre bars in terms of this much more stable 'quantum' reference. Eventually, it was possible metaphorically to throw away the 1889 metre and redefine it in 1960 as '1650 763.73 wavelengths in vacuum of the radiation corresponding to the transition between the levels $2p_{10}$ and $5d_5$ of the *kr 86* atom' (Fig. 1).

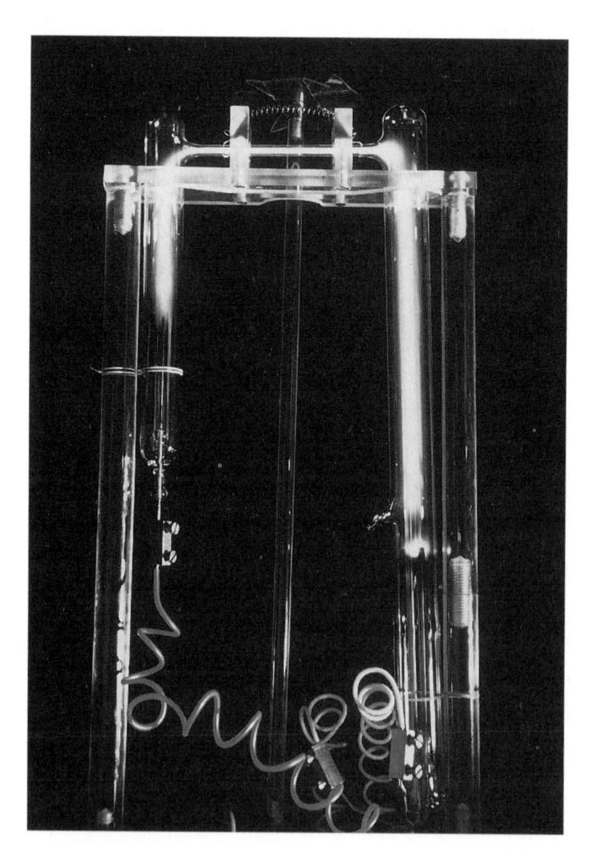

Fig. 1 The Krypton 86 lamp used to define the standard of length between 1960 and 1983.

Krypton lamps

There was a price to pay for this greater accuracy and the 'universality' of the standard because the equipment became more and more complex. Although it was much more accurate and stable or reproducible than the metre bar, the krypton lamp was much more tricky to deal with and had to be operated under carefully controlled conditions. Fortunately, very soon after the international definition of the metre was changed the laser was invented and like so many advanced developments, was put to immediate use in precision metrology.

Iodine lasers—the current length standard

The advantages of light from a laser were that it could be, by its very nature, much narrower than the krypton line and that it was much brighter. Today, laser standards have superseded the krypton lamp.

The wavelength of light from a laser can be tuned by making changes to the distance between the mirrors. The challenge was to find some stable reference point within the tuning range and keep the laser tuned to it. The solution lay with the discovery that iodine vapour absorbed light from the helium–neon laser. Some careful spectroscopy and calculations showed that the centres of 10 iodine absorption lay within the turning range. However, just like the krypton lamp the iodine molecules were all moving round so even these absorption lines had a 'width'. The trick was to put a tube of iodine inside the laser, which meant that the laser light could be made to have a special interaction with the iodine atoms (saturated absorption) and could pick off only those molecules that moved at right angles to the laser beam. As a result the system could detect super-narrow lines. By keeping the laser adjusted so that its wavelength was at the centre of one of these absorption lines it was possible to create today's length standard reference—about a thousand times narrower than the krypton line (Fig. 2).

Electric standards—the Josephson volt

Until the late 1980s, the most reliable and stable voltage standards were Weston standard cells—small chemical batteries which, treated well, could be remarkably stable. But the search for better electrical standards was similar to that for better length ones. Metrologists wanted standards which did not vary with time or with the material from which they are made. Today's voltage standard achieves these objectives and is based on the Josephson effects in superconductivity. Devices, rather like a sandwich of two superconducting metals separated by a thin barrier of insulating material, have very interesting properties when kept at the

Fig. 2 The iodine stabilized laser used to realize today's definition of the metre.

temperature of liquid helium: –279°C. Although there is a physical barrier, two superconducting electrons pair up and can tunnel through between the two metals without developing a voltage between them. If a voltage is applied to the Josephson junction, the electron pairs tunnel to a state of different energy. Energy has, however, to be conserved and the electron pair has to lose some energy in the form of microwaves. The Josephson voltage standard reverses this process and shines microwave energy on to the junction. What actually happens is that the two effects mix with each other and result in very accurate quantized voltages which depend only on the applied frequency and various fundamental physical constants—e, the charge on the electron and h, Planck's constant. Because frequency is a very accurate quantity, we can now have an almost equally accurate voltage.

The voltage produced by one junction is extremely small, only about one ten-thousandth of a volt, so is itself hard to measure and, in practice, between 3000 and 20 000 individual junctions are put together so as to build up voltages of up to 10 V. The arrays are now made commercially and used in several industrial laboratories where high accuracy is needed, as well as in the majority of National Standards laboratories.

Time

One of the world's great clocks was Harrison's chronometer. In the mid-eighteenth century the government offered a prize for a marine clock. The target they set was for a loss of less than 2 minutes over a 6-week period. Harrison's clock lost only 15 seconds over 5 months. Today's time standard is, however, a caesium atomic clock. It works rather like

Fig. 3 Atomic clocks used at NPL to set the UK's time scale.

the laser length standard but we actually use microwaves rather than light to make the caesium atom jump between two of its energy levels. These clocks are the most accurate of all today's measurement standards—an uncertainty of the equivalent of only 1 second in 300 000 years. The beauty of this standard is that its easily made available to anybody with a radio receiver through what is called the 'Greenwich time signal' but which actually comes from NPL's clocks. (Fig. 3)

Temperature standards—the kelvin

Temperature is another of the 'base units'. Until 25 or so years ago most scientists thought 100°C was the boiling point of water and 0°C the freezing point. But as in other cases, better experiments in the science of thermodynamics started to challenge these 'practical' measurements and as a result the boiling point of steam turned out to be only 99.97°C! The international temperature scale had to be redefined because physicists are happiest with a sound thermodynamics scale even though to do 'real' measurements there are set fixed points.

The most common fixed point is known as a triple point. At a triple point a solid, liquid and a gas can coexist—a unique thermodynamic situation, which can be described by theories and made consistent with all the other measurements and standards. When a carefully prepared pure reference mixture is cooled the temperature drops with time until it reaches a stable 'plateau' when all the three phases exist. The international standard for the kelvin is defined using a triple point cell of highly pure water in an evacuated phial and is 1/273.16 of the thermodynamic temperature of pure water at its triple point. (Fig. 4)

For higher or lower temperatures, the international temperature scale uses several other triple and freezing or meeting points of the various

Fig. 4 Cell used to define the Kelvin at the triple point of water.

metals or mixtures used to set fixed points for very low to very high temperatures.

Mass standards—the kilogram

Next—and definitely *not* an atomic standard—is the kilogram. Quite remarkably, it is still based on its nineteenth century predecessor—a small cylinder of platinum iridium. There is a curious mystique about the kilogram. *The* kilogram is in Paris at the BIPM and has been since 1889. Each country has a copy which has been weighed carefully against it (Fig. 5). All are checked—three times in the last 120 years—to see how they drift around. Usually kilograms gain weight by absorbing gases or pollutants from the air—they are kept in air rather than in a vacuum because once a polluting layer has built up on the metal surface, it remains extremely stable.

Metrologists can compare kilograms to one-thousandth of a microgram—good enough for industrial and commercial measurement but a tiresome and painstaking measurement. There are, therefore, several attempts to relate the kilogram to the more accurate electrical standards, which look like promising replacements for the kilogram within the next 10 years.

Summary

We have concentrated—for practical reasons—on five of the seven international standards. The most accurate is the atomic clock closely

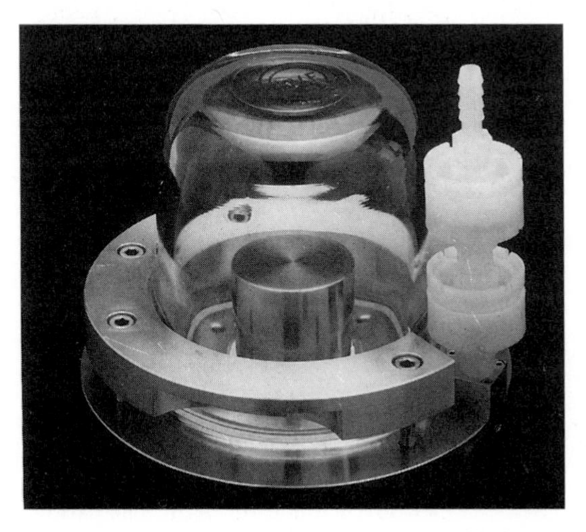

Fig. 5 The UK's copy of the international kilogram.

followed by the electrical and laser length standards. In nearly all cases industrial and scientific demands for improvements are increasing and so in some areas metrologists are already on the look out for new or improved ways of measuring.

Applications

I began by claiming that accurate measurement was at the heart of society and its industrial and commercial life. A few final examples can help illustrate just a few of today's measurement challenges.

Force measurement relates directly to the kilogram and metre (weight per unit area) and is one of the areas where industry is pressing national standards laboratories most vigorously for improved accuracy and range of measurement. Figure 6 shows NPL's 1.2 MN force machine (120 tonnes). The weights used to produce the forces are directly traceable to the kilogram and each is known to 3 p.p.m. In building it they had to be

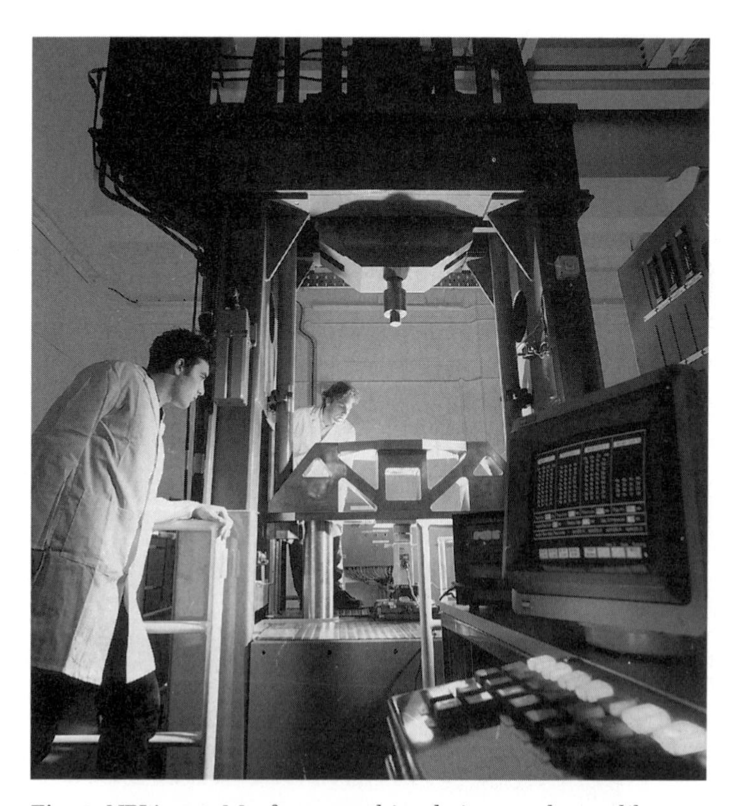

Fig. 6 NPL's 1.2 Mn force machine being used to calibrate a load cell.

aligned to 1 mm and 'little g' (the acceleration due to gravity) measured over its 30 m height. The accuracy is about 10 p.p.m. and is needed:

(1) for measuring jet engine thrust—big engines have about 30 tonne thrusts, which have to be measured to half a tenth of a per cent so as to conform with customers' specifications;

(2) for weighing oil rigs, such as the Conoco plant in the North Sea, which has to be weighed so that the correct forces can be applied to keep it tethered to the sea bed; and

(3) to calibrate thrust transducers incorporated in 'Thrust 2' the car currently attempting to set a new world land speed record.

Length measurement is increasingly being applied to enable others to make precise measurements in industry and science. Figure 7 shows recent results from a Cambridge group which employed up to date interferometric technologies to make for the first time milli-arc second resolution measurements of the rotating double star, Capella.

Medical applications now demand carefully controlled ball/socket clearance of 10 μm in artificial hip joints.

In *time* measurement satellite-based clocks can be used to locate a position on the earth to about 3 m through the 'global positioning system' (GPS), which is also used to help control the Conoco oil rig

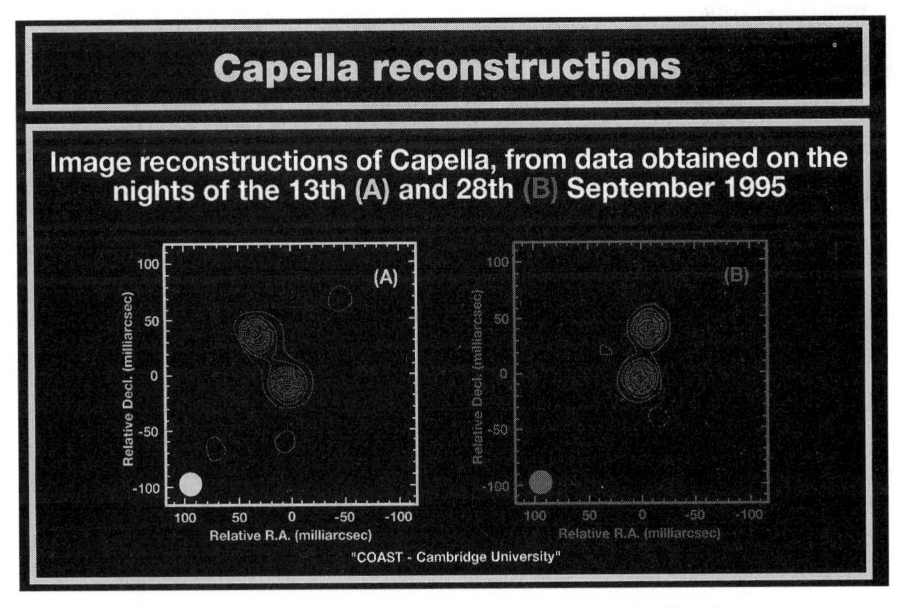

Fig. 7 Image reconstruction of the double star 'Capella' showing the rotation of the two stars with milli-arc second resolution.

discussed in the second point on force measurement. Advances in technology mean that it is now possible to buy hand-held GPS receivers which show latitude, longitude, and height.

Advances in satellite-based time measurement and transfer will mean that the twenty-first century aeroplanes will land and navigate by improved GPSs.

Conclusions

Measurement is, indeed, all around us and is subject to ever-increasing demands for improved accuracy. This trend is likely to continue and we expect to be challenged by new applications of measurement in a variety of different fields.

Readers of this record may like to know that the lecturer, knowing the importance of precise timing at the Friday Evening Discourses, presented the Royal Institution with a radio-controlled clock so that NPL's 'atomic' time signal could be used to calibrate the lecture theatre's clocks.

ANDREW WALLARD

Born 1945 and educated at the University of St Andrews. He joined the National Physical Laboratory in 1968 to work on interferometry and the use of frequency stabilized lasers as standards of length. 1978–1990 he was employed in a number of Whitehall positions including the Central Policy Review Staff and policy support groups for Ministers at the Department of Trade and Industry (DTI). He was responsible for several DTI research programmes in electronics and optoelectronics and for the European Union ESPRIT programme which complemented the UK's 'Alvey' IT programme. In 1990 he returned to NPL as Deputy Director, a post which he continues to hold following the privatization of NPL in 1995.

Pondering on Pisa

JOHN BURLAND

Introduction

In 1989 the civic tower of Pavia collapsed without warning, killing four people. The Italian Minister of Public Buildings and Works appointed a commission to advise on the stability of the Pisa tower. The commission recommended closure of the tower to the general public and this was instituted at the beginning of 1990. There was an immediate outcry by the Mayor and citizens of Pisa who foresaw the damage that the closure would inflict on the economy of Pisa, heavily dependent on tourism as it is. In March 1990 the Prime Minister of Italy set up a new commission, under the chairmanship of Professor Michele Jamiolkowski, to develop and implement measures for stabilizing the tower. It is the fifteenth commission this century and its membership covers a number of disciplines including structural and geotechnical engineering, architecture, architectural history, archaeology, and restoration.

It is instructive to imagine a tower, founded on jelly and slowly inclining to the point at which it is about to fall over. Any support would also have to rest on the jelly. Worse still, the masonry composing the tower is so fragile that it could explode at any time. This is a reasonable description of the state of the Leaning Tower of Pisa, and helps to explain why stabilizing it represents the ultimate civil engineering challenge.

Details of the tower and ground profile

Figure 1 shows a cross-section through the tower. It is nearly 60 m high and the foundations are 19.6 m in diameter. The weight of the tower is 14 500 t. At present the foundations are inclined due south at about 5.5° to the horizontal. The average inclination of the axis of the tower is somewhat less due to its slight curvature as will be discussed later. The seventh cornice overhangs the first cornice by about 4.5 m.

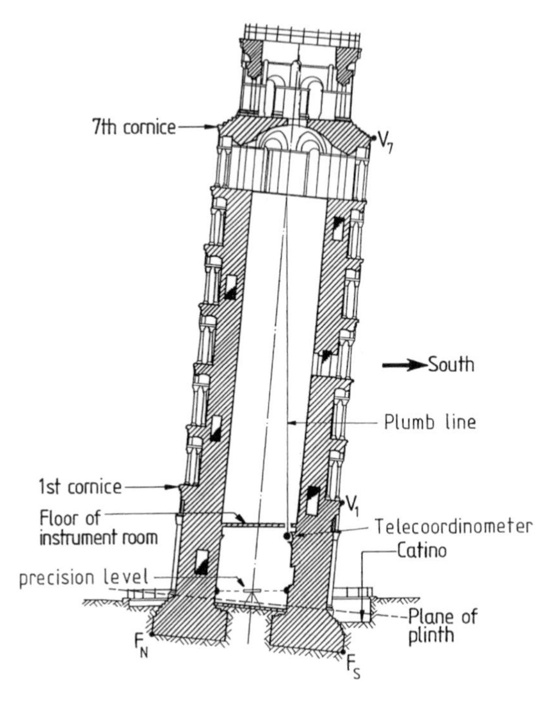

Fig. 1 Cross-section through tower.

Construction is in the form of a hollow cylinder. The inner and outer surfaces are faced with marble and the annulus between these facings is filled with rubble and mortar within which extensive voids have been found. The spiral staircase winds up within the annulus.

Figure 2 shows the ground profile underlying the tower. It consists of three distinct layers. Layer A is about 10 m thick and primarily consists of estuarine deposits laid down under tidal conditions. As a consequence the soil types consist of rather variable sandy and clayey silts. At the bottom of layer A is a 2 m thick, medium dense, fine sand layer (the upper sand). Based on sample descriptions and cone tests, the material to the south of the tower appears to be more clayey than to the north and the sand layer is locally much thinner. Therefore, to the south layer A could be expected to be slightly more compressible than to the north.

Layer B consists of soft, sensitive, normally consolidated marine clay, which extends to a depth of about 40 m. The upper clay, known as the Pancone Clay, is very sensitive to disturbance which causes it to lose strength. The lower clay is separated from the Pancone Clay by a sand layer (the intermediate sand) overlain by a layer of stiffer clay (the intermediate clay). The Pancone Clay is laterally very uniform in the vicinity

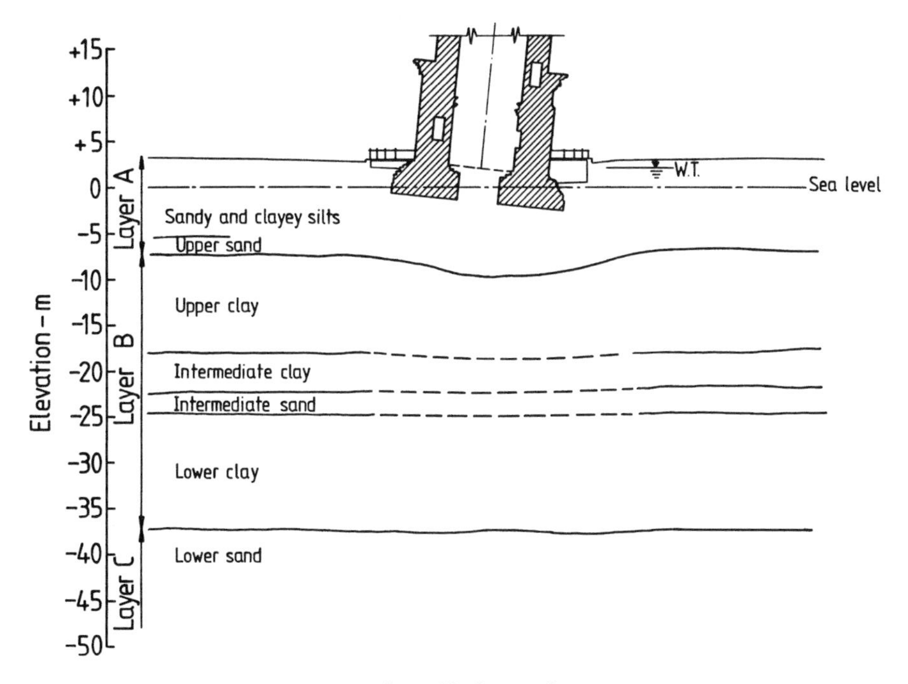

Fig. 2 Soil profile beneath tower.

of the tower. Layer C is a dense sand which extends to considerable depth (the lower sand).

The water table in horizon A is between 1 and 2 m below ground surface. Pumping from the lower sand has resulted in downward seepage from layer A with a vertical pore pressure distribution through layer B, which is slightly below hydrostatic.

The many borings beneath and around the tower show that the surface of the Pancone Clay is dished beneath the tower from which it can be deduced that the average settlement is between 2.5 and 3 m.

History of construction

The tower is a campanile for the cathedral, construction of which began in the latter half of the eleventh century. Work on the tower began on 9 August 1173 (by the modern calendar). By about 1178 construction had progressed to about one-quarter of the way up the fourth storey when work stopped. The reason for the stoppage is not known but had it continued much further the foundations would have experienced a bearing capacity failure within the Pancone Clay. The work recommenced in about 1272, after a pause of nearly 100 years, by which time the strength of the clay had increased due to consolidation under the weight of the tower. By about 1278 construction had reached the seventh cornice when work again stopped—possibly due to military action. Once again there can be no doubt that, had work continued, the tower would have fallen over. In about 1360 work on the bell chamber was commenced and was completed in about 1370—nearly 200 years after commencement of the work.

It is known that the tower must have been tilting to the south when work on the bell chamber was commenced as it is noticeably more vertical than the remainder of the tower. Indeed, on the north side there are four steps from the seventh cornice up to the floor of the bell chamber while on the south side there are six steps. Another important historical detail is that in 1838 the architect Alessandro Della Gherardesca excavated a walk-way around the foundations. This is known as the catino and its purpose was to expose the column plinths and foundation steps for all to see as was originally intended. This activity resulted in an inrush of water on the south side, as here the excavation is below the water table, and there is evidence to suggest that the inclination of the tower increased by as much as half a degree. As is described later, Gherardesca left us an unpleasant surprise.

History of tilting

One of the first actions of the commission was to undertake the development of a computer model of the tower and the underlying ground that could be used to assess the effectiveness of various possible remedial measures. Calibration of such a model is essential and the only means of doing this is to attempt to simulate the history of tilting of the tower during and subsequent to its construction. Hence it became apparent very early on that we needed to learn as much as possible about the history of the tilt of the tower. In the absence of any documentary evidence all the clues to the history of tilt lie in the adjustments made to the masonry layers during construction and in the shape of the axis of the tower.

Over the years a number of measurements of the dimensions of the tower have been made and many of them are conflicting. The Polvani Commission measured the thickness of each of the masonry layers and its variation around the tower.[1] This information has proved extremely valuable in unravelling the history of tilt.

Figure 3 shows the shape of the axis of the tower deduced from the measured relative inclinations of the masonry layers assuming that con-

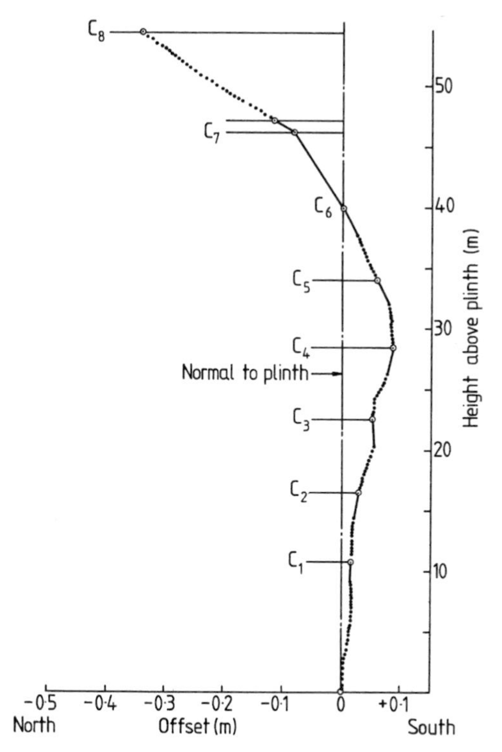

Fig. 3 Shape of axis of tower deduced from relative inclinations of masonry layers.

struction proceeded perpendicular to each masonry layer. This shape compares favourably with other independent measurements at a few locations up the tower. It can be seen that the axis is curved. For years the tower has been unkindly referred to as having a banana shape. I prefer to call it a question mark (?) so as to reflect the enigma of the tower.

Some important observations can be made from the measurements on the masonry layers. For most of the storeys, construction took place using parallel sided blocks of masonry. With one or two notable exceptions adjustments only took place close to each floor using tapered blocks. The most important exception can be seen in Fig. 3 where there is an obvious kink one-quarter the way up the fourth storey. It will be recalled that the construction remained at this level for about 100 years. Evidently, the tower was tilting significantly when work recommenced and the masons made adjustments to correct it.

We see that the history of the tilting of the tower is tantalizingly frozen into the masonry layers. If only we knew the rules that the masons followed in adjusting for the tilt we would be able to unravel the history.

We have to put ourselves in the place of a mason or architect in the twelfth or thirteenth century and ask ourselves: 'What is the most practical thing to do when you arrive at a given floor and find that the tower is out-of-plumb?' A widely accepted hypothesis is that the masons would always try to keep the masonry layers horizontal and the Polvani Commission adopted this. Although this seems reasonable for a low aspect ratio building like a cathedral, it does not make sense for a tower as it would tend to perpetuate the overall out-of-plumb. After a few trials, a child building a tower of bricks on a carpet will soon learn to compensate for any tilt by attempting to place successive bricks over the centre of the base of the tower, i.e. by bringing the centre of the tower back vertically over the centre of the foundations (or possibly even further, away from the direction of tilt). Therefore, an alternative hypothesis is one in which the masons aimed to bring the centre line of the tower back, vertically over the centre of the foundations at the end of each storey. The architectural historians on the commission are satisfied that the masons would have had the technology to make such an adjustment, particularly as the stones for each storey were carved and assembled on the ground prior to hoisting into position.

Figure 4 shows the re-constructed history of inclination of the foundations of the tower using the alternative hypothesis.[2] In this figure the

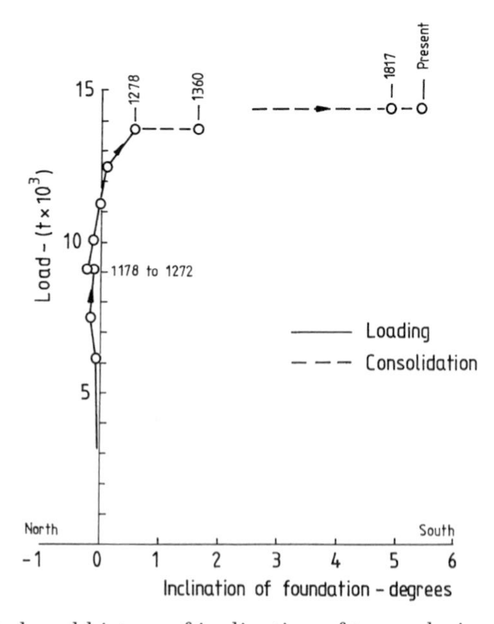

Fig. 4 Deduced history of inclination of tower during and subsequent to construction.

weight of the tower at any time is plotted against the deduced inclination. It can be seen that initially the tower inclined slightly to the north amounting to about 0.2° in 1272 when construction recommenced. As construction proceeded the tower began to move towards the south at an increasing rate. In 1278, when construction had reached the seventh cornice, the tilt was about 0.6°. During the 90-year pause, the tilt increased to about 1.6°. After the completion of the bell chamber in about 1370 the inclination of the tower increased dramatically. The point dated 1817 is based on measurements made by two British architects Cressy and Taylor using a plumb line. A further measurement was made by the Frenchman Ruhault de Fleury in 1859, which showed that the excavation of the catino by Gherardesca in 1838 caused a significant increase of inclination. The history of tilting depicted in Fig. 4 has been used to calibrate numerical and physical models of the tower and underlying ground.

Computer modelling of the movements of the tower

The analysis was carried out using a suite of finite element geotechnical computer programs developed at Imperial College and known as ICFEP.[3] The constitutive model is based on critical state concepts[4] and is nonlinear elastic work-hardening plastic. Fully coupled consolidation is incorporated so that time effects due to the drainage of pore water out of or into the soil skeleton are included.

It must be emphasized that the prime objective of the analysis was to develop an understanding of the mechanisms controlling the behaviour of the tower.[5] Accordingly, a plane strain approach was used for much of the work and only later was three-dimensional analysis used to explore certain detailed features.

The layers of the finite element mesh matched the soil sublayering that had been established from numerous extensive soil exploration studies. Figure 5 shows the mesh in the immediate vicinity of the foundation. In layer B (see Fig. 2) the soil is assumed to be laterally homogeneous. However, a tapered layer of slightly more compressible material was incorporated into the mesh for layer A1 as shown by the shaded elements in Fig. 5. This slightly more compressible region represents the more clayey material found beneath the south side of the foundation as discussed in the second section. In applied mechanics terms the insertion of this slightly more compressible tapered layer may be considered to act as an 'imperfection'. The overturning moment generated by the lateral movement of the centre of gravity of the tower was incorporated into the model as a function of the inclination of the foundation as shown in Fig. 5.

The analysis was carried out in a series of time increments in which the loads were applied to the foundation to simulate the construction history of the tower. The excavation of the catino in 1838 was also simulated in the analysis. Calibration of the model was carried out by adjusting the relationship between the overturning moment generated by the centre of gravity and the inclination of the foundation. A number of runs were carried out with successive adjustments being made until good agreement was obtained between the actual and the predicted present-day value of the inclination.

Figure 6 shows a graph of the predicted changes in inclination of the tower against time, compared with the deduced historical values. It is important to appreciate that the only point that has been predetermined in the analysis is the present-day value. The model does not simulate the initial small rotation of the tower to the north. However, from about 1272 onwards there is remarkable agreement between the model and the historical inclinations. Note that it is only when the bell chamber was added in 1360 that the inclination increases dramatically. Also of considerable interest is the excavation of the catino in 1838 which results in a predicted rotation of about 0.75°. It should be noted that the final imposed inclination of the model tower is 5.44°, which is slightly less than the present-day value of 5.5°. It was found that any further increase in the final inclination of the model tower resulted in instability—a clear indication that the tower is very close to falling over.

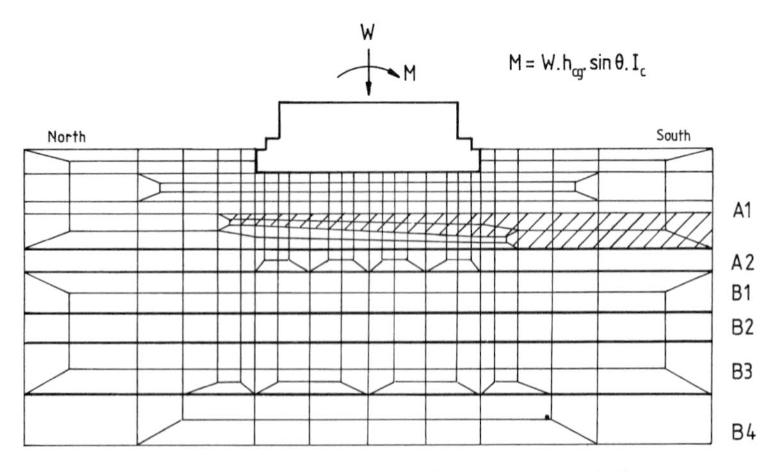

Fig. 5 Finite element mesh in the vicinity of the tower foundation.

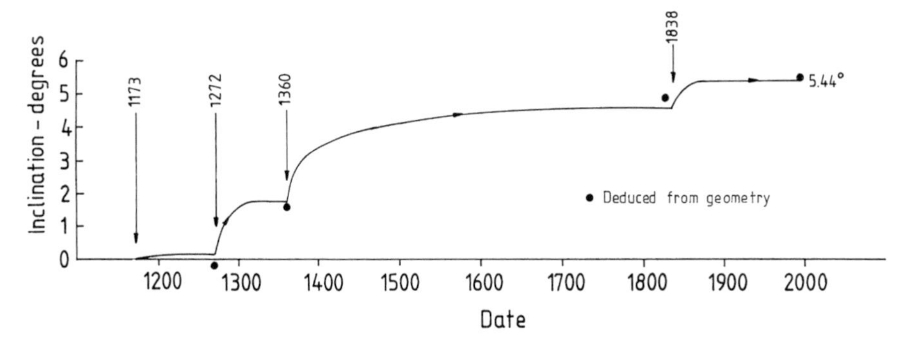

Fig. 6 Relationship between time and inclination for the computer simulation of the history of the Pisa Tower.

Burland and Potts[5] concluded from a careful study of the computer model that the impending instability of the tower foundation is not due to a shear failure of the ground but can be attributed to the high compressibility of the Pancone Clay. This phenomenon was called 'leaning instability' by the late Edmund Hambly[6] who used it to explain the lean of the Pisa tower. No matter how carefully the structure is built, once it reaches a critical height the smallest perturbation will induce leaning instability. As pointed out by Hambly: '... leaning instability is not due to lack of strength of the ground but is due to insufficient stiffness, i.e. too much settlement under load'. Children building brick towers on a soft carpet will be familiar with this phenomenon!

In summary, the finite element model gives remarkable agreement with the deduced historical behaviour of the tower. It is important to emphasize that the predicted history of foundation inclinations and overturning moments were self-generated and were not imposed externally in a predetermined way. The only quantity that was used to calibrate the model was the present-day inclination. The analysis has demonstrated that the lean of the tower results from the phenomenon of settlement instability due to the high compressibility of the Pancone Clay. The role of the layer of slightly increased compressibility beneath the south side of the foundations is to act as an 'imperfection'. Its principal effect is to determine the direction of lean rather than its magnitude. The main limitation of the model is that it is a plane strain one rather than fully three-dimensional. Also, the constitutive model does not deal with creep so that no attempt has been made to model the small time-dependent rotations that have been taking place this century and which are described in the next section. Nevertheless, the model provides important insights into the basic mechanisms of behaviour and has proved valuable in assessing the effectiveness of various proposed stabilization measures. Its role in evaluating the effectiveness of the temporary counterweight solution is described later.

Observed behaviour of the tower this century

Change of inclination

For most of this century the inclination of the tower has been increasing. The study of these movements has been important in developing an understanding of the behaviour of the tower and has profoundly influenced the decisions taken by the commission. It is important to appreciate that the magnitudes of the movements are about three orders of magnitude less than the movements that occurred during construction. Thus changes in inclination are measured in arc seconds rather than degrees (one arc second equals 1/3600th of a degree).

Figure 7 is a plan view of the *Piazza dei Miracoli* showing the location of the baptistry, cathedral, and tower. Since 1911 the inclination of the tower has been measured regularly by means of a theodolite. The instrument is located at the station marked E and the angles between station D and the first cornice (V1 in Fig. 1) and between station D and the seventh cornice (V7) are measured. The difference between these two angles is used to calculate the vertical offset between the seventh and first cornice and hence the inclination of the tower.

In 1928 four levelling stations were placed around the plinth level of the tower and were referred to a bench mark on the baptistry. Readings

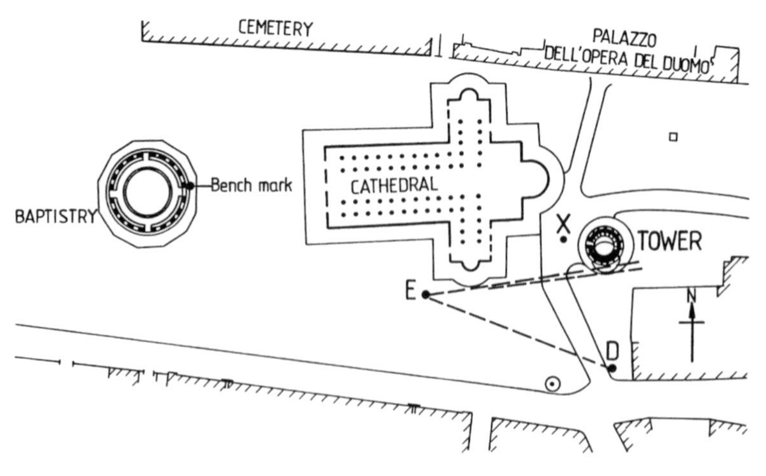

Fig. 7 Piazza dei Miracoli.

were taken in 1928 and 1929 but not again until 1965. In 1965, 15 levelling points were installed around the tower at plinth level and about 70 surveying monuments were located around the piazza.

In 1934 a plumb line was installed in the tower, suspended from the sixth floor and observed in an instrument room whose location is shown in Fig. 1. The instrument was designed by the engineers Girometti and Bonechi and is known as the GB pendulum. Also in 1934 a 4.5 m long spirit level was installed within the instrument room. The instrument rests on brackets embedded in the masonry and can be used to measure both the north–south and the east–west inclination of the tower. The instrument was designed by the officials of the Genio Civile di Pisa and is known as the GC level.

Figure 8 shows the change of inclination with time since 1911. From 1934 to 1969 the GC level was read regularly once or twice a year except during the Second World War. For some reason readings with the GC level ceased in 1969, but fortunately precision levelling on the 15 points around the tower began in 1965 and continued regularly until 1985. In 1990 Professor Carlo Viggiani and I read the GC level again and found that the inclination agreed to within a few seconds of arc with that derived from the precision levelling around the plinth.

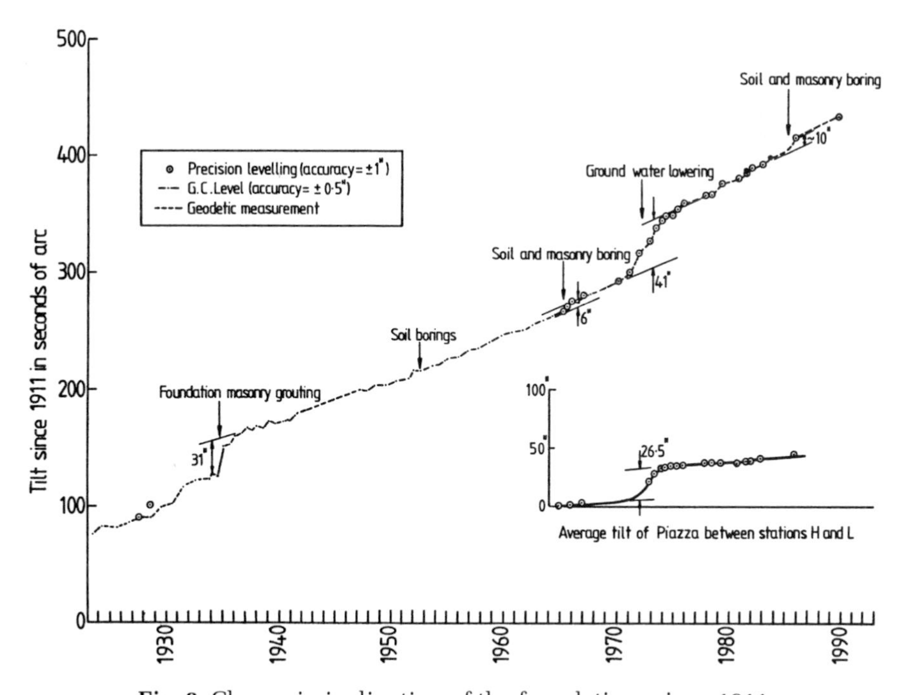

Fig. 8 Change in inclination of the foundations since 1911.

It can be seen from Fig. 8 that the inclination–time relationship for the tower is not a smooth curve but contains some significant 'events'. In 1934, Girometti drilled 361 holes into the foundation masonry and injected about 80 t of grout with a view to strengthening the masonry. This activity caused a sudden increase in tilt of 31 arc seconds. Like Gherardesca, Girometti also left us an unpleasant surprise as is described later. In 1966 some soil and masonry drilling took place and caused a small but distinct increase of tilt of about 6 arc seconds. Again, in 1985 an increase in tilt of 10 arc seconds. resulted from masonry boring through the foundations. In the late 1960s and early 1970s pumping from the lower sands caused subsidence and tilting towards the south-west of the piazza. This induced a tilt of the tower of about 41 arc seconds. When pumping was reduced the tilting of the tower reduced to its previous rate. It is clear from these events that the inclination of the tower is very sensitive to even the smallest ground disturbance. Hence any remedial measures should involve a minimum of such disturbance. The rate of inclination of the tower in 1990 was about 6 seconds of arc per annum or about 1.5 mm at the top of the tower.

The motion of the tower foundation

Previously, studies have concentrated on the changes of inclination of the tower. Little attention has been devoted to the complete motion of the foundations relative to the surrounding ground. The theodolite and precision levelling measurements help to clarify this. It will be recalled from Fig. 7 that angles were measured relative to the line ED. Hence it is possible to deduce the horizontal displacements of the tower relative to point D. Figure 9 shows a plot of the horizontal displacement of point

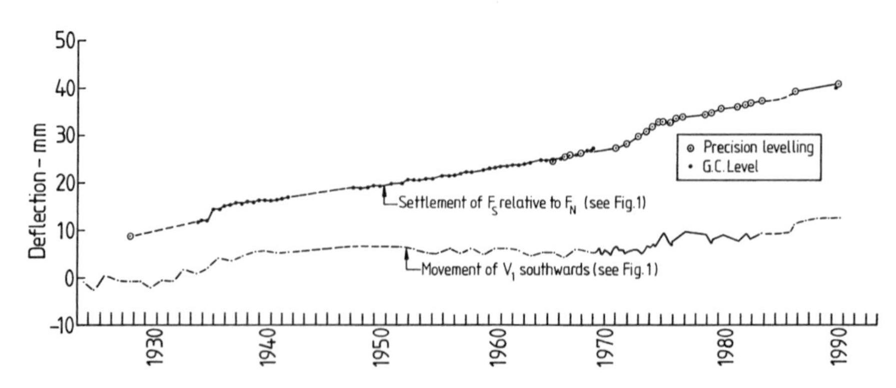

Fig. 9 Horizontal displacement of V1 on the first cornice since 1911.

V_1 on the first cornice relative to point D since 1911. Also shown, for comparison, is the relative vertical displacements between the north and south sides of the foundation (points F_N and F_S). It can be seen that up to 1934 the horizontal movement of V_1 was very small. Between 1935 and 1938, following the work of Girometti, point V_1 moved southwards by about 5 mm. No further horizontal movement took place until about 1973 when a further southward movement of about 3–4 mm took place as a result of the groundwater lowering. A further small horizontal movement appears to have taken place in about 1985 as a result of masonry drilling at that time. These observations reveal the surprising fact that during steady-state creep-rotation point V_1 on the first cornice does not move horizontally. Horizontal movements to the south only take place when disturbance to the underlying ground takes place.

Study of the precision levelling results shows that between 1928 and 1965 the centre of the foundations at plinth level rose by 0.3 mm relative to the baptistry—a negligible amount. Between 1965 and 1986 the relative vertical displacement between the centre of the plinth and a point a few metres away from the tower was again negligible. Thus, not only does point V_1 not move horizontally during steady-state creep, but also negligible average settlement of the foundations has taken place relative to the surrounding ground.

The observations described above can be used to define the rigid-body motion of the tower during steady-state creep-rotation as shown in Fig. 10. It can be seen that the tower must be rotating about a point approximately located level with point V_1 and vertically above the centre of the foundation. The direction of motion of points F_N and F_S are

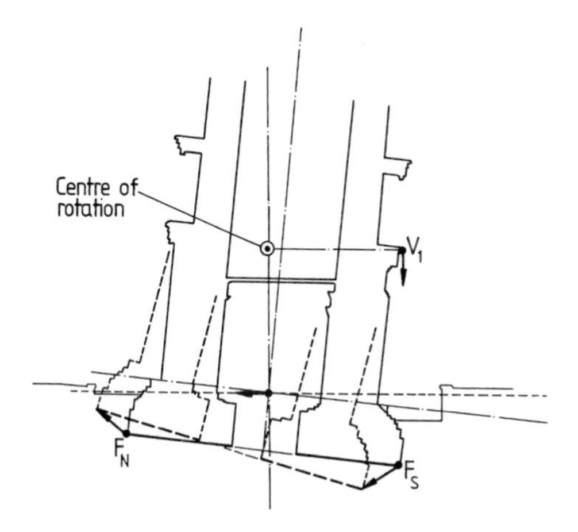

Fig. 10 Motion of the tower during steady-state creep-rotation.

shown by vectors and it is clear that the foundations are moving northwards with F_N rising and F_S sinking.

Conclusions from the observed motion of the tower foundations

The discovery that the motion of the tower is as shown in Fig. 10 has turned out to be a most important finding in a number of respects. Previously, it had been believed that the foundations were undergoing creep settlements with the south side settling more rapidly than the north. However, the observation that the north side had been steadily rising led to the suggestion that the application of load to the foundation masonry on the north side could be beneficial in reducing the overturning moment.[8]

The form of foundation motion depicted in Fig. 10 leads to the very important conclusion that the seat of the continuing long-term tilting of the tower lies in horizon A and not within the underlying Pancone Clay as had been widely assumed in the past. It can therefore be concluded that this stratum must have undergone a considerable period of ageing since last experiencing significant deformation. Thus, in developing the computer model, it is reasonable to assume that the clay has an increased resistance to yield subsequent to the excavation of the catino in 1838. This conclusion has proved of great importance in the successful analysis of the effects of applying the lead counterweight.[5]

The continuing foundation movements tend to be seasonal. Between February and August each year little change in the north–south inclination takes place. In late August or early September the tower starts to move southward and this continues through till December or January amounting to an average of about 6 arc seconds. In the light of the observed motion of the tower foundations, the most likely cause of these seasonal movements is thought to be the sharp rises in groundwater level that have been measured in horizon A resulting from seasonal heavy rainstorms in the period September to December each year. Thus, continuing rotation of the foundations might be substantially reduced by controlling the water table in horizon A in the vicinity of the tower.

Temporary stabilization of the tower

There are two distinct problems that threaten the stability of the tower. The most immediate one is the strength of the masonry. It can be seen from the cross-section in Fig. 1 that at first floor level there is a change in cross-section of the walls. This gives rise to stress concentrations at the south side. In addition to this, the spiral staircase can be seen to pass through the middle of this change in cross-section giving rise to a significant magnification in the stresses. The marble cladding in this location shows signs of cracking. It is almost impossible to assess accurately the margin of safety against failure of the masonry, but the consequences of failure would be catastrophic. The second problem is the stability of the foundations against overturning.

The approach of the commission to stabilization of the tower has been a two-stage one. The first stage has been to secure an increase in the margin of safety against both modes of failure as quickly as possible by means of temporary measures. Having achieved this, the second stage is to develop permanent solutions recognizing that these would require time to carry out the necessary investigations and trials. Significant progress has been made with the first stage. It is a prerequisite of restoration work that temporary works should be non-destructive, reversible, and capable of being applied incrementally in a controlled manner.

Temporary stabilization of the masonry

The masonry problem has been tackled by binding lightly pre-stressed plastic covered steel tendons around the tower at the first cornice and at intervals up the second storey as shown in Fig. 11. The work was carried out in the summer of 1992 and was effective in closing some of the

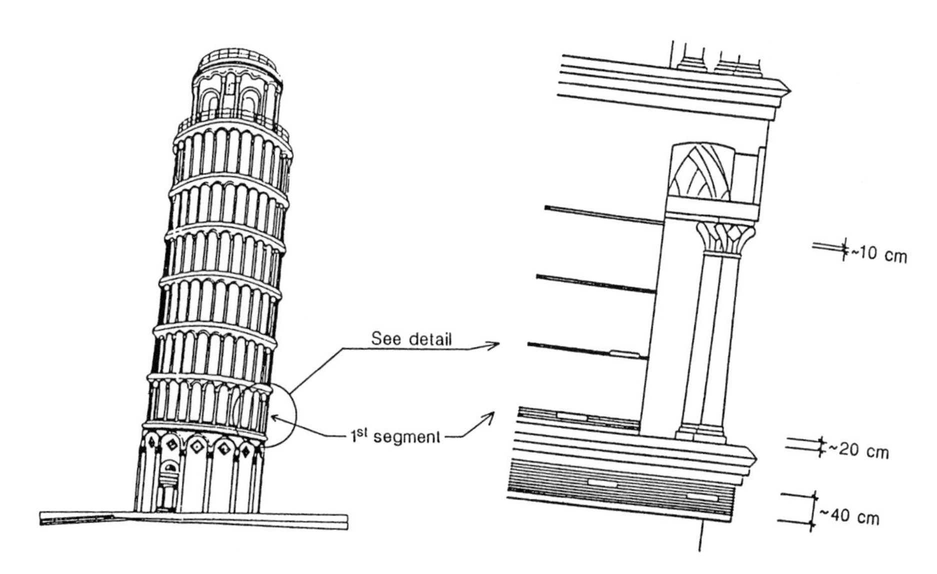

Fig. 11 Temporary stabilization of the masonry with light circumferential pre-stressing.

cracks and in reducing the risk of a buckling failure of the marble cladding. The visual impact has proved to be negligible.

Temporary stabilization of the foundations

As mentioned previously, the observation that the northern side of the foundation had been steadily rising for most of this century led to the suggestion that application of load to the foundation masonry on the north side could be beneficial in reducing the overturning moment. Clearly, such a solution would not have been considered if it had not been recognized that leaning instability rather than bearing capacity failure was controlling the behaviour of the tower or if the north side of the foundation had been settling.

Before implementing such a solution it was obviously essential that a detailed analysis should be carried out. The purpose of such an analysis was twofold: (i) to ensure that the proposal was safe and did not lead to any undesirable effects, and (ii) to provide a best estimate of the response against which to judge the observed response of the tower as the load was being applied. A detailed description of the analysis is given by Burland and Potts[5] who found that a satisfactory result was only forthcoming if the effects of ageing of the underlying Pancone Clay was incorporated in the computer model. The justification for such ageing lay in the observed motion of the foundations depicted in Fig. 10 as described in the previous section. The computer analysis indicated that it was safe to apply up to a maximum of 1400 t load to the north side of the foundation masonry. Above that load there was a risk that the underlying Pancone Clay would begin to yield resulting in a southward rotation of the tower and excessive settlement of the foundations.

Accordingly, a design was developed by Professors Leonhardt and Macchi for the application of a north counterweight and the details of construction are shown in Fig. 12. It consists of a temporary pre-stressed concrete ring cast around the base of the tower at plinth level. This ring acts as a base for supporting specially cast lead ingots, which were placed one at a time at suitable time intervals. The movements experienced by the tower are measured with a highly redundant monitoring system consisting of the following: (i) precision inclinometers and levellometers installed on the wall of the ground floor room; (ii) high precision levelling of eight survey stations mounted on the wall of the above room; and (iii) external high precision levelling of 15 bench marks located around the tower plinth and 24 bench marks located along north–south and east–west lines centred on the tower. All the levels are related to a deep datum installed in the *Piazza dei Miracoli* by the commission.

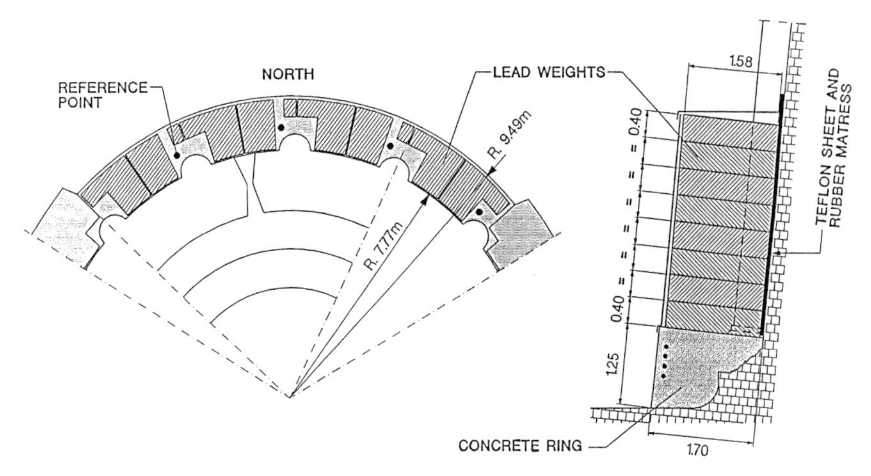

Fig. 12 Details of the north counterweight.

Observed response

Burland *et al.*[7] describe the response of the tower to the application of the counterweight. Construction of the concrete ring commenced on 3 May 1993 and the first lead ingot was placed on 14 July 1993. The load was applied in four phases with a pause between each phase to give time to observe the response of the tower. The final phase was split in two either side of the Christmas break. The last ingot was placed on 20 January 1994.

Figure 13 shows the change of inclination of the tower towards the north during the application of the lead ingots as measured by the internal high precision levelling and the inclinometer placed in the north–south plane. The agreement between the two independent monitoring systems is excellent. (Note that Fig. 15 does not include the inclination induced by the weight of the concrete ring, which amounted to about 4 arc seconds.) It can be seen that the amount of creep between the phases of load is small. However, subsequent to completion of loading, time-dependent northward inclination has continued. On 20 February 1994 (1 month after completion of loading) the northward inclination was 33 arc seconds. By the end of July 1994 it had increased to 48 arc seconds giving a total of 52 arc seconds including the effect of the concrete ring. On 21 February 1994 the average settlement of the tower relative to the surrounding ground was about 2.5 mm.

Comparison between predictions and observations

Figure 16 shows a comparison of the predictions from the computer model and measurements of: (i) the changes in inclination, and (ii) the

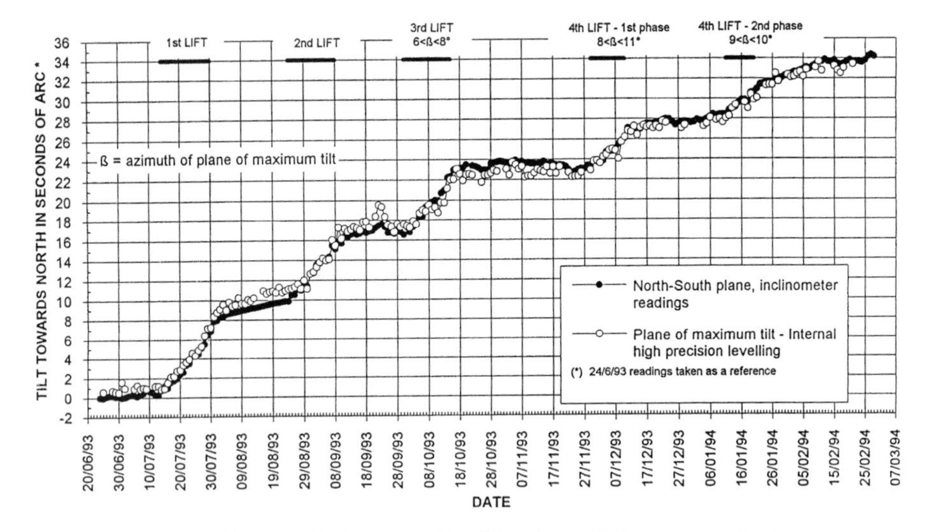

Fig. 13 Observed change of inclination of the tower during application of the counterweight.

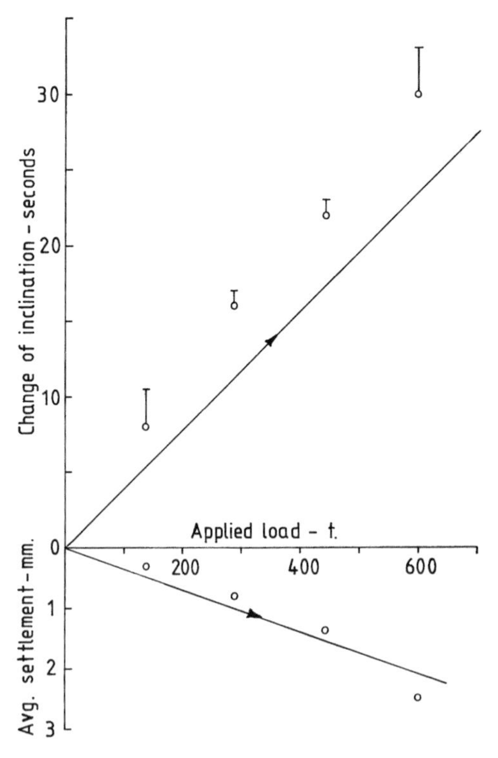

Fig. 14 Predicted and observed response of the tower to the counterweight.

average settlements of the tower relative to the surrounding ground during the application of the lead ingots. The points in the upper part of Fig. 16 represent the measured rotations at the end of each phase of loading and the vertical lines extending from them show the amount of creep movement between each phase. For the final phase the creep after 1 month is shown. It can be seen that the predictions of the computer model give changes in inclination, which are about 80% of the measured values. However, the predicted settlements are in excellent agreement with the measurements.

It is perhaps worth emphasizing that the purpose of the computer model was to clarify some of the basic mechanisms of behaviour and it was calibrated against inclinations measured in degrees. The use of the model in studying the effects of the counterweight was to check that undesirable and unexpected responses of the tower did not occur. In this respect the model has proved to be very useful. It has led to a consideration of the effects of ageing and it has drawn attention to the importance of limiting the magnitude of the load so as to avoid yield in the underlying Pancone Clay. It is perhaps expecting too much of the model for it to make accurate quantitative predictions of movements that are three orders of magnitude less than those against which it was calibrated and the fact that it has done as well as it has is remarkable. The observed movements due to the application of the counterweight have been used to refine the model further.

Permanent stabilization

For bureaucratic and financial reasons work on the temporary stabilization of the tower has taken longer than had been hoped. In parallel with these operations the commission has been exploring a variety of approaches to permanently stabilizing the tower. The fragility of the masonry, the sensitivity of the underlying clay and the very marginal stability of the foundations has already been referred to. Because of these severe restraints, any measures involving the application of concentrated loads to the masonry or underpinning operations beneath the south side of the foundation have been ruled-out. Moreover, aesthetic and conservation considerations require that the visible impact of any stabilizing measures should be kept to an absolute minimum.

The commission has decided to give priority to so-called 'very soft' solutions aimed at reducing the inclination of the tower by up to half a degree by means of induced subsidence beneath the north side of the foundation without touching the structure of the tower. Such an approach allows the simultaneous reduction of both the foundation instability and the masonry overstressing with a minimum of work on the tower fabric itself.

Some of the key requirements of stabilization by reducing inclination are as follows.

1. The method must be capable of application incrementally in very small steps.
2. The method should permit the tower to be 'steered'.
3. It must produce a rapid response from the tower so that its effects can be monitored and controlled.
4. Settlement at the south side must not be more than 0.25 of the settlement of the north side. This restriction is required to minimize damage to the catino and disturbance to the very highly stressed soil beneath the south side.
5. There must be no risk of disturbance to the underlying Pancone Clay, which is highly sensitive and upon whose stiffness, due to ageing, the stability of the tower depends.
6. The method should not be critically dependent on assumed detailed ground conditions.
7. The impact of possible archaeological remains beneath the tower must be taken into account.
8. Before the method is implemented it must have been clearly demonstrated by means of calculation, modelling, and large-scale trials that the probability of success is very high indeed.
9. It must be demonstrated that there is no risk of an adverse response of the tower.
10. Any preliminary works associated with the method must have no risk of impact on the tower.
11. Methods which require costly civil engineering works prior to carrying out the stabilization work are extremely undesirable for a number of reasons.

After careful consideration of a number of possible approaches the commission chose to study three in detail.

1. The construction of a ground pressing slab to the north of the tower, which is coupled to a post-tensioned concrete ring constructed around the periphery of the foundations.
2. Consolidation of the Pancone Clay by means of carefully devised electro-osmosis.

3. The technique of soil extraction as postulated by Terracina[9] for Pisa and widely used in Mexico City to reduce the differential settlements of a number of buildings due to regional subsidence and earthquake effects. This technique involves the controlled removal of small volumes of soil from the sandy silt formation of horizon A beneath the north side of the foundation.

All three approaches have been the subject of intense investigation. Numerical and centrifuge modelling of the north pressing slab have shown that the response of the tower is somewhat uncertain and, if positive, is small while the induced settlements are large. Full-scale trials of the electro-osmosis showed that the ground conditions at Pisa are not suited to this method. Both these methods require costly civil engineering works prior to commencement of the stabilization work. Work on the method of soil extraction is proving much more positive but before describing it a major set-back took place in September 1995 and first this will be described.

A set-back

It is not widely appreciated that the decree establishing the commission has never been ratified. The position in Italian law is that a decree has to be ratified by the Italian Parliament within 2 months of publication or else it fails. Thus, every 2 months, the decree relating to the commission has to be renewed. On a number of occasions the commission has been suspended because of delays in the renewal of the decree. Such an arrangement makes the commission very vulnerable to media and political pressures. Moreover, long-term planning becomes very difficult.

Shortly after the successful application of the temporary counterweight a view emerged that the commission was politically vulnerable and that something needed to be done that would clearly demonstrate the effectiveness of the work so far. There was also considerable concern among some members that, should the commission cease to exist, the unsightly lead counterweight would be left in position for many years. Therefore, a scheme was developed to replace the lead weights with 10 tensioned cables anchored in the lower sands at a depth of about 45 m as shown in Fig. 17. Additional benefits of this proposal were seen to be that the increased lever-arm would give a slightly larger stabilizing moment than the lead counterweight and tensions in the anchors could be adjusted to 'steer' the tower during implementation of induced subsidence. It is important to appreciate that this 10-anchor solution was always intended to be temporary.

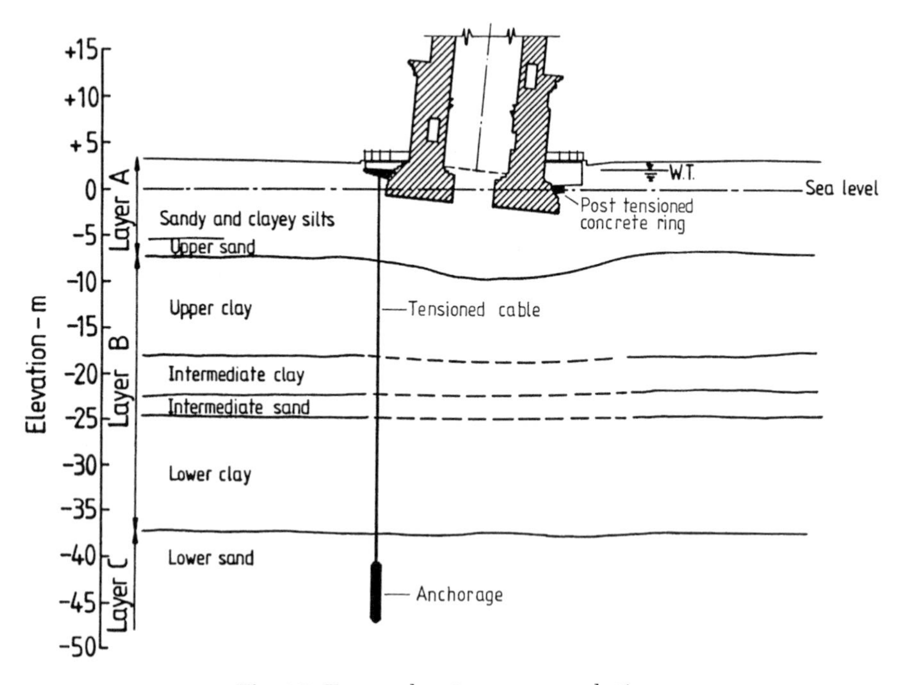

Fig. 15 Ten anchor temporary solution.

The major problem with the 10-anchor solution is that the anchors have to react against a post-tensioned concrete ring around the tower foundation and this involves excavation beneath the catino at the south side—an operation of the utmost delicacy as it is below the water table. Various schemes for controlling the water were considered and it was decided to employ local ground freezing immediately beneath the catino floor but well above foundation level. The post-tensioned concrete ring was to be installed in short lengths so as to limit the length of excavation open at any time.

Shortly before commencement of the freezing operation exploratory drilling through the floor of the catino revealed the existence of an 80 cm thick ancient concrete (conglomerate) layer, which had evidently been placed by Gherardesca in 1838. There are no archaeological records of this conglomerate and its discovery came as a complete surprise. A key question was whether it was connected to the tower. Exploratory drilling was carried out to investigate the interface between the conglomerate and the masonry foundation. A circumferential gap was found all around the foundation and it was concluded that the conglomerate was not connected to the masonry. Work then started on installing the post-tensioned concrete ring.

Freezing commenced on the north side and the northern sections of the ring were successfully installed. The freezing operations consisted of 36 hours of continuous freezing using liquid nitrogen followed by a maintenance phase when freezing was carried out for 1 hour/day so as to control the expansion of the ice front. Some worrying southward rotation of the tower did take place during freezing at the north but this was recovered once thawing commenced. Of far greater concern was the discovery of a large number of steel grout-filled pipes connecting the conglomerate to the masonry foundation. These were installed by Girometti in 1934 when the foundation masonry was grouted. In none of the engineering reports of the time is there any reference to these grout pipes or of the conglomerate.

In September 1995 freezing commenced on the south-west and south-east sides of the foundation. During the initial 36 hours of continuous freezing no rotation of the tower was observed. However, as soon as the freezing was stopped for the maintenance phase the tower began to rotate southward at about 4 arc seconds per day. The operation was suspended and the southward rotation was controlled by the application of further lead weights on the north side. The resulting southward rotation of the tower was small, being about 7 arc seconds, but the counterweight had to be increased to about 900 t. The main concern was the uncertainty about the strength of the structural connection between the conglomerate and the masonry formed by the steel grout pipes. In view of this uncertainty the freezing operation was abandoned and work on developing the permanent solution was accelerated.

Induced subsidence by soil extraction

Figure 18 shows the proposed scheme whereby small quantities of soil are extracted from layer A below the north side of the tower foundation by means of an inclined drill. The principle of the method is to extract a small volume of soil at a desired location leaving a cavity. The cavity gently closes due to the overburden pressure causing a small surface subsidence. The process is repeated at various chosen locations and very gradually the inclination of the tower is reduced.

Two key questions had to be addressed.

1. Given that the tower is on the point of leaning instability, is there a risk that extraction of small quantities of soil from beneath the north side will cause an *increase* in inclination?
2. Is the extraction of small volumes of ground in a controlled manner feasible, will the cavities close and what is the response at the soil/ foundation interface?

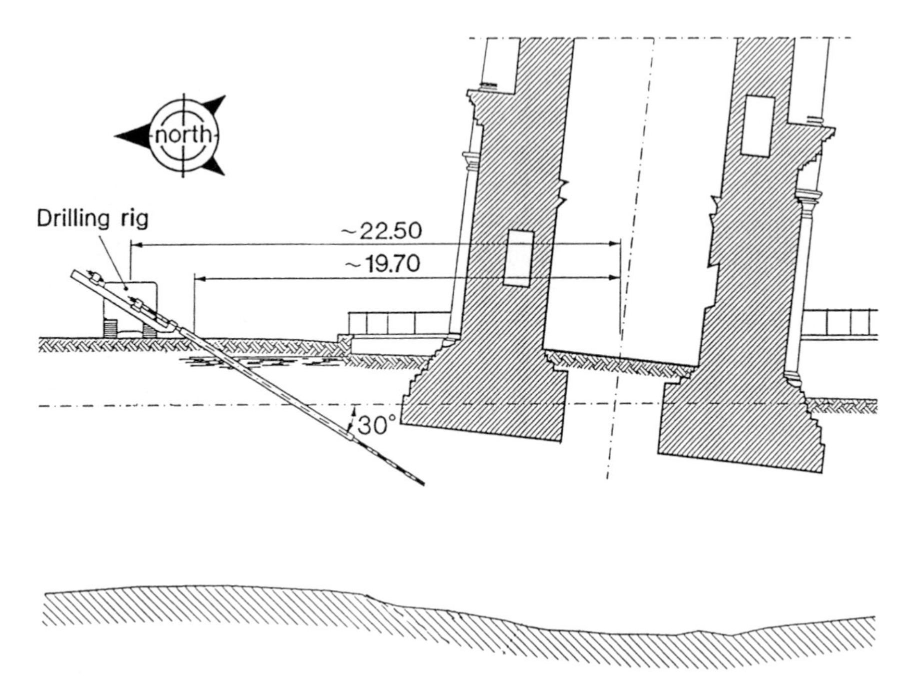

Fig. 16 Induced subsidence by soil extraction.

The first issue has been studied in great detail using two independent approaches—numerical modelling and physical modelling on the centrifuge. The numerical model described previously was used to simulate the extraction of soil from beneath the north side of the foundation. Even though the tower was on the point of falling over it was found that, provided extraction takes place north of a critical line, the response is always positive. Moreover, the changes in contact stress beneath the foundations were small. Advanced physical modelling was carried out on a centrifuge at ISMES in Bergamo. As for the numerical modelling, the ground conditions were carefully reproduced and the model was calibrated to give a reasonably accurate history of inclination. The test results showed that soil extraction always gave a positive response.

The results of the modelling work were sufficiently encouraging to undertake a large-scale development trial of the drilling equipment. For this purpose a 7 m diameter eccentrically loaded instrumented footing was constructed in the piazza north of the baptistry as shown in Fig. 19. The objectives of the trial were:

(1) to develop a suitable method of forming a cavity without disturbing the surrounding ground during drilling;

Fig. 17 Soil extraction trial showing 7 m diameter eccentrically loaded footing and inclined drill.

(2) to study the time involved in cavity closure;

(3) to measure the changes in contact stresses and pore water pressures beneath the trial footing;

(4) to evaluate the effectiveness of the method in changing the inclination of the trial footing;

(5) to explore methods of 'steering' the trial footing by adjusting the drilling sequence; and

(6) to study the time effects between and after the operations.

It must be emphasized that the trial footing was not intended to represent a scale model of the tower.

The results of the trial have been very successful. Drilling is carried out using a hollow-stemmed continuous flight auger inside a contra-rotating casing. When the drill is withdrawn to form the cavity an instrumented probe located in the hollow stem is left in place to monitor its closure. A cavity formed in the horizon A material has been found to close smoothly and rapidly. The stress changes beneath the foundation were found to be small. The trial footing was successfully rotated by about 0.25° and directional control was maintained even though the ground conditions were somewhat non-uniform. Rotational response to soil extraction was rapid taking a few hours. Very importantly, an effective system of communication, decision taking, and implementation was developed.

Concluding remarks

Both numerical and physical modelling of the response of the tower to soil extraction has proved positive. The large-scale trials of the drilling technology has shown that the method works for the soil in layer A. A decision has been taken by the commission to carry out preliminary soil extraction beneath the north side of the tower itself. The objective is to observe the response of the tower to a limited and localized intervention. Before this work is undertaken a safeguard structure is to be constructed in the form of a horizontal cable stay attached to the tower at the third storey. If the preliminary soil extraction proves successful it is estimated that it will take about 2 years to reduce the inclination of the tower by about half a degree, which will be barely visible. It is also anticipated that it will be necessary to stabilize the groundwater level in the vicinity of the tower and to carry out some strengthening of the masonry at the south side of the second storey.

At the time of writing the operation of the commission has again been suspended and it is unlikely that work will recommence on the tower before the middle of 1998.

References

1. Ministero dei Lavori Pubblici. (1971). *Ricerche e studi sulla Torre di Pisa ed i fenomeni connessi alle condizione di ambiente.* 3 Vol., I.G.M., Florence.
2. Burland, J.B. and Viggiani, C. (1994). Observazioni sulcomportamento della Torre di Pisa. *Rivista Italiana di Geotecnica*, **28** (3), 179–200.
3. Potts, D.M. and Gens, A. (1984). The effect of the plastic potential in boundary value problems involving plane strain deformation. *International Journal for Numerical and Analytical Methods in Geomechanics*, **8**, 259–86.
4. Schofield, A.N. and Wroth, C.P. (1968). *Critical state soil mechanics.* McGraw-Hill Book Co., London.
5. Burland, J.B. and Potts, D.M. (1994). Development and application of a numerical model for the Leaning Tower of Pisa. *International symposium on pre-failure deformation characteristics of geo-materials.* IS-Hokkaido '94, Japan, Vol. 2, 715–38.
6. Hambly, E.C. (1985). Soil buckling and the leaning instability of tall structures. *The Structural Engineer*, **63A** (3), 77–85.
7. Burland, J.B., Jamiolkowski, M., Lancellotta, R., Leonards, G.A., and Viggiani, C. (1994). Pisa update—behaviour during counterweight application. *ISSMFE News*, **21**, (2).
8. Burland, J.B. (1990). Pisa Tower. A simple temporary scheme to increase the stability of the foundations. *Unpublished report to the Commission.*
9. Terracina, F. (1962). Foundations of the Tower of Pisa. *Geotechnique*, **12** (4), 336–9.

JOHN BURLAND

Born 1936, educated in South Africa and studied civil engineering at the University of Witwatersrand. Returned to England in 1961, working with Ove Arup & Partners, Consulting Engineers. Obtained a PhD at Cambridge University and joined the Building Research Station in 1966, becoming Head of the Geotechnics Division in 1972 and Assistant Director in charge of the Materials and Structures Department in 1979. In 1980 was appointed to the Chair of Soil Mechanics at Imperial College of Science, Technology and Medicine. Was responsible for the design of the underground car park at the Palace of Westminster and the foundations for the Queen Elizabeth II Conference Centre. Is a member of the Italian Prime Minister's Commission for the stabilization of the Tower of Pisa, and of the international Board of consultants advising on the underpinning of the Metropolitan Cathedral of Mexico City. Was London Underground's expert witness for the Parliamentary Select Committees on the Jubilee Line Extension and CrossRail. Was awarded the Kelvin Gold Medal for outstanding contributions to engineering in 1989.

An arts/science interface: medieval manuscripts, pigments, and spectroscopy

ROBIN J.H. CLARK

Introduction

Much of our cultural heritage is enshrined in manuscripts, paintings, pottery, china, enamels, faience, and other artwork and has been preserved for hundreds and, in some cases, thousands of years. The analysis and identification of pigments on or in such artefacts and the interpretation given to the presence of particular pigments on items from a particular place at a particular date encompass many disciplines spanning both the arts and the sciences. An important role of the scientific disciplines is in the identification of the techniques and materials used in the production of an artefact and the bearing which this may have on our understanding of the development and spread of chemical technology, of artistic styles and techniques, and on the whereabouts of trading routes. Being at the arts/science interface, this subject attracts as much interest from art historians, librarians, and museum scientists as from conservationists and research scientists.

There have been enormous advances in the techniques of analytical chemistry over the past 50 years, to the extent that the identification of any pigment may now be straightforward. However, most techniques currently in use are either intrinsically destructive of a pigment, or cannot be applied *in situ*. Such techniques clearly cannot be used to identify pigments on important manuscripts held throughout the world in galleries and libraries as these operate non-sampling policies. This situation has had the consequence that many of the most significant, beautifully illustrated, and otherwise well-documented artworks currently held in public and private collections have, to date, been poorly characterized in respect of the pigments with which they have been illuminated. Accordingly, a great deal of interesting sociological and technological detail about these works of art has yet to be revealed. It is now considered that Raman microscopy is the best single technique for the identification of pigment grains on manuscripts, paintings, and other artefacts in that it combines the required attributes of being reliable, sensitive, non-destructive, largely immune to interference (from other pigments, binder, and fluorescence), and is applicable to *in situ* studies.[1] Moreover, as the excitation lines usually used lie in the visible region of the spectrum, the spatial resolution possible is $\leq 1\ \mu m$. The situation has the consequence that adjacent pigment grains of this size can separately be studied, which is important where the components of a pigment mixture are to be identified.

Despite the above attributes of Raman microscopy its development as an analytical technique for the above purposes has been slow until recently. This has been partly due to the intrinsic weakness of the Raman effect[2,3] in the absence of resonance effects[4] and partly to the moderate detection capabilities (semiconductor or diode), which were available. Many of these problems have, however, been solved with modern spectrometers and many new features, especially as regards CCD (charge-coupled device) detectors, are continually being developed.

The aims of this article are to outline the purposes behind pigment studies, to indicate the nature of the pigments used throughout time on manuscripts, paintings, and other artefacts, to convey an idea as to how the technique works, to review a number of recent applications of great interest, and to point the way to future developments in the area.

Purpose of the examination of pigments

The purposes behind examining pigments from works of art are matters to do with characterization, restoration, conservation, authentication, and dating.[5] Thus, they are:

1. To identify the pigment, its crystal form and its place of origin; and to determine whether it is uniform in composition and particle size at different depths or whether it is layered.
2. To establish whether restoration of damage is feasible, given the identity of the pigments to be replaced.
3. To consider proper measures to conserve a work of art from the effects of heat, light and gaseous pollutants.

4. To consider whether knowledge of the pigments present on a work of art may give an indication as to the date of the work and hence to authenticity. Although minerals are difficult or impossible to date, the year in which a synthetic pigment was first made is usually well established and provides a marker for an indirect form of dating.

The artists' concerns in a pigment are with texture, permanence of colour, compatibility with other pigments, fastness in media, wettability, miscibility, oil absorption, stability of consistency, tinting strength, hiding power, transparency, drying effects, toxicity, and possible presence of adulterants or impurities.[1,7] The analyst, restorer, and authenticator will be concerned to learn whether the pigment identified is one normally found on a work of art executed at the particular time and place to which the artwork is attributed and whether the same pigment is used in other areas of the work.

Pigments—inorganic

Inorganic pigments have the advantage over organic ones of generally being the more stable photochemically. Most of those used before the eighteenth century were minerals, and many required little treatment to produce the required artists' pigment. Nevertheless, certain very long established pigments, e.g. Egyptian blue and lead(II) antimonate, are synthetic and required not inconsiderable chemical knowledge for their synthesis.[5–9] The colours of inorganic pigments arise from ligand-field, charge-transfer, intervalence charge-transfer transitions and/or from specular reflectance, i.e. mirror-like reflectance from the top surfaces of crystallites. As was well known to early artists, the depth of colour of any pigment is related to the size of its particles, as this affects the balance between the diffuse reflectance (controlled by the absorption coefficient and the bandwidth) and the specular reflectance (controlled by complementary factors).[10,11] The fact that deep blue crystals of $CuSO_4.5H_2O$ become much paler when ground to a powder is a well known illustration of this effect.

By way of example, some of the materials which have been used as blue pigments at different periods are listed alphabetically in Table 1 together with, in each case, the chemical formula, the basis of the colour absorption, and (if synthetic) the date of first manufacture. Similar tables are available for black, orange-brown, green, red, white, and yellow pigments and more extensive information can be had from a wide range of books[1,5–9] and pigment company literature. Although the identification

Table 1. Blue inorganic pigments

Pigment	Chemical name	Formula	Date[a]	Transition[b]
azurite	basic copper(II) carbonate	$2CuCO_3.Cu(OH)_2$	min.	LF
cerulean blue	cobalt(II) stannate	$CoO.nSnO_2$	1821	LF
cobalt blue	cobalt(II)-doped alumina glass	$CoO.Al_2O_3$	1775	LF
Egyptian blue (cuprorivaite)	calcium copper(II) silicate	$CaCuSi_4O_{10}$	3rd millenium BC	LF
lazurite (from lapis lazuli)	sulphur radical anions in a sodium aluminosilicate matrix	$Na_8[Al_6Si_6O_{24}]S_n$	min./1828	S_3^- CT
manganese blue	barium manganate(VII) sulphate	$Ba(MnO_4)_2 + BaSO_4$	1907	CT
phthalocyanine blue (Winsor blue)	copper(II) phthalocyanine	$Cu(C_{32}H_{16}N_8)$	1936	π–π^*
posnjakite	basic copper(II) sulphate	$CuSO_4.3Cu(OH)_2.H_2O$	min.	LF
Prussian blue	iron(III) hexacyanoferrate(II)	$Fe_4[Fe(CN)_6]_3.14–16H_2O$	1704	IVCT
smalt	cobalt(II) silicate	$CoO.nSiO_2$ (+ K_2O + Al_2O_3)	~1500	LF
verdigris	basic copper(II) acetate	$Cu(O_2CCH_3)_2.2Cu(OH)_2$	min.	LF

[a] The pigment is either specified to be a mineral (min.), and/or the date of its first manufacture is listed. Listing taken from ref. 1, which also contains listings of common black, orange-brown, green, red, white and yellow pigments.

[b] LF = ligand field transition; CT = charge transfer transition; IVCT = intervalence charge transfer transition; π–π^* = electric-dipole–allowed charge transfer transition of the phthalocyanine ring system.

of any one pigment will not, of itself, provide precise information as to the date of a piece of artwork, if this information is coupled with that gained by analogous studies on pigments of all other hues, it is then probable that a worthwhile comment on this matter can be made. Clearly, this is of great importance in the valuations of artworks.

Organic pigments and dyes

Among the most common dyes extracted from plants by medieval dyers were indigo from woad for blue, alizarin and madder for red, luteolin from weld for yellow, and crocetin from saffron, also for yellow; all had been established by the Middle Ages, and were widely used on manuscripts. Other organic dyes were extracted from marine molluscs (Tyrian purple, i.e. 6,6′-dibromoindigotin),[12] carmine from scale insects (kermes or cochineal),[8] sepia from cuttlefish, Indian yellow from cow urine,[8] and yet others from many different lichens, etc. Many of the commonly used organic dyes[1] fluoresce and/or are prone to photochemical degradation. An interesting consequence of degradation of one component (say, weld) of a mixture of pigments (e.g. lapis lazuli with weld) is that the colour of the mixture will change with time, posing awkward decisions for conservators as to the extent of restoration which is appropriate.

Raman microscopy

Many different techniques have been used to try to identify pigments or dyes on manuscripts, paintings and other artefacts, notably scanning electron microscopy (SEM), X-ray fluorescence (XRF), X-ray diffraction (XRD), particle-induced X-ray emission (PIXE), particle-induced gamma-ray emission (PIGE), infra-red spectroscopy, ultraviolet/visible spectroscopy, and optical microscopy. Some of these techniques are specific to the elements present, some to the chemical groups present, some to the compounds themselves, and some to the pigment material in bulk. The most recent to be applied is Raman microscopy which is a variant on Raman spectroscopy. The effect, which was first recognized by C.V. Raman in Calcutta in 1928, and which earned the discoverer the Nobel Prize in Physics for 1930, is a consequence of the irradiation of a sample (gaseous, liquid, or solid) with a monochromatic (i.e. single colour) beam of light (nowadays from a laser) and the analysis of the inelastic components of the light scattered by the sample, i.e. the components scattered with frequencies (wavenumbers) different from that of the excitation line.[2,3] The inelastic (the so-called Raman) scattering gives rise to many bands, up to $3N-6$ of them for a non-linear molecule, where N is the number of atoms in the molecule. These bands—collectively referred to as a Raman spectrum—are highly specific in wavenumber, intensity, and bandwidth to the sample, and thus constitute a unique fingerprint of the latter. Indeed Raman spectroscopy is a remarkably effective analytical technique; it has been argued that, when coupled to a microscope (as in Raman microscopy), it is the ideal technique for the identification of pigment grains on manuscripts, paintings, and other artefacts, for the reasons given in the Introduction. In a typical set-up (Fig. 1), a microscope is coupled to a spectrometer with a sensitive diode array or CCD detector. The laser beam (frequency ν_0) is brought to a focus on each pigment grain in turn by use of the microscope objective (×50 or ×100). The Raman scattering at frequencies $\nu_0 \pm \nu_i$, where the ν_i are the frequencies of the internal transitions of the scattering species, retraces the path of the incident beam, being collected by the same objective before being directed by a beam splitter to the monochromator and then the detector. The data are then processed and displayed on a screen or as hard copy. Low laser powers (< 1 mW) are used in order to eliminate the possibility of pigment degradation. Many different laser lines of different frequencies can be used in order to obtain the optimum Raman spectrum, including infra-red (1064 nm) excitation in the case of Fourier transform (FT) Raman spectroscopy.

Case studies

Two spectroscopic studies are now outlined to reveal how, in the one case, the identities of the chromophores in a well known pigment were established and, in the other, how research in the nineteenth century in synthetic chemistry in the area of materials science led to the development of a whole range of closely related pigments with colours which could be changed systematically with change of composition.

The intense royal blue colour of the mineral lapis lazuli, has been highly prized for at least 5000 years; its synthetic equivalent, ultramarine blue has been known since 1828. Not only was the semi-precious gemstone admired as such, but the pigmentary properties of the material have been highly valued since the sixth century, and especially during and after the Middle Ages. The origin of the colour was not understood until a combination of resonance Raman and electron spin resonance studies led to the identification of the chromophores (the parts of the pigment responsible for the light absorption) as sulphur radical anions

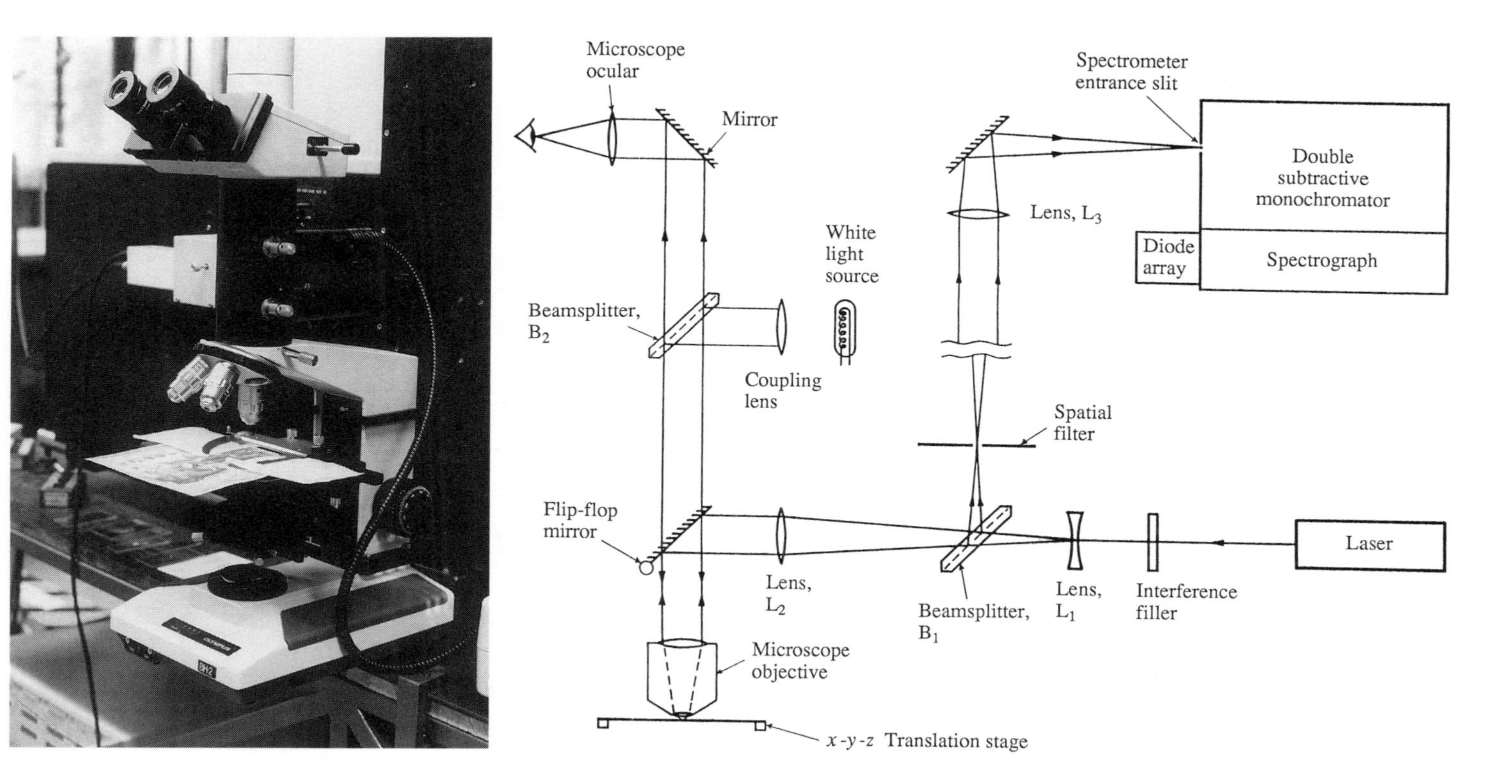

Fig. 1 A Raman microscope together with a schematic representation of the optical cofiguration.

trapped (effectively matrix isolated) within the cubic holes of the host aluminosilicate (sodalite) cage, albeit present in very low (≤ 1%) proportions.[13] The chromophores present were shown to be S_3^- (λ_{max} = 610 nm, ω_1 = 550.3 cm^{-1}) together with some S_2^- (λ_{max} = 380 nm, ω_e = 590.4 cm^{-1}). The key Raman observations on this pigment were the long resonance Raman progressions observed in the totally symmetric stretching mode of each radical anion, features which are characteristic of S_3^- and S_2^- when substituted into alkali halide matrices. Not all pigments, of course, have such intense and highly characteristic Raman spectra as ultramarine blue, but none the less this study illustrates the importance of Raman spectroscopy in characterizing certain minerals and other materials in an easy and highly effective manner.

Cadmium sulphide was first recommended as a yellow artists' pigment in 1818, but the native minerals (greenockite and hawleyite) could not be much used owing to their scarcity. It was not until about 1846 that cadmium sulphide became available commercially, whereupon it soon became popular among impressionist painters, e.g. see Claude Monet's 'Bordighera', 1884, and Charles Demuth's 'Gladioli No. 4', 1925, etc.[8] In addition, it was discovered that selenium could be made to substitute for sulphur, leading to the formation of cadmium selenosulphides with colours ranging from orange through to maroon depending upon the selenium content. This gave the impressionists a wide continuous range of pigment hues to draw upon, and substantially increased the size of their palettes.

Several recent case studies are now discussed in order to illustrate the way in which Raman microscopy has been used to probe the identity of pigments—particularly the inorganic ones—on medieval manuscripts, paintings, and other works of art. Such work has resolved many ambiguities in the minds of art historians in a quick and definitive manner.

Lucka Bible c. 1270

Plate 3 shows the historiated initial letter 'I' from 'In principio ...' at the beginning of the book of Genesis in a bible known as the *Lucka Bible*, which is one of many that were made in Paris at c. 1270.[14] Eight pigments have been identified on this bible by Raman microscopy. Six of them were much used (white lead, vermilion, red lead, lapis lazuli, azurite, and orpiment), while two others (realgar and malachite) were found to be present only in trace amounts, possibly as unintended components of orpiment and azurite, respectively. The Raman spectra of the eight pigments (Fig. 2), all inorganic, are distinctive and thus allow clear identification in each case.[14]

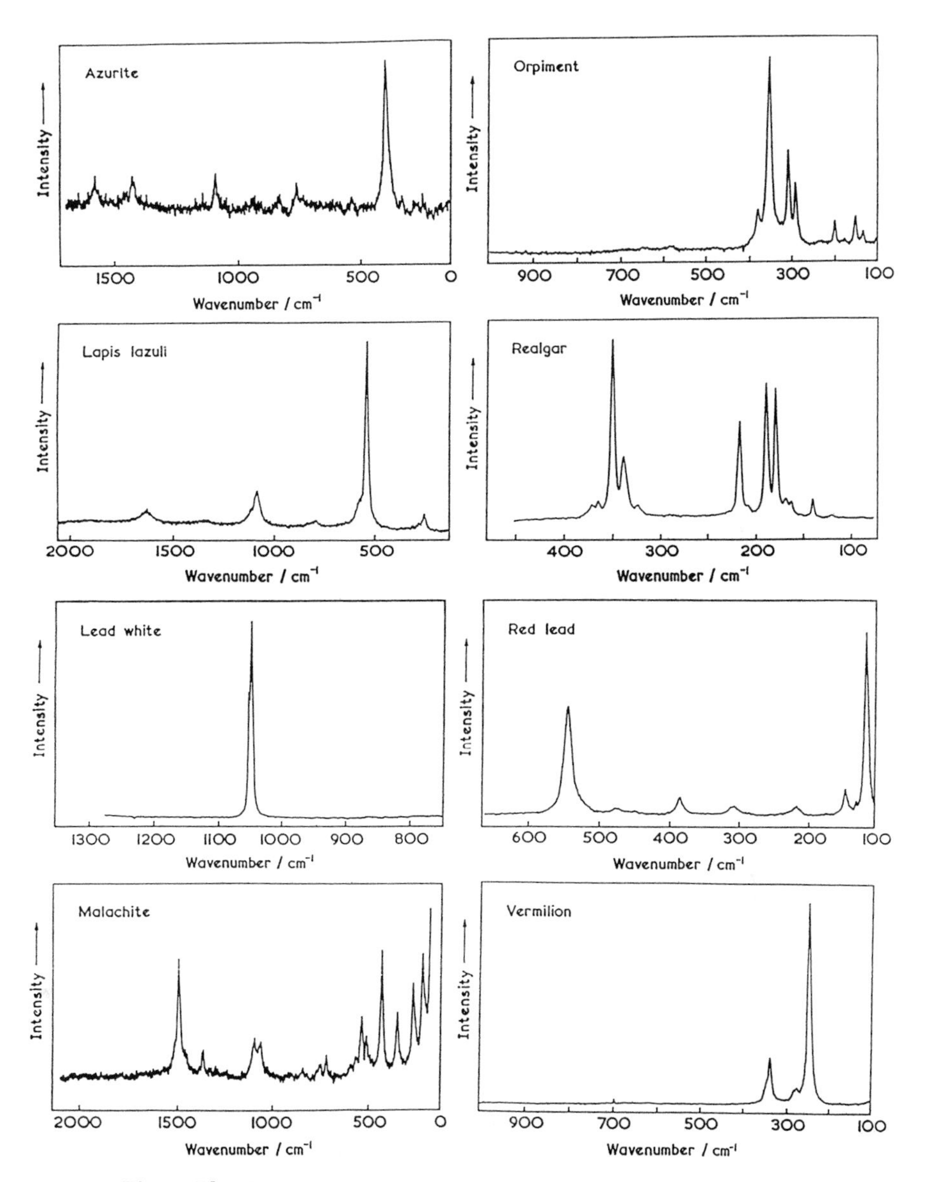

Fig. 2 The Raman spectra of pigments identified on the historiated letter 'I' (see Plate 3).[14]

Skarð copy of the Icelandic Law Book, the Jónsbók c. 1360

Three historiated initials and c. 15 illuminated initials with associated background painting and embellishments from the Skarð copy of the Icelandic Law Book, the *Jónsbók*, have been examined.[15] Six pigments were identified by Raman microscopy, namely vermilion, orpiment, realgar, red ochre, azurite, and bone white. The pigments responsible for some of the green and blue colours could not be identified unambiguously, although the diffuse reflectance spectra of the lighter green and blue pigments appear to arise from verdigris or variants thereof. The dark green colours arise from verdigris, possibly mixed with green earth. Neither red lead nor white lead, pigments commonly used in northern Europe, were found on this manuscript; whether this was due to lack of availability (despite the existence of trading routes, neither pigment being native to Iceland) or to artistic preference is not clear.

German manuscript, sixteenth century

The elaborately historiated initial letter 'R' on a sixteenth century German manuscript has been extensively studied, and eight pigments—azurite, lead tin yellow type I, malachite, vermilion, white lead, red lead, carbon, and massicot—have been identified thereon.[1] It has been shown by Raman microscopy that in this case the two shades of blue on the garments arise from the same pigment, azurite, rather than two different ones, the illuminator having used less azurite and more binder to produce the lighter shade. The deeper blue arises from coarse grains of pigment (~ 30 μm diameter) whereas the lighter blue arises from fine grains (~ 3 μm diameter); this effect, as discussed earlier, arises from the different relative importance of diffuse and specular reflectance with differently sized grains.[10,11]

It is of particular interest that the dark grey colour of the pillar top was not obtained by use of a single pigment but by colour subtraction of a mixture of at least the seven pigments, white lead, carbon, azurite, vermilion, red lead, massicot, and lead tin yellow type I. The mixture is evident at ×1000 magnification in Plate 4 and, by Raman microscopy, each different pigment grain (some down to 1 μm across) may be identified. Mixing of pigments was a common practice of artists in order to obtain hues unavailable from a single pigment.

Studies on Persian, Latin, German, and Chinese manuscripts and on a Qur'an

Similar Raman studies have now been completed on a wide range of Persian,[16] Latin,[17] German (Plate 5),[17] and Chinese[18] manuscripts, and on a Qur'an (thirteenth century, from Iran or Central Asia),[19] and have led to the establishment of the palette in each case. By way of illustration, one of the German manuscripts studied—a fifteenth century Book of Prayers

(MS Ger 4) written in Alsatian dialect and held in the DMS Watson Library, University College London—contains many skilfully illuminated initials with rich floral decorations (Plate 5). The pigments used to decorate the initials include vermilion (red), azurite (blue), and white lead (both in its own right and to temper other colours), as well as gold and silver. Unusually, the green pigment present was not malachite, a copper-based pigment, but a mixture of lapis lazuli (blue) and lead tin yellow type I (Pb_2SnO_4); this result could readily be established by Raman microscopy.[17]

Eight pre-tenth century manuscript fragments and one textile fragment from in or near Dunhuang, an oasis town on the Silk Road in Chinese Central Asia, were shown to be illuminated in each case with vermilion, HgS, on the basis of the unique and intense Raman spectrum of this mineral.[18]

Study and identification of degradation products: Byzantine/Syriac Gospel lectionary, thirteenth century

There is great potential for Raman spectroscopy to provide information which would assist conservators to preserve manuscripts and to identify degradation products of pigments. By way of illustration, a study of a very rare, early thirteenth century Byzantine/Syriac Gospel lectionary has been carried out in which unambiguous proof was given as to the nature of a serious conservation problem (the apparent transformation of a white pigment to a black compound) affecting virtually all of the 60 illuminations in this massive volume (Plate 6).[20] Analysis by Raman microscopy proved unambiguously that the white pigment was lead white (basic lead(II) carbonate) and that the black compound was lead(II) sulphide. The possible cause of the degradation of the white pigment was discussed, as were the merits of a proposed treatment with hydrogen peroxide to 'reverse' the process. The artist's palette was also characterized and, in addition to lead white and lead(II) sulphide, a further five pigments were identified unambiguously: vermilion (mercury(II) sulphide), lapis lazuli, orpiment (arsenic(III) sulphide), realgar (arsenic(II) sulphide), and pararealgar. Pararealgar, a yellow pigment, is a light-induced transformation product of the orange/red pigment realgar, As_4S_4, and had not previously been identified on any manuscript.

The thermal degradation of white lead in static and flowing air and in nitrogen and in oxygen atmospheres is very complicated, and has recently been studied by Raman, thermoanalytical and X-ray diffraction techniques in order to define the conditions under which the mineral pigments Pb_3O_4 (orange-red), tetragonal PbO (litharge, yellow-orange) and orthorhombic PbO (massicot, yellow) are formed.[21]

Other degradation studies

Many other studies concerned with degradation and conservation problems are currently being addressed. These include:

(1) the study of the interconversion of HgS (red) to HgS (black) and its possible reversal;

(2) the study of the photochemical decomposition of orpiment (As_2S_3);

(3) the study of the wavelength dependence of the photochemical conversion of α- or β-realgar (As_4S_4, both orange) to para-realgar (yellow);

(4) the study of the evolvement of acetic acid from verdigris and basic verdigris, which leads to changes in the colour from blue-green to brown-green and possible degradation of the paper, and

(5) the study of malachite/azurite interconversion.

Authenticity studies

The dating aspect of the technique is, of course, of interest to auction houses such as Sotheby's. As indicated earlier, the information gained relates solely to whether or not synthetic pigments of known first date of manufacture are, or are not, present on a manuscript. Clearly, the identification of a twentieth century pigment on a purportedly much earlier manuscript would be a cause of great concern—greatly affecting the value—and several studies of this sort have been carried out. One currently under discussion is aimed at finally establishing whether or not the famous Vinland map is authentic. This purportedly pre-Columbian map of the world includes, significantly, the north-east coastline of the USA. Forty years after its discovery in a secondhand bookstore in 1957 it is still unclear whether or not it is a fake.[22] The difficulty hinges on two scientific investigations: the first, by polarized light microscopy on 20–30 microsamples removed from the surface of the map and found in the ink layer, indicated the presence of titanium dioxide (anatase); this material could not used as a white pigment before 1923, when it was first able to be manufactured pure. The second, by PIXE (a technique whereby elemental composition can be determined over sizeable areas of the map but with much poorer (1 mm) spatial resolution) led to no evidence for abnormal levels of titanium. The implied incompatibility of these results could almost certainly be resolved by Raman microscopy, taking advantage of the ability with this technique to range over the whole map at very high (1 μm) resolution.

Pigment sections from paintings

There are several advantages in taking pigment samples for analysis. These are: (i) sampling does not require the removal of an artefact from its permanent location; (ii) well-planned sampling is a one-off operation that needs never to be repeated and so involves much less handling of the artefact; (iii) only the sample, not the entire artefact, is subjected to irradiation; and (iv) samples may be able to be taken in such a way that the lacunae left by excision of the pigment particles are not discernible to the naked eye; they may also be obtainable, in the case of manuscripts, from offsets transferred on to the opposite page. However, although pigment sampling may be permitted from a painting, it is almost never the case from a manuscript owing to the fragility of the latter.

A recent case of interest in which samples from paintings were, indeed, made available for Raman microscopy involved the study and characterization of lead tin yellow. This pigment has a long and complicated history owing in part to the fact, not originally appreciated, that it exists in two distinct forms, type I which is Pb_2SnO_4 and which has a tetragonal structure isostructural with Pb_3O_4, and type II which is $PbSn_{0.76}Si_{0.24}O_3$ and has a defect pyrochlore structure, 24% of the tin sites having been substituted at random by silicon in the lattice.[23] Both types are synthetic and have been in vogue at different periods of time over the past 2000 years. Owing to the similarities in their colours they have sometimes been confused with one another and on other occasions with lead(II) antimonate, $Pb_2Sb_2O_7$ (Naples yellow). The Raman spectra of these pigments are quite different from one another and hence the pigments may readily be distinguished on this basis. A recent Raman study of two late sixteenth century paintings held in the National Gallery has revealed very clearly that the yellow pigment on one of them—'Death of Acteon', Titian, yellow bush in foreground—consists of type I, whereas that on the other—'Allegory of Love IV', Paolo Veronese, man's cloak—consists of type II.[23] Raman studies on samples or cross-sections of pigment layers taken from paintings are comparatively easy to carry out and have widespread analytical application.

Applications to ceramics, pottery, faience, etc.

The technique of Raman microscopy has recently been applied for the first time to the study of pigments on ceramics, pottery, and faience. The first such study was made of the pigments used in the glazes of fragments of medieval items of pottery dating back to the latter half of the thirteenth century and found buried beneath a church in the abandoned village of Castel Fiorentino, near Foggia in southern Italy. This research led for the first time to the identification of lapis lazuli as the blue pigment in a pottery glaze (Plate 7).[24] Brown-black pigments which were also present were identified to be manganese oxides, probably MnO_2.[24]

Another very recent investigation was concerned with the identification of pigments present in ancient Egyptian faience fragments from El-Amarna in the Nile valley (Plate 8). These fragments, which date from the eighteenth dynasty (1350–1334 BC), i.e. from the period of Akhenaten's rule, were discovered by Sir William Flinders Petrie in 1891–1892. The faience pottery, which consists essentially of silica, SiO_2, was found to have been coloured red with red ochre (essentially iron(III) oxide) and yellow with lead(II) antimonate, $Pb_2Sb_2O_7$, on the basis of the unique Raman spectra given by these two inorganic pigments.[25] For reasons not wholly understood, Raman studies of glazed pottery are not successful if the laser beam is directed through the glaze on to the pottery, but only when directed at cross-sections in which the glaze is broken or chipped. So far, red (647.1 nm) excitation appears to be the best to use for Raman studies of pottery.

Future developments

With the development of very much faster detectors (i.e. CCDs), notch filters, and better spectrometer designs, it is now possible to construct Raman spectrometers which need, for their operation, less powerful lasers than previously; the latter may be air cooled and therefore mobile, rather than large, water-cooled, and static. The consequence is that the size of the spectrometer system can now be greatly reduced, which makes it possible to take, if necessary, the complete unit to a gallery or museum for in-house studies rather than the artefact to the laboratory. The optimization of the optical and electronic components of spectrometers for different exciting lines can now also be carried out readily so that potential problems to do with sample heating on irradiation and with sample fluorescence can be overcome. Further important developments involve the use of fibre-optic probes connected to a miniaturized colour video camera in a remote head ('superhead'), which enable *in situ* studies of pigments on wall paintings to be carried out. Moreover, the use of two-dimensional motorized stages to permit both the mapping of surfaces as well as depth profiling *in situ* in order to establish the composition of pigment layers is likely to provide uniquely valuable further information.

Conclusions

Raman microscopy is now clearly established as a major technique for the rapid identification of pigments on manuscripts, paintings, and other artefacts. Its relevance to the identification of the components of pigment mixes at high ($\leq 1\ \mu m$) spatial resolution is unparalleled. The main difficulties arise with certain organic pigments which either fluoresce (or their supports or binders do), are photosensitive, or fail to yield a Raman spectrum owing to small particle size and/or high degree of dilution, e.g. in a lake. The technique of Raman spectroscopy has, of course, a vast number of technological applications other than to the identification of pigments, namely, to the identification of contaminants in microelectronics, of inhomogeneities formed during the crystallization of polymers, the detection of inclusions in minerals, the monitoring of the curing of polymer resins, and of pesticides in tissues of organisms in the food chain, together with other applications in medicine, jewellery studies, polymer science, and forensic science.[26–28] It should be emphasized, however, that the most effective studies on pigment identification as well as in the other areas of science mentioned are usually carried out by use of Raman microscopy in conjunction with one or more other techniques. In this way the weaknesses of the one may be complemented by the strengths of another.

Acknowledgements

I am most grateful to my recent co-workers in the area who have engendered much interest and displayed great skill in the successful prosecution of this work. These include Drs I.M. Bell, D.A. Ciomartan, and P.J. Gibbs, and L. Burgio, M.L. Curri, M.A.M. Daniels, A. Hardy, and K. Huxley. Support for the work has come from the Leverhulme Trust, the EPSRC, and the ULIRS.

References

1. R.J.H. Clark, *Chem. Soc. Rev.*, 1995, **24**, 187–96.
2. D.A. Long, *Raman spectroscopy*, McGraw-Hill, New York, 1977.
3. R.J.H. Clark and R.E. Hester (ed.), *Advances in spectroscopy*, Wiley, Chichester, Vols 1–26, 1975–98.
4. R.J.H. Clark and T.J. Dines, *Angew. Chem., Internat. Ed. Engl.*, 1985, **25**, 131–58.
5. D.V. Thompson, *The materials and techniques of Medieval painting*, Dover, New York, 1956.

6. R.J. Gettens and G.L. Stout, *Painting materials*, Dover, New York, 1966.
7. K. Wehlte, *The materials and techniques of painting*, Van Nostrand Reinhold Co. Ltd, New York, 1975.
8. R.L. Feller (ed.), *Artists' pigments*, Cambridge University Press, Cambridge, Vol. 1, 1986.
9. A. Roy (ed.), *Artists' pigments*, Oxford University Press, Oxford, Vol. 2, 1993.
10. R.J.H. Clark, *J. Chem. Educ.*, 1964, **41**, 488–92.
11. W.M. Wendlandt and H.G. Hecht, *Reflectance spectroscopy*, Interscience, New York, 1966.
12. R.J.H. Clark, C.J. Cooksey, M.A.M. Daniels, and R. Withnall, *Endeavour*, 1993, **17**, 191–9.
13. R.J.H. Clark, T.J. Dines, and M. Kurmoo, *Inorg. Chem.*, 1983, **22**, 2766–72 and references therein.
14. S.P. Best, R.J.H. Clark, M.A.M. Daniels, and R. Withnall, *Chem. Brit.*, 1993, **29**, 118–22.
15. S.P. Best, R.J.H. Clark, M.A.M. Daniels, C.A. Porter, and R. Withnall, *Studies Conservation*, 1995, **40**, 31–40.
16. D.A. Ciomartan and R.J.H. Clark, *J. Brazil. Chem. Soc.*, 1996, **7**, 395–402.
17. L. Burgio, D.A. Ciomartan, and R.J.H. Clark, *J. Raman Spectrosc.*, 1977, **28**, 79–83; *J. Mol. Struct.*, 1997, **405**, 1–11.
18. R.J.H. Clark, P.J. Gibbs, K.R. Seddon, N.M. Brovenko, and Y.A. Petrosyan, *J. Raman Spectrosc.*, 1997, **28**, 91–4.
19. R.J.H. Clark and K. Huxley, *Science and technology for cultural heritage*, 1996, **5**, 95–101.
20. R.J.H. Clark and P.J. Gibbs, *Chem. Commun.*, 1997, 1003–4. (The manuscript, valued at c. £1 million, was exhibited at the Metropolitan Museum of Art, New York, at an exhibition of Byzantine art in March 1997.)
21. D.A. Ciomartan, R.J.H. Clark, L.J. McDonald, and M. Odlyha, *J. Chem. Soc. Dalton*, 1996, 3639–45.
22. K.M. Towe, *Acc. Chem. Res.*, 1990, **23**, 84–7.
23. R.J.H. Clark, L. Cridland, B.M. Kariuki, K.D.M. Harris, and R. Withnall, *J. Chem. Soc. Dalton*, 1995, 2577–2582.
24. R.J.H. Clark, M.L. Curri, and C. Laganara, *Spectrochim. Acta*, 1997, **53A**, 597–603; R.J.H. Clark, L. Curri, G.S. Henshaw, and C. Laganara, *J. Raman Spectrosc.*, 1997, **28**, 105–9.
25. R.J.H. Clark and P.J. Gibbs, *J. Raman Spectrosc.*, 1997, **28**, 99–103. (The items concerned are held in the Petrie Museum, University College London.)
26. J. Corset, P. Dhamelincourt, and J. Barbillat, *Chem. Brit.*, 1989, 612–16.
27. G. Turrell and J. Corset (ed.), *Raman microscopy*, Academic Press, London, 1996.
28. A. Paipetis, C. Vlattas, and C. Galiotis, *J. Raman Spectrosc.*, 1996, **27**, 519–26.

ROBIN J.H. CLARK

Born in 1935, and educated at the Universities of Canterbury and Otago before being awarded a PhD in Inorganic Chemistry at University College London in 1961. He joined the academic staff at that Institution shortly thereafter, becoming a Professor in 1982, Dean of Science 1988–89, and Sir William Ramsay Professor and Head of Department in 1989. His research in inorganic chemistry and spectroscopy has led to the publication of nearly 400 scientific papers, three books, and 36 edited books. He has held many visiting professorships and has lectured at over 250 universities and institutions in 31 countries throughout the world. He has given the Royal Society of Chemistry's Tilden, Nyholm, Thomas Graham and Harry Hallam Lectures and has served on many national and international committees, including the Royal Society and Royal Institution Councils. He is also Chairman of the Advisory Council and Trustee of the Ramsay Memorial Fellowships Trust. He was elected an Honorary Fellow of the Royal Society of New Zealand in 1989, a Fellow of the Royal Society and a Member of the Academia Europaea in 1990, a Fellow of the Royal Society of Arts in 1992, and a Fellow of University College London in 1993.

Disappearing species: problems and opportunities

NORMAN MYERS

Summary

The chapter[1] opens with a brief account of the nature and extent of biodiversity. It outlines some values of biodiversity to humankind, notably bio-ecological and economic values. It goes on to review the scope and scale of the biotic crisis unfolding, with potential to precipitate a mass extinction of species matching most such episodes in the prehistoric past. The assessment focuses on tropical moist forests as the biome harbouring by far the greatest number of species and undergoing the most rapid depletion. The chapter engages in a detailed examination of the part played by human populations in the crisis, differentiating between the role of developing and developed nations. It emphasizes the fast-growing numbers of displaced peasants as the predominant cause of tropical deforestation, and of negligent and wasteful consumption in industrialized countries as the principal cause of pollution-derived degradation of wildland habitats. The chapter concludes with a selective appraisal of policy response to safeguard biodiversity.

What is biodiversity?

Biodiversity comprises the sum total of life forms at all levels of organization in biological systems. The most readily recognized manifestation lies with species, and they are the chief focus of this chapter. So far as we know, there are at least 10 million species sharing the Earth with us. But this is only a very crude estimate. I was once a couple of hours late meeting an entomologist friend in a remote forest in Kenya. No problem, he assured me, he had spent the time finding a new insect species. He hadn't named the bug after himself as he'd discovered dozens in just the past year. Another friend, this one in California, told me that even in that highly explored part of the world it would not be difficult to track down a new insect or three.

Each year scientists add thousands to the 1.4 million species documented. Probably another 9 million await their attention in the great 'out there' of biodiversity that remains absurdly uninvestigated; the total could well be 30 million, conceivably 100 million. In other sectors of science we display greater knowledge and understanding of the world about us. We can measure the distance from a given point on the Earth's surface to a given point on the Moon's surface at a given point in time with an accuracy of less than 1 cm. Yet when it comes to measuring the phenomenon that makes our planet unique in the known universe, namely life in all its abundance and variety, we remain grotesquely unaware of what's what. If a space probe were to pick up signs of life on Jupiter—likely a slime mould or some ultra-rudimentary form of life compared with the advanced species on Earth—the find would make television screens and newspaper headlines. Sometimes there seems little sense to modern science and the frontiers it chooses to explore.

Yet as we continue to eliminate unique manifestations of life with ever-greater energy and ingenuity, we also continue to expand the limits of our knowledge, albeit at snail's pace (if that doesn't insult a snail). We are gaining a preliminary grasp of the abundance and variety of species, including the lesser known categories. It seems there could be 1.8 million micro-organisms, including 1.5 million fungi as opposed to a total of 69 000 fungi documented. The other 300 000 micro-organisms comprise bacteria and viruses, although we have named only 14 000 bacteria. There could also be 1 million each of nematodes and mites, although only 12 000 nematodes and 30 000 mites are known to us. Could it be that God's inordinate fondness was not for beetles but for creatures generally smaller than the smallest beetle (which is about the size of a full stop)? One acre of English pasture has been found to contain an estimated 71 million beetle individuals, 249 million springtails, and 666 million mites, plus 135 million assorted aphids, bristletails, and miscellaneous other arthropods (creatures with jointed appendages).

In addition, in turns out there are spectacular assemblies of species on the deep-sea floor, a world with no light, near-freezing temperatures, and many atmospheres of pressure. The marine realm has traditionally been viewed as biotically depauperate in comparison with the terrestrial realm, with an estimated 1 million species or so. But recent research by Dr Fred Grassle of Rutgers University in the United States asserts the true total in just the ocean benthos could be as high as 10 million, mostly made up of molluscs, crustaceans, and polychaete worms. In an area 2.5 miles down and no bigger than two tennis courts, Grassle has found 90 677 small invertebrates living on or in the sediments, many of them no larger than half a millimetre. They comprised 798 species, of which 460 were new to science. More important still, they represented 171 families and 14 phyla—a higher-taxon diversity that could not remotely be matched on land. Whereas terrestrial habitates feature 11 of the supercategories known as phyla (only one of which is limited to land), the seas are home to 28 phyla, 13 of them found nowhere else.

The seas have even revealed a new species of whale. Discovered off Peru's coast in 1976, *Mesoplodon peruvianus* is the smallest of 13 known species in its genus, reaching a length of no more than 4 m and with a remarkably diminutive cranium. The same year also produced the largest shark species, the megamouth, so distinctive that it warrants a family all to itself.

Other prominent categories of species are producing newcomers to science. According to Dr Orlando Garrido of Cuba's National Museum, the island has thrown up three formerly unknown geckos, four snakes, five frogs, and seven lizards in just the past few years. Ichthyologists discover more than 100 fish species every year, virtually all of them in tropical freshwater systems. No group of animals is better known than birds, yet 43 new species have been discovered during the 1980s, and during the 1990s there have come to scientific light a shrike in Somalia, a parrotlet in Peru, a rail in the Solomon Islands, a babbler in the Philippines, and a bishopbird in Tanzania—the latter discovered in the form of four specimens in a cage bound for export, two of them dead and the other two expiring a few days later.

Even more striking has been the discovery of several new species of primate. Four years ago there surfaced a new lemur on Madagascar and a guenon in Gabon. Two years later a lion tamarin showed its black face in a small forested island off the coast of Brazil less than 300 km from the 18-million city of São Paulo, and surviving in the shape of only a few dozen individuals. Just 2 years ago, a pocket-sized marmoset made its scientific appearance in Brazil too, this time in central Amazonia. Occasionally, a new species turns up in unlikely circumstances, like the ant discovered in a plant pot in a Washington DC office.

In big-picture terms, invertebrates dominate biodiversity, making up perhaps 9 million of the putative global total of 10 million species. Bugs are not going to inherit the Earth—they own it already. (Bacteria too are more important than we might think: they make up one-tenth of our body weight.) If, as is on the cards, we lose about half of all mammal species within the foreseeable future—primates, carnivores, and the like—that will not be a fraction so drastic as if we lose half of all invertebrates. Dr David Pimentel and his student colleagues at Cornell University have calculated that more than 40 US crops are absolutely dependent upon insects for pollination, with a farm-gate value of $30 billion per year. Because of the decline in wild populations of bees, many American farmers now have to rent domestic bees.[2] About one-third of the human diet is made up of insect-pollinated fruits, vegetables, and legumes. We could get by materially without the panda or the blue whale, but the demise of the honeybee would cause us ecological and economic hiccups aplenty.

Biodiversity occurs at subspecies level too. If, as is likely, an average species comprises hundreds of genetically distinct populations, there could be billions of such populations worldwide.[3] Because of their ecological differentiation, some populations will be better equipped than others to adapt to the swiftly changing conditions of our environmental future. We may well find that if the depletion of biodiversity continues unabated, the loss of species' populations will eventually induce adverse biospheric repercussions on a scale to rival those arising from the loss of species themselves.

Given the large numbers of species' subunits, each species features much genetic variability. According to the Indian geneticist, Dr Madhav Gadgil, there is a theoretical planetary total of 10^{600} possible genes making up the global genome. In practical terms, however, the average species probably comprises a rough complement of only 1 million organisms, and each organism probably features only between 10^4 and 10^5 genes (viruses may possess as few as three genes). If we accept there is a minimum total of 10^7 organisms on Earth, albeit some sharing a considerable part of their genetic material, the complete stock of Earth's genetic diversity ostensibly features 10^9 genes—this being a very small fraction of the potential that could eventually be generated if we allowed evolution to continue its creative course.

While intraspecies genetic differences generally appear slight, they can be quite pronounced. Consider, for example, the variability manifested in the many races of dogs or the specialized types of wheat. Yet even this gives only a crude picture. A typical bacterium may contain about 1000 genes, certain fungi 10 000, and many flowering plants and a few animals 400 000 or even more. A typical mammal organism such as a mouse may harbour 'only' 100 000 genes, a total found in each and every one of its cells. As has been graphically expressed by Dr Edward Wilson of Harvard University, and a doyen of biodiversity,

> Each of the cells (of the housemouse) contains four strings of DNA, each of which comprises about one billion nucleotide pairs organized into a hundred thousand structural genes. If stretched out fully, the DNA would be roughly one yard long. But this molecule is invisible to the naked eye because it is only 20 angstroms in diameter. If we magnified it until its width equalled that of a wrapping string to make it plainly visible, the fully extended molecule would be 600 miles long. As we travelled along its length, we would encounter some 20 nucleotide pairs to the inch. The full information contained therein, if translated into ordinary-sized printed letters, would just about fill all 15 editions of the Encyclopedia Britannica published in 1768.

Biodepletion

Biodiversity is the key characteristic of our planet, possibly unique in the universe. Yet this distinctive phenomenon is being depleted at ever-more rapid rates. Moreover, the biodiversity problem, or rather the biodepletion problem, is different from all other environmental problems. Soil loss, desertification, deforestation, ozone-layer thinning, global warming, and the rest could eventually be made good over a period of a few centuries at most. Biodepletion, by contrast, is irreversible within conventional time horizons: it will impoverish the biosphere for millions of years. Hence the issue deserves special attention.

Biodepletion stems almost entirely from loss of habitat. In turn, this derives from the expansion of human activities—partly due to growing human numbers, partly to growing human aspirations, and partly to growing human techno-capacity to exploit environments and their natural resources. Note in particular that the human species is already appropriating 40% of net plant growth each year, leaving the other 60% for millions of other species.[4] What when human numbers reach twice as many as today, as is projected within a few decades' time—and when, given recent trends of humans' intensifying exploitation of plant growth, the ratio may become still more disproportionately skewed? Note too that one-third of the Earth's vegetated surface has undergone human-induced degradation through reduction of its potential biotic productivity; and that the degradation rate is accelerating almost everywhere, certainly in the most populous nations (G.C. Daily, personal communication and ref. 5)

Of course, population—human numbers and their growth rate—is central to biodepletion. But is it prominent, predominant, or overwhelming? The issue is subject to much debate, as this chapter reveals.

Before we consider this vital question, let us review some ways in which biodiversity 'matters'. What does it do for us?

Biodiversity's values to humankind

Biodiversity supplies abundant values to humankind, hence its decline is regrettable on several counts. It supplies a host of environmental services such as controlling the gaseous mix of the atmosphere, generating and maintaining soils, controlling pests, and running biogeochemical cycles, among other biospheric functions for the transfer of matter, nutrients, and energy—all of which are vital to human societies everywhere.[6]

The values of biodiversity for humankind are immediately apparent through the myriad contributions of species and their genetic resources to plant-based medicines, drugs, and pharmaceuticals.[7,8] The cumulative commercial value of these products, mostly derived from tropical plants of developing nations, is estimated to amount during the 1990s to $500 billion (1984 dollars) for the developed nations alone.[9] To put the figure of $500 billion in perspective, it is equivalent to half the gross domestic product of Britain. Moreover, the calculation applies only to the products' commercial value as reflected through sales. Their economic value, including the benefits of reduced morbidity and mortality, worker productivity maintained, and the like, is several times larger. In the absence of greatly expanded conservation efforts, we could well witness an extinction on average every 2 years of a plant species with the potential to produce a drug. The cumulative retail-market loss from each such plant extinction through the year 2050 will amount to $12 billion in the United States alone.[9]

A number of plants generate anticancer drugs, along the lines of vincristine and vinblastine used against childhood leukaemia and derived from Madagascar's rosy periwinkle. Each year about 30 000 lives are saved in the United States thanks to plant-derived anticancer drugs. Using Organization for Economic and Cooperative Development indices the economic value of anticancer plant materials in the United States already amounts to $372 billion in terms of lives saved, let alone reduced morbidity and the like. We can realistically triple these figures to obtain values for all developed nations.[9] Most anticancer plants are to be found in the tropics, meaning for the most part the developing nations. This too is where the great bulk of extinctions is occurring.

Similar considerations apply in the field of agriculture. During the late 1970s a wild species of maize was discovered in a montane forest of south-central Mexico, being the most primitive known relative of modern maize.[10] At the time of its discovery, it was surviving in three forest patches covering a mere 4 ha, a habitat threatened with imminent destruction by agricultural settlement. The wild species is a perennial, unlike all other forms of maize. Cross-breeding with established commercial varieties of maize has opened up the prospect that maize growers (and maize consumers) can be spared the seasonal expense of ploughing and sowing, as the plant springs up again of its own accord like daffodills. Even more important, the wild maize offers resistance to four of eight major viruses and mycoplasmas that have hitherto baffled maize breeders.[11] These four diseases cause at least a 1% loss to the world's maize harvest each year, worth more than $500 million in 1980 dollars. Equally to the point, the wild maize grows at elevations above 2500 m, and is thus adapted to habitats cooler and damper than established maize-growing lands. This offers scope for expanding the cultivation range of maize by as much as one-fifth. The overall commercial benefits supplied by this wild plant, surviving when first found in the form of no more than a few thousand stalks, could be worth at least $4 billion per year (1980 dollars).[12]

In summary, species and their genetic constituents supply many market-place benefits to humans, even though pharmacognosy scientists and agronomic experts have undertaken detailed examination of only one plant species in 100, and fewer than one animal species in 5000, to assess their potential for medicine and agriculture. Further, the scope for species' contributions to industry remains largely unexplored even though we already enjoy multiple applications in the form of fibres, exudates, oils, waxes, acids, sterols, esters, phenols, elastomers, and polystyrenes. All in all, we can reasonably assert that the planetary spectrum of species represents the most abundant and diverse, the least exploited, and the most rapidly declining stock of natural resources with which humankind can tackle many of its most fundamental challenges.

Decline of biodiversity

Mass extinction of species

There is much evidence that we are into the opening phase of a mass extinction of species.[13–16] A mass extinction can be defined as an exceptional decline in biodiversity that is substantial in size and generally global in extent, and affecting a broad range of taxonomic groups over a short period of time.[17] In this sense, the present mass extinction—if it remains unchecked by conservation action of appropriate scope and scale—will rival and conceivably surpass the numbers of species lost during the mass-extinction episodes of the prehistoric past.[14]

Earth's stock of species is widely estimated[18,19] to total a minimum of 10 million. Some scientists[20,21] believe the true total could well be 30 million, possibly 50 million, and conceivably 100 million. Of the conservative estimate of 10 million species, about 90% are usually considered to be terrestrial; and of the terrestrial species, roughly 80% or more than 7 million are believed to occur in the tropics, with roughly 5 million in tropical moist forests.[22] As we shall see, these tropical forests are not only the richest biome biotically but constitute the biome where habitat depletion is occurring fastest—'depletion', including not only outright destruction but gross disruption (fragmentation etc.), with loss of many ecosystem attributes including food webs. So this is the prime locus of the mass extinction underway, and it is the main focus of the present analysis.

Consider again the higher estimates for the planetary spectrum of species, between 30 million and 100 million[16,20] (for critiques, see refs 18 and 23). As many if not most of the additional species are believed to occur in tropical forests, the true planetary total is not only a matter for speculation. If the real total is 30 million species, the extinction rate will be much higher than the rate postulated on the basis of only 10 million species. But for the sake of being cautious and conservative, let us accept a total of 10 million species. As noted, at least half of these species live in tropical forests, even though remaining forests cover only 6% of the Earth's land surface.

Tropical forests are being destroyed at a rate of at least 150 000 km^2 per year[24–26] (for a slightly lower estimate, see ref. 27). In addition to this outright destruction, an expanse as large again is being grossly disrupted through over-heavy logging and slash-and-burn cultivation, with much degradation and impoverishment of ecosystems and species' life-support systems. But in the interests of being cautious and conservative again, let us consider only forest destruction. The current annual loss represents 2% of remaining forests; and the amount increased by 89% during the 1980s. If present patterns and trends of forest destruction persist, leading to still more acceleration in the annual rate, the current destruction of 2% of remaining forests may well double again during the 1990s.

How shall we translate a 2% annual rate of forest destruction into an annual species extinction rate? An analytic approach is supplied by the theory of island biogeography,[28] being a well established theory with much empirical evidence from on-ground analyses around the world[29–31] (for a dissident view, see ref. 32). The theory states that the number of species in an area rises or falls as approximately the fourth root of the increase or decline in the area, with actual values varying between the third and fifth roots. So when a habitat loses 90% of its original extent, it can generally support no more than 50% of its original species.

The calculation is cautious and conservative. It depends critically upon the status of the remaining 10% of habitat. If this relict expanse is split into many small pieces (as is often the case with remnant tracts of tropical forests), a further 'islandizing effect' comes into play, reducing the stock of surviving species still more. It is not clear how severe this additional depletion can be. Informed estimates[16] suggest the 50% can readily be reduced to 40%, more generally to 30%, sometimes to 20%, and occasionally even to 10%. Similarly, if most species concerned occur in small, local endemic communities, the percentage loss of species can readily approach the percentage loss of area. Moreover, isolated remnants of forest become prone to additional depauperizing processes such as 'edge effects', also to dessicating effects due to local climate change.[33] So the 90/50 calculation for island biogeography extinctions should be viewed as a minimum estimate.

The most broad-scale application of island biogeography to tropical deforestation and species extinctions has been presented by one of the authors of the original theory, Professor Edward O. Wilson of Harvard University.[16] He estimates a current annual loss of 27 000 species in the forests. As he emphasizes repeatedly, this estimate is extremely optimistic. If we employ a more 'realistic' reckoning by qualitatively incorporating a number of other bioecological factors such as alien introductions, over-hunting, and diseases, the annual total will become larger. (Indeed an alternative analysis[6] proposes that 60 000 to 90 000 species are lost in tropical forests each year.) In addition a good number of species, albeit undetermined in even preliminary terms, is presumably disappearing in other parts of the world. Suppose we accept a bare minimum estimate of 30 000 species eliminated worldwide per year. This means we are witnessing a rate at least 120 000 times higher than the 'natural' rate of extinctions before the advent of the human era, considered to be perhaps one species every 4 years.[34]

True, the tropical-forest calculation is a broad-brush affair, viewing the forests as a single homogeneous expanse even though they are highly heterogeneous. So the island biogeography approach needs to be complemented by a local-scale assessment, available in the form of a 'hot spots' analysis. Hot spots are areas that: (i) feature exceptional concentrations of species with exceptional levels of endemism, and (ii) face exceptional threats of imminent habitat destruction. There are 14 such hot spots in tropical forests and another four in Mediterranean-type zones (plus others, unappraised as yet, in coral reefs, wetlands, montane environments, and many islands among other localities rich in endemic species). Collectively, these 18 hot spots contain 50 000 endemic plant species, or 20% of all the Earth's plant species confined to 0.5% of its land surface. So far as we can determine, they contain an even larger proportion of the Earth's animal species with highly localized distributions. Most of these hot spots have already lost the great bulk of their biodiversity habitats, and it is reasonable to suppose that many thousands of species are disappearing annually in hot-spot areas alone.[35,36]

So much for the current extinction rate. What of the future? Through detailed analysis backed by abundant documentation, Wilson[16] considers we face the prospect of losing 20% of all species within 30 years and 50% or more thereafter. Another long-experienced expert, Raven,[15] calculates that half of all species exist in those tropical forests that, in the absence of adequate conservation measures, will be reduced to less than one-tenth of their present expanse within the next three decades. In accord with island biogeography, Raven concludes—and he stresses this is a conservative prognosis—that one-quarter of all species are likely to be eliminated during the next 30 years. Worse, 'fully half of total species may disappear before the close of the 21st century.' Another two biodiversity analysts, Paul and Anne Ehrlich,[6] assert that 'If the current accelerating trends [of habitat destruction] continue, half of Earth's species might easily disappear by 2050. These estimates are in line with those of several other biodiversity specialists.[14,23,37] If we accept, then, that half of all species are likely to be eliminated within the foreseeable future, this means in turn that two-fifths of the problem is confined to the 18 hot-spot localities.

Consequences for evolution

The loss of large numbers of species will be far from the only outcome of the present biotic debacle, supposing it proceeds unchecked. There is likely to be a significant disruption of certain basic processes of evolution.[38,39] The forces of natural selection and speciation can work only with the 'resource base' of species and subunits available[34,40]—as we have seen, this crucial base is being grossly reduced. To cite the graphic phrasing of Soule and Wilcox,[41] 'Death is one thing; an end to birth is something else.' Given what we can discern from the geologic record, the recovery period, i.e. the interval until speciation capacities generate a stock of species to match today's in abundance and variety, will be protracted. After the late Cretaceous crash, between 5 and 10 million years elapsed before there were bats in the skies and whales in the seas.

Following the mass extinction of the late Permian when marine invertebrates lost roughly half their families, as many as 20 million years were needed before the survivors could establish even half as many families as they had lost.[42]

The evolutionary outcome this time around could prove yet more drastic. The critical factor lies with the likely loss of key environments. Not only do we appear set to lose most if not virtually all tropical forests. There is progressive depletion of tropical coral reefs, wetlands, estuaries, and other biotopes with exceptional abundance and diversity of species and with unusual complexity of ecological workings. These environments have served in the past as pre-eminent 'powerhouses' of evolution, meaning they have thrown up more species than other environments. Virtually every major group of vertebrates and many other large categories of animals have originated in spacious zones with warm, equable climates, notably the Old World tropics and especially their forests.[43,44] The rate of evolutionary diversification—whether through proliferation of species or through emergence of major new adaptations—has been greatest in the tropics, especially in tropical forests. Tropical species, notably tropical forest species, appear to persist for only brief periods of geologic time, which implies a high rate of evolution.[45,46]

As extensive environments are eliminated wholesale, moreover, the current mass extinction applies across most if not all major categories of species. The outcome will contrast sharply with the end of the Cretaceous, when not only placental mammals survived (leading to the adaptive radiation of mammals, eventually including humans), but also birds, amphibians, and crocodiles among many other non-dinosaurian reptiles. In addition, the present extinction spasm is eliminating a large portion of terrestrial plant species, by contrast with mass-extinction episodes of the prehistoric past when terrestrial plants have survived with relatively few losses[47]—and have thus supplied a resource base on which evolutionary processes could start to generate replacement animal species forthwith. If this biotic substrate is markedly depleted within the foreseeable future, the restorative capacities of evolution will be the more diminished.

All this will carry severe implications for human societies extending throughout the recovery period estimated to last at least 5 million years, possibly several times longer. Just 5 million years would be 20 times longer than humankind itself has been a species. The present generation is effectively imposing a decision on the unconsulted behalf of at least 200 000 follow-on generations. It must rank as the most far-reaching decision ever taken on behalf of such a large number of people in the course of human history. Suppose that the Earth's population maintains an average of 2.5 billion people during the next 5 million years, and that the generation time remains 25 years. The total affected will be 500 trillion people.* [48]

Biodiversity and population

What are the main driving forces behind the mass extinction underway? We have already noted the role of population—both present numbers of people and their rate of increase. Let us now consider this factor in more detail.

The role of developing countries

The broad-scale decline of biodiversity has been underway since roughly 1950, i.e. the time when there began a major increase in human encroachment on to wildland environments. Since that time too, the population of developing countries has expanded from 1.7 billion to 4.6 billion people, more than a 170% increase. But this is not to say that more people must necessarily mean fewer species habitats. Many other variables are at work, notably poverty, inefficient agriculture, poor land-use planning, inadequate technology, and deficient policy strategies among other prominent factors of countries concerned; plus adverse exogenous factors of aid, trade, debt, investment, and South/North relationships generally. There are abundant linkages that make the picture far more complex than a simple population/biodiversity equation.[49–51]

Let us establish a rough analytic framework within which to evaluate the impact of population growth on wildland environments. The sudden upsurge in human numbers has often exacted a toll in terms of the environmental underpinnings of human societies, as witness tropical deforestation, desertification, soil erosion, and the like. But population growth has not been the only factor at issue. Soil erosion is as severe in Indiana as in India, even though Indiana's population pressures *per se* are only a fraction of those of India. One of the fastest desertifying coun-

* The statistic 'trillion' is much bandied about by economists, politicians and others. How far do we realize what the statistic means? How can we wrap our minds around such a large number? Reader, try a thought experiment. Make a guesstimate of how long a period is represented by 1 trillion seconds. Do it quickly, don't calculate. If you claim a basic understanding of the world you live in, you should be roughly on target. Then check with your pocket calculator. When this experiment was tried on a gathering of hot-shot scientists in Geneva a while ago, most of them were out by an order of magnitude (this writer by twice as much).

tries is Botswana, with only 1.5 million people in a territory the size of Spain or Texas—and a country where the bulk of desertifying processes is due to the activities of a few hundred large-scale cattle ranchers who make up only 5% of the livestock-owning community.

So population is no more than one among many variables, even though it can often rank as *primus inter pares*. Important too are technology types, energy supplies, economic systems, trade relations, political persuasions, policy strategies, and a host of other factors that either reduce or aggravate the impact of population growth. All the same, population can be a prominent if not pre-eminent factor when it exceeds the capacity of a country's natural-resource base to sustain it, and likewise exceeds the capacity of development planners to accommodate it.[52,53] In these circumstances it causes—or at the very least it makes a major contribution to—a spillover of human communities on to biodiversity habitats.

Consider the case of Kenya, a country with exceptional biodiversity. The 1996 population of 28 million people is projected to expand to well over 100 million by the time zero growth is attained in the twenty-second century. Yet even if the nation were to employ western Europe's high-technology agriculture, it could support no more than 52 million people off its own lands;[54] and even if Kenya were to achieve the two-child family forthwith, its population would still double because of demographic momentum (48% of Kenyans are under the age of 15, meaning that large numbers of future parents are already in place). So Kenya will have to depend on steadily increasing amounts of food from outside to support itself. Regrettably, and in large part because of its high population growth rate (about 4% during the 1980s), its per-capita economic growth since 1970 has been well under 2%.[55] Worse, Kenya's terms of trade have long been declining until they have recently been negative in more years than not, meaning the country faces the prospect of diminishing financial reserves to purchase food abroad. Its export economy will have to flourish permanently in a manner far better than it has ever achieved to date if the country is to buy enough food to meet its fast-growing needs. Worst of all, the country will have to undertake this challenge with a natural-resource base from which forests have almost disappeared, watershed flows for irrigation agriculture are widely depleted, and much topsoil has been eroded away.[56]

We must anticipate, then, that Kenya will continue to feature growing throngs of impoverished peasants who will encroach on to whatever environments are available for them to gain their subsistence livelihoods. As that will usually mean more cutting down of forests, digging up of grasslands and cultivation of steep slopes (the phenomenon of 'marginal people in marginal environments'), these peasant farmers have the capacity to induce exceptionally pervasive injury of Kenya's wildlands throughout the foreseeable future. The country's forests, being the most species-rich habitats, covered 12% of national territory in 1960, when the population was only 6 million. As the forests often occupy fertile lands where property rights are vague at best, they have been a prime focus for agricultural settlement on the part of landless peasants, until today they occupy only 2% of Kenya's land area.

Tropical forests and the shifted cultivator

Let us now move on to a larger-scope perspective and examine the generic case of tropical forests. As we have seen, these forests contain at least 50% of all species in only 6% of the Earth's land surface; and they are being depleted faster than any other large-scale biome. This is where the mass extinction envisaged will largely occur—or be contained. If we allow the forests to be virtually eliminated within the next few decades, there will be a massive loss of species regardless of our conservation efforts in the rest of the world; and if we manage to save most of the forests while allowing other major environments to be widely depleted, we shall surely avoid a mass extinction when viewed in proportion to the planetary complement of species. Tropical forests are uniquely critical to the biodiversity prospect writ large, hence they receive special emphasis here. Of current deforestation, only one-fifth is due to excessive logging, which has hardly increased since the early 1980s. Roughly one-tenth is due to cattle ranching, which has actually declined somewhat. One-seventh is due to road building, dam construction, commercial agriculture, fuelwood gathering, and other peripheral activities, which have likewise shown scant increase. Over three-fifths is attributable to the slash-and-burn farmer, with fast-expanding impact.[57] This principal agent of deforestation operates not so much as a shifting cultivator of tradition. Rather he is a displaced peasant who finds himself landless in established farming areas of countries concerned and feels obliged to migrate to the last unoccupied public lands available, tropical forests.[58,59] Driven significantly by population growth and resultant pressures on traditional farmlands (albeit often cultivated with only low or medium levels of agrotechnology, hence cultivated in extensive rather than intensive fashion), slash-and-burn agriculture is the main source of deforestation in such leading tropical-forest countries as Colombia, Ecuador, Peru, Bolivia, Ivory Coast, Zaire, Madagascar, India, Thailand, Indonesia, and Philippines.[24] Countries with acute population pressures include nine of the 14 tropical-forest hot spots, with endemic plant species comprising 9% of the planetary total.

Being powerless to resist the factors that drive him into the forest, the shifted cultivator is no more to be blamed for deforestation than a soldier is to be held responsible for starting a war. His life-style is determined by a host of factors—economic, social, legal, institutional, political—of which he has scant understanding and over which he has virtually no control.[60,61] But he advances on forest fringes in ever-growing numbers, pushing deeper into the forests year by year. Behind come more multitudes of displaced peasants. By contrast with the shifting cultivator and his sustainable use of forest ecosystems, the shifted cultivator is unable to allow the forest any chance of regeneration.

To reiterate, many other factors are at work in the shifted cultivator phenomenon. They include peasant poverty, maldistribution of established farmlands, inequitable land-use systems, lack of property rights and tenure regimes, low-level agrotechnologies, insufficient policy support for subsistence farming, deficient rural infrastructure, and inadequate development overall. But population growth is likely to be a prominent if not the predominent factor in the progressive expansion of shifted cultivator numbers foreseeable, and hence in the accelerating rates of deforestation and of species extinctions.

Again, we must be careful not to over-state the case. In the island of New Guinea, population pressures are slight to date (8 million people in 827 000 km^2), and appear unlikely to become unduly significant within the foreseeable future. Much the same applies to the countries of the Zaire Basin in central Africa; to the countries of the Guyana shield, namely Guyana, Suriname, and French Guiana; and to the western sector of Brazilian Amazonia. All in all, these areas comprise roughly 3 million km^2 of tropical forest, or two-fifths of the remaining biome.

As a measure of the scope for increasing streams of shifted cultivators to migrate into the forests, consider three salient factors. First, tropical forest countries will not only account for the bulk of population growth (an extra 3 billion people worldwide within the next 40 years). More significantly, the numbers of shifted cultivators have recently been growing far faster than national populations, usually at rates between 4 and 16% annually,[26] meaning their projected doubling time is between 17.25 and 4.3 years.

Secondly, alternative forms of livelihood for landless peasants are becoming still more limited by unemployment problems. Developing countries need to generate 40 million new jobs each year throughout the next 20 years simply to accommodate new entrants into the labour market, let alone to relieve present unemployment and underemployment which often amount to 30–40% of the work force. To put the 40 million figure in perspective, note that the United States, with an economy half as large again as all the developing world's often has difficulty in generating an additional 2 million jobs per year. In Brazil, 1.7 million new people seek jobs each year, over half of them failing to find enough employment to support themselves, and many of them joining the migratory surge toward Amazonia.

Thirdly, there is a pervasive problem of farmland shortage in many if not most developing countries, where land provides the livelihood for about three-fifths of populations and where the great bulk of the most fertile and accessible land has already been taken. A full 200 million farmers have too little land for subsistence needs of food and fuel.[62] For some sample figures for leading tropical forest countries, note that of Indonesia's and Thailand's rural populations in the late 1980s, 15% were landless; of Peru's, 19%; of Myanmar's, 22%; and of the Philippines', 34%. This translated into a total in Thailand of 6.4 million people landless, in Myanmar, 6.7 million, in the Philippines, 11.9 million, and in Indonesia, 19.2 million.[63]

Many rural poor are increasingly encroaching on to tropical forests among other low-potential lands, where they have no option but to overexploit environmental resource stocks in order to survive. Worse, farmland shortages continue to spread among the rural poor at rates between 3 and 5% a year, i.e. faster than population growth rates—although much of the problem has been set up by population growth in the past.[63]

These two last considerations also apply to wildland environments and hence to species habitats in zones other than tropical forests—zones where population pressures already cause much decline in biodiversity. According to Harrison,[64] population growth in developing countries accounted for: (i) 72% of expansion of arable lands during 1961–85, leading to the decline of many natural environments, not only through deforestation but desertification and other forms of land degradation; and (ii) 69% of the increase in livestock numbers during 1961–85, leading to soil erosion and desertification. Several other analyses (e.g. refs 27, 49, 52, 53) supply similar findings, albeit with marginally different statistical conclusions.

Of course these calculations can appear oversimplified, even simplistic. A situation with abundant complexities is the opposite of simple. The multiple linkages are not all direct in their operation, still less are they exclusively causative. But they are there. Just because they cannot be demonstrated and documented in detail does not mean they are any less real than more readily defined linkages. They exert their effect, and they are growing stronger and more numerous. The fact that it is difficult to perceive all the linkages in question may tell us less about the linkages than about our limited capacity to think concisely and systematically about them in their full scope. To the extent that environmental scientists are sometimes reluctant to assert a population–deforestation linkage because they lack information of conventional quantity and quality, they might reflect that, all too unwittingly, an unduly cautious approach to linkage analysis could prove unduly erroneous.

This relates particularly to the response of the policy maker, who may feel that absence of evidence about a problem infers evidence of absence of a problem—whereupon he or she may decide to do nothing, even though that is to decide to do a great deal in a world that is not standing still. The asymmetry of evaluation means that we should be wary of being preoccupied with what can be counted if that is to the detriment of what ultimately counts. As in other situations of uncertainty where a negative outcome carries exceptionally adverse consequences, it will be better for us to find we have been roughly right than precisely wrong.

The role of developed countries

Population in developed countries, in conjunction with related factors of excessively consumerist and wasteful life-styles, also contributes to biodepletion. Consider first these countries' role in tropical deforestation. There are the well known linkages of the hamburger and cassava connections, also the tropical timber trade, whereby developed countries' consumerist life-styles, working in conjunction with inequitable trading patterns, generate market-place pressures that induce deforestation.[65] Not so well known is the debt connection. Tropical forest countries owe roughly half of international debt totalling $1.7 trillion. Debt-burdened countries often feel inclined to overexploit their forest resources in order to raise foreign exchange. More importantly, debt leaves them with fewer financial resources to tackle problems of peasant poverty, landlessness and unemployment. So strong is this linkage in certain instances that a $5-billion reduction in a country's debt can lead to a reduction of anywhere between 250 and 1000 (occasionally more) km^2 of annual deforestion.[66,67] Debt relief could be a potent factor in the case of, for example, Peru with its $23 billion debt and annual deforestation of 3500 km,^{2} and Philippines with debt of $39 billion and annual deforestation of 2700 km.^{2}

At the same time, much biodiversity is threatened within developed countries. In the United States, for instance, more than 2250 plant species, 12% of the total, face extinction within less than 10 years (many species are restricted to a single population).[68] At least half the country's freshwater animal species are severely endangered through disruption of rivers, lakes, and other wetlands.[69] Two Mediterranean-type hot spots are located in California and south-western Australia, where almost 5000 endemic plant species and a still larger number of endemic animal species are confined to habitats undergoing progressive depletion.[36] (To put this plant figure in perspective, note that the British Isles' flora amounts to only some 1600 species, fewer than 20 of which are endemic.) There are many such instances of species concentrations threatened in developed countries, albeit with not nearly such large totals as in tropical forests.

Present threats to biodiversity worldwide are not to be compared, however, with those that could lie ahead through climate change. Global warming, mainly induced to date through profligate consumption of fossil fuels in developed countries, would likely cause widespread depletion of habitats in many parts of the Earth.[70,71] Indeed, burners of fossil fuel could eventually rank second to burners of tropical forests as a source of mass extinction.

Here too lies a role for population growth. While developed-country population growth rates are generally low, they count. Britain, for example, with a 0.2% annual population growth rate, features a net increase through natural growth of 117 000 persons per year. By contrast, Bangladesh, with a 2.0% annual growth rate, features a net increase of 2.4 million persons, 20 times larger. But because the fossil-fuel consumption of each new British person is more than 30 times that of a Bangladeshi, Britain's population growth contributes more than twice as much carbon dioxide to the global atmosphere, and hence to global warming, as does Bangladesh's. (The average British family comprises less than two children, but when we factor in resource consumption and pollution impacts, and then compare the British life-style with the global average, the 'real world' size of a British family is more like 10 children.) Ironically, Britain—a country where family planning benefits exceed costs by a ratio of 5:1[72]—could achieve zero population growth by the simple expedient of eliminating half of all unwanted births.

Population growth in all developed countries will contribute half as much again of the build-up of carbon dioxide in the global atmosphere during the 1990s as will population growth in developing countries. Much the same applies to emissions of ozone-depleting chemicals and toxic substances that likewise degrade extensive sectors of the biosphere. Yet not a single developed country has formulated a population policy within a context of environmental linkages overall.

Policy responses

Fortunately, there are many policy responses available. Consider in particular the scope to tackle the leading biodiversity problem through

Table 1. Population growth and socio-economic factors in leading tropical forest countries

Country	Population (millions)		Population growth rate, 1996 (%)	Population in rural areas, 1996 (%)	Population projected (millions)		Projected size of stationary population (millions)	Per-capita GNP, 1994 (US$)	Foreign debt 1994 (billions) (US$)
	1950	1996			2000	2025			
Latin America									
Brazil	53	161	1.7	24	175	230	285	3370	151
Colombia	12	38	2.1	23	38	49	63	1620	19
Ecuador	3	12	2.3	41	13	18	25	1310	15
Peru	8	24	2.1	30	26	37	48	1890	23
Venezuela	5	22	2.1	16	24	35	41	2760	37
Asia									
India	362	950	1.9	74	1026	1392	1886	310	99
Indonesia	77	201	1.6	69	213	276	354	880	97
Malaysia	6	21	2.4	49	22	32	43	3520	25
Papua New Guinea	2	4	2.3	85	5	8	12	1160	3
Philippines	20	72	2.1	51	75	105	140	960	39
Thailand	20	61	1.4	81	62	74	105	2210	61
Africa									
Cameroon	5	14	2.9	59	15	29	56	680	7
Congo	1	3	2.3	42	3	6	15	640	5
Gabon	n/a	1	1.5	27	2	3	7	3550	4
Ivory Coast	3	15	3.5	54	17	37	67	510	18
Madagascar	5	15	3.2	74	17	34	49	230	4
Zaire	14	47	3.2	71	51	105	172	220	

Adapted from refs 57 and 58.

measures to reduce tropical deforestation on the part of the shifted cultivator. Unless there is a reduction of population growth together with a resolution of landlessness (the latter prospect appears distinctly unpromising),[63] it is difficult to see that much forest will remain in just a few decades' time in most countries in question. Yet the shifted cultivator problem is complex and taxing. Whereas it would be fairly straightforward to relieve deforestation pressures from the commercial logger (by growing timber in plantations on deforested lands) and from the cattle rancher (by expanding sustainable beef production on established pasturelands), no such 'easy fix' is available for the shifted cultivator. His needs can be addressed only through a broad-scale effort to bring him into the development process through socio-economic advancement, redistribution of existing farmlands, reform of land-tenure systems, build-up of agricultual extension services, improvement of rural infrastructure, and provision of agrotechnologies that enable the impoverished peasant to practise more intensive and sustainable farming in traditional agricultural areas, namely the areas from which he feels obliged to migrate.

To date, there seems limited prospect of the shifted-cultivator problem being resolved within a time horizon that will assist tropical forests, unless much more attention is directed by governments concerned, also international development agencies, to the particular development challenges he poses. It is a measure of how far his cause is systematically neglected that we have no sound idea of how numerous his communities are, beyond estimates that range from 300 to 600 million. If the latter total is correct (it may well be an underestimate), the shifted cultivator accounts for over 1 in 10 of humankind. Yet he remains a forgotten figure. Although the issue was identified as long ago as the late 1970s, the amount done to tackle the problem through policy reform of appropriate scope and scale is scant at best.

Most measures to tackle deforestation seek to relieve problems within the forests, even though these measures respond to less than half the problem overall, located as it is in the shifted cultivator issue originating in lands well outside the forests. Many anti-deforestation measures, e.g. more protected areas, reflect efforts to tackle symptoms rather than sources of problems, given that—to reiterate the principal point—the ultimate source of the biggest problem lies in lands far outside the forests. This dimension to the deforestation issue remains beyond the policy purview of most tropical forest planners.

The shifted cultivator issue demonstrates that there is limited scope for biodiversity protection in the traditional strategy of parks and reserves. This is not to say there is no great need for many more such areas. Ecologists and biogeographers consider that in tropical forests alone, we should at least triple the expanse protected—while recognizing that one-third of existing parks and reserves are already subject to agricultural encroachment and other forms of human disruption. Additional forms of degradation are likely to stem in the future from pollution: as much as 1 million km^2 or almost 15% of remaining forests could soon become subject to acid precipitation.[73] There could eventually be further and more widespread depletion from global warming.[74]

This means that protected areas are becoming far less of the sufficient conservation response they were once considered to be. The time has arrived when, as a bottom-line conclusion, we must recognize we can ultimately safeguard biodiversity only by safeguarding the biosphere—with all that entails for agriculture, industry, energy, and a host of other sectors, especially the growth of both population and consumption. Increasingly too it is apparent we may ultimately find we inhabit a world where there are no more protected areas: either because they have been over-run by landless peasants and grand-scale pollution, or because we have learned to manage all our landscapes in such a manner that there is automatic provision for biodiversity.[75]

A final thought on the policy front. The Global Environment Facility has provided half a billion dollars over 3 years to assist biodiversity. This is quite the largest such dispensation ever made. But compare it with what is at stake. Every year, just the commercial value to just the rich nations of just the present array of plant-derived pharmaceuticals is 100 times greater. More significant still, the worldwide amount spent annually on 'perverse' subsidies, i.e. subsidies that inadvertently foster over-loading of croplands, over-grazing of rangelands, profligate burning of fossil fuels, wasteful use of water, over-cutting of forests and over-harvesting of fisheries (to cite but a few examples of activities that also reduce biodiversity) is 30 times greater again. Note too that the shortfall in spending to support those 130 million developing-world women who possess the motivation to reduce their fertility but lack the family-planning facilities, is $3 billion. If we were to take care of these unmet needs—which should be catered for on humanitarian grounds even if there were no population problem at all—we would reduce the ultimate world population size by 2 billion people.

A closing reflection

Biodiversity is indeed a profound problem. It is also a glorious opportunity. As everyone knows, species extinction is irreversible. And we face the prospect of eliminating species in their millions, perhaps half of all that share this planet with us, by virtue of what we do (and don't do) during the immediate future. Time is of the super-essence. Suppose we allow ourselves until, say, the year 2010 to take the vital decisions that will affect our planetary home for hundreds, thousands, and even millions of years. After the year 2010, the processes of habitat destruction will surely have worked up so much momentum that they will be hard to slow down, let alone to halt or even turn around. Till 2010 there are roughly 5000 days. We lose 1% of our manoeuvring room every 7 weeks. Exciting times to be alive!

If we switch from environmental breakdown to breakthrough, we shall be acclaimed way beyond the year 2010. People will look back and say 'Those people of the 1990s, when they realized what was going on, did they ever get to grips with the greatest challenge in history—and didn't they make themselves giants of the human condition till they must have felt ten feet tall.' We live in much more than exciting times.

I am sometimes asked if I am hopeful about our environmental outlook. There are occasions, I admit, when I am reminded of the person defining an optimist and a pessimist. The optimist proclaims this is the best of all possible worlds—to which the pessimist responds that that is probably true. But then I snap out of my downside mood by reflecting that I am a member of what is truly a priveleged generation. No other human generation could ever enjoy a challenge so creative as ours. We have it in our hands to save species at a time when millions of them face terminal threat. And—here's the clincher—we are the sole generation to face such an extreme yet glorious prospect. People in the past have never enjoyed our chance: today's problems have simply never arisen before. Nor will any generation of the future have our chance: if we do not get on with the job, our descendants will be left with nothing but to pick up the pieces.

It is up to us alone. Shall we not delight that we are alive at a time of unprecedented challenge? Are we not fortunate beyond dreams to be scientists, conservationists, and citizens of embattled Earth at this momentous stage in the human enterprise?

Acknowledgements

It is a pleasure to recognize back-up support in the form of many illuminating ideas from my Research Associate, Jennifer Kent.

References

1. Adapted from papers published by the author in professional journals, namely, N. Myers. Mass extinctions: what can the past tell us about the present and the future? *Global and Planetary Change*, 1990, **82**, 175–85; N. Myers, Tropical deforestation: population, poverty and biodiversity. In T.M. Swanson, ed., *The economics and ecology of biodiversity decline* 111–36, Cambridge University Press, Cambridge, 1995; N. Myers, Population and biodiversity. In F. Graham-Smith, ed., *Population: the complex reality*: 117–36, The Royal Society, London, 1996.
2. D. Pimentel *et al.* Conserving biological diversity in agricultural/forestry systems. *BioScience* 1992, **42**, 354–62.
3. P.R. Ehrlich and G.C. Daily. Population extinction and saving biodiversity. *Ambio*, 1993, **22**, 64–8.
4. P.M. Vitousek, P.R. Ehrlich, A.H. Ehrlich, and P.A. Matson. Human appropriation of the products of photosynthesis. *BioScience*, 1986, **36**, 368–73.
5. G.M. Woodwell. When succession fails. In M.K. Wali, ed., *Ecosystem rehabilitation*, Vol. 1: *Policy issues*: 27–35. Academic Publishing, SPE The Hague, Netherlands, 1992.
6. P.R. Ehrlich and A.H. Ehrlich. The value of biodiversity. *Ambio*, 1992, **21**, 219–26.
7. N. Myers. *A wealth of wild species: storehouse for human welfare.* Westview Press, Boulder, Colorado, 1983.
8. M.L. Oldfield. *The value of conserving genetic resources.* Sinauer Associates, Sunderland, Massachusetts, 1989.
9. P. Principe, P. Monetizing the pharmocological benefits of plants. In M.J. Balick, *et al.*, eds., *Tropical forest medical resources and the conservation of biodiversity.* Columbia University Press, New York, 1996.
10. H.H. Iltis, J.F. Doebley, R.M. Guzman, and B. Pazy. *Zea diploperennis* (Gramineae), a new teosinte from Mexico. *Science*, 1979, **203**, 186–88.
11. L.R. Nault and W.R. Findley. Primitive relative offers new traits for corn improvement. *Ohio Report*, 1981, **66** (6), 90–2.
12. A.C. Fisher. *Economic analysis and the extinction of species*, Department of Agriculture and Resource Economics, University of California, Berkeley, California, 1982.
13. P.R. Ehrlich and A.H. Ehrlich. *Extinction: the causes and consequences of the disappearance of species.* Random House, New York, 1981.
14. N. Myers. Mass extinctions: what can the past tell us about the present and the future? *Global and Planetary Change*, 1990, **82**, 175–85.
15. P.R. Raven. The politics of preserving biodiversity. *BioScience*, 1990, **40** (10), 769–74.
16. E.O. Wilson. *The diversity of life.* Harvard University Press, Cambridge, Massachusetts, 1992.
17. D. Jablonski. Causes and consequences of mass extinctions: a comparative approach. In D.K. Elliott, ed., *Dynamics of Extinction*: 183–229. Wiley, New York, 1986.
18. K.J. Gaston. The magnitude of global insect species richness. *Conservation Biology*, 1991, **5**, 283–96.

19. R.M. May. How many species are there on Earth? *Science*, 1988, **241**, 1441–1449.
20. T.L. Erwin. How many species are there?: revisited. *Conservation Biology*, 1991, **5**, 330–3.
21. P.R. Ehrlich and E.O. Wilson. Biodiversity studies: science and policy. *Science*, 1991, **253**, 758–62.
22. R. Ricklefs and D. Schluter, eds. *Species diversity: geographical and historical aspects*. University of Chicago Press, Chicago, Illinois, 1992.
23. R.M. May. How many species inhabit the Earth? *Scientific American*, 1992, **267**, 42–8.
24. N. Myers. *Deforestation rates in tropical forests and their climatic implications*. Friends of the Earth, London, 1989.
25. N. Myers. *Future operational monitoring of tropical forests: an alert strategy*. Joint Research Centre, Commission of the European Communities, Ispra, Italy, 1992.
26. N. Myers. *The primary source: tropical forests and our future*. Norton, New York, 1992.
27. Food and Agriculture Organization. *Tropical forest resources assessment*. Food and Agriculture Organization, Rome, Italy, 1993.
28. R.H. MacArthur and E.O. Wilson. *The theory of island biogeography*. Princeton University Press, Princeton, New Jersey, 1967.
29. R. Ricklefs. *Ecology* (3rd edn). W.H. Freeman, San Francisco, California 1990.
30. C.L. Shafer. *Nature reserves: island theory and conservation practice*. Smithsonian Institution Press, Washington DC, 1991.
31. D.S. Wilcove. From fragmentation to extinction. *Natural Areas Journal*, 1987, **7**, 23–9.
32. B.L. Zimmerman and R.O. Bierregaard. The relevance of the equilibrium theory of island biogeography and species-area relations to conservation with a case from Amazonia. *Journal of Biogeography*, 1986, **13**, 133–43.
33. N. Myers, ed. *Tropical forests and climate*. Kluwer, Dordrecht, 1992.
34. D.M. Raup. *Extinction: bad genes or bad luck?* Norton, New York, 1991.
35. N. Myers. Threatened biotas: 'hot spots' in tropical forests. *The Environmentalist*, 1988, **8**, 187–208.
36. N. Myers. The biodiversity challenge: expanded hot-spots analysis. *The Environmentalist*, 1990, **10**, 243–56.
37. J.M. Diamond. The present, past and future of human-caused extinction. *Philosophical Transactions of the Royal Society of London B*, 1989, **325**, 469–78.
38. N. Myers. *Tackling mass extinction of species: a great creative challenge*, Horace M. Albright Lecture in Conservation, University of California, Berkeley, California, 1986.
39. N. Myers. Questions of mass extinction. *Biodiversity and Conservation*, 1993, **2**, 2–17.
40. N. Eldredge. *The miner's canary*. Prentice-Hall, New York, 1991.
41. M.E. Soule and B.A. Wilcox. Conservation biology: its scope and its challenge. In M.E. Soule and B.A. Wilcox, eds, *Conservation biology: an evolutionary–ecological perspective*: 1–8. Sinauer Associates, Sunderland, Massachusetts, 1980.

42. D. Jablonski. Extinctions: a palaeontological perspective. *Science*, 1991, **253**, 754–7.
43. P.J. Darlington. *ZooGeography: the geographical distribution of animals.* Wiley, New York, 1957.
44. E. Mayr. *The growth of biological thought: diversity, evolution and inheritance.* Harvard University Press, Cambridge, Massachusetts, 1982.
45. D. Jablonski. The tropics as a source of evolutionary novelty through geological time. *Nature*, 1993, **364**, 142–4.
46. S.M. Stanley. *The new evolutionarey timetable.* Basic Books, New York, 1991.
47. A.H. Knoll. Patterns of extinction in the fossil record of vascular plants. In M.H. Nitecki, ed., *Extinctions*: 21–68, University of Chicago Press, Chicago, Illinois, 1984.
48. N. Myers. Biodiversity and the precautionary principle. *Ambio*, 1993, **22**, 74–9.
49. K. Davis and M.S. Bernstam, eds. *Resources, environment and population: present knowledge, future options*, Oxford University Press, New York 1991.
50. N. Myers. *Scarcity or abundance: a debate on the environment.* Norton, New York 1994.
51. R. Repetto. *Population, resources, environment: an uncertain future.* Population Reference Bureau, Washington DC, 1987.
52. P.R. Ehrlich and A.H. Ehrlich. *The population explosion.* Simon and Schuster, New York, 1990.
53. G.D. Ness, W.D. Drake, and S.R. Brechin, eds. *Population–environment dynamics: ideas and observations.* University of Michigan Press, Ann Arbor, Michigan, 1993.
54. Food and Agriculture Organization. *Potential population supporting capacities of lands in the developing world.* Food and Agriculture Organization, Rome 1984.
55. World Bank. *World development report 1996*, Oxford University Press, New York 1996.
56. World Resources Institute. *World Resources 1996–97*, Oxford University Press, New York 1996.
57. N. Myers. Tropical deforestation: population, poverty and biodiversity. In T.M. Swanson, ed., *The economics and ecology of biodiversity decline*: 111–36. Cambridge University Press, Cambridge, 1995.
58. W.J. Peters and L.F. Neuenschwander. *Slash and burn farming in Third World forests.* University of Idaho Press, Moscow, Idaho, 1988.
59. W.C. Thiesenhusen, ed. *Searching for agrarian reform in Latin America.* Unwin Hyman, Boston, Massachusetts 1989.
60. M. Colchester and L. Lohmann, eds. *The struggle for land and the fate of the forests.* Zed Books, London, 1992.
61. J. Westoby. *Introduction to world forestry.* Basil Blackwell, Oxford 1989.
62. R.M. El-Ghonemy. *The political economy of rural poverty: the case for land reform.* Routledge, London, 1990.
63. I. Jazairy, M. Alamgir, and T. Panuccio. *The state of world rural poverty: an inquiry into its causes and consequences.* Intermediate Technology Publications, London, 1992.

64. P. Harrison. *The third revolution: environment, population and a sustainable world.* Tauris, London, 1992.
65. N. Myers. The question of linkages in environment and development. *BioScience*, 1993, **43**, 302–10.
66. J.R. Kahn and J.A. McDonald. *Third World debt and tropical deforestation.* State University of New York, Binghampton, New York, 1992.
67. R.E. Gullison and E.C. Losos. The role of foreign debt in deforestation in Latin America. *Conservation Biology*, 1993, **7**, 140–7.
68. L. Robert. Extinction imminent for native plants. *Science*, 1988, **242**, 1508.
69. J.E. Williams, J.E. Johnson, D.A. Hendrickson, D.A., S. Contraras-Balderas, J.D. Williams, M. Navarro-Mendoza, D.E. McAllister, and J.E. Deacon. Fishes of North America endangered, threatened or of special concern, 1989. *Fisheries*, 1989, **14**(6), 2–38.
70. R.L. Peters and T.E. Lovejoy, eds. *Consequences of the greenhouse warming to biodiversity.* Yale University Press, New Haven, Connecticut, 1992.
71. G.M. Woodwell, ed. *The earth in transition: patterns and processes of biotic impoverishment.* Cambridge University Press, New York, 1990.
72. V. Estaugh and J. Wheatly. *Family planning and family well-being.* Family Policy Studies Centre, London, 1991.
73. H. Rodhe and R. Herrera, eds. *Acidification in tropical countries.* Wiley, Chichester, 1988.
74. G.S. Hartshorn. Ten possible effects of global warming on the biological diversity in tropical forests. In R.L. Peters and T.E. Lovejoy, eds, *Global warming and biological diversity*: 137–46. Yale University Press, New Haven, Connecticut, 1992.
75. J.A. McNeely. The future of national parks. *Environment*, 1990, **32**(1) 16–20, 36–41.
76. C. Haub and M. Yanagishita. *World population data sheet, 1996.* Population Reference Bureau, Washington DC, 1996.

NORMAN MYERS

Born 1934 in Yorkshire. Graduating from Oxford and Berkeley, he works as an independent consultant in environment and development, focusing on a variety of issues such as mass extinction of species, tropical forests, environmental economics, population, global warming and environmental security, all within a context of sustainable development. He has received various honours and awards for his efforts on behalf of the environment, including the Sasakawa Environment Prize and the Volvo Environment prize, and is a Pew Scholar in Conservation and Environment. He has had more than 200 papers published in professional journals, and is the author of 250 popular articles in newspapers and magazines and twelve books with sales of one million copies in eleven languages. In 1998 he was appointed a Companion of the Order of St Michael and St George for 'services to the global environment.'

Self-assembly: nature's way

KUNIAKI NAGAYAMA

Introduction

What do you visualize from the word 'nature': the Lake District loved by William Wordsworth or the Woods of Walden Pond where Henry David Thoreau lived? For me as a Japanese person, 'nature' is first of all beautiful green mountains and rivers. This probably reflects the environment where I grew up. However, the nature I am going to discuss here is not nature on such a large scale. It is that part of nature which is internal to us and intrinsically of a very small scale. We cannot stop marvelling at the ingenious behaviour of very small creatures such as ants and butterflies. However, it seems that behind this marvelling there is an unconscious prejudice based on our own size as humans.

To help you understand how big a human is as a life form, let us borrow the wisdom of 'Gulliver'[1] and compare the size of a human with creatures both in the small and large worlds. For example, the ratio between a 10 mm ant and a 1 μm cell is the same as the ratio between a city 10 km wide and a 1 m human. And the ratio between a 1 μm cell and a 1 m human is the same as the ratio between a 1 m human and Great Britain which is about 1000 km long. As cells are the basic unit of life in small-scale nature, this means that our human life form has the same complexity as that of Great Britain, including the city of London in it. In this sense a human is actually monstrously large beyond our imagination.

Now we are at the door to a wonderland of nature developing in the small scale. This chapter will give you a key to open it and step inside. In the first part, in addition to scale I will address force, which is responsible for assembling materials, and next information, which is necessary to produce complex materials. In the second part several examples with marbles, coins, colloid particles, and protein molecules will be shown. You will understand how scale, force, and information works cooperatively to produce complex forms. In the last part I will summarize and try to make a contrast between our own technology for assembling materials and nature's way of manufacturing.

Nature's way to assemble materials in complex forms

Scale is a question of geometry but we have to address physical force when we are dealing with a concrete substance. When considering the difference between the forces which work in nature at both small and large scale, the ratio of the surface area to the volume of objects is very important. In the large-scale world, the dominant force is gravity. In fact, the structure of the universe and our civilization are severely influenced by gravitational force. On the other hand, in the small world, for life forms and tools in our daily life, the dominant factor is the intermolecular force. The intermolecular force bonds matter. Why did these two forces, gravity and the intermolecular force, find different niches? This is because the intermolecular force is proportional to the surface area whereas gravity is proportional to the volume. The force ratio (N) between the intermolecular force and gravity can be rewritten by the ratio of the surface area to the volume. The ratio of the surface area to the volume is reciprocal to the size (s) and therefore the smaller the object, the larger the ratio. Thus in the small scale world, the intermolecular force becomes dominant. Two worlds, of large and small scale nature, are divided by a straight line representing the scale and force relation ($\log N = -\log s + c$) and only the intermolecular force drives self-assembly, which is a phenomenon occurring in the small world.

Before moving on to illustrate self-assembly phenomena on a small scale, let us, as a reference, consider how gravity works to assemble matter. Suppose some marbles are sandwiched between a pair of transparent plastic plates. Obviously they wouldn't self-assemble. But if the plate is tilted, they assemble because of gravity. In order to make a hexagonal lattice, a border with a specific shape has to be designed, and the marbles line up in an orderly fashion. Although there is no force between the marbles, a hexagonal lattice results from cooperation between gravity and the borders. Gravity is widely used in industry for assembling things. Remember, too, that some rocks such as sandstone and sedimental rock are also made by gravity during a geological era on the earth.

A living entity makes itself, but the intermolecular force alone is not enough to build such a complex structure. If an object is built solely by the intermolecular force, then we can only reach the level of sand and rocks. Another decisive factor for making a complex structure is required, and that is information. When you make a complex tool, you need a design drawing, and this is what I call information.

Figure 1 shows what is actually controlled in life by the genetic information inherited from parents. We should also bear in mind the fact that all the genetic information is used to define the sequence of amino acid polymers. When this amino acid sequence is defined, a single unique three-dimensional structure of the protein is then determined.

Each of thousands of proteins in our body has a specific structure,[2] which gives rise to specific functions and activities. Particularly important among these is molecular recognition, based on the intermolecular force. When they freely move in aqueous solution, as revealed by Brownian motion, proteins recognize each other and build up complex structures. The general function of molecular recognition is to assemble proteins together into an architecture more complicated than that of the constituent proteins. Of course other biomolecules such as nucleic acids and lipids are occasionally required to obtain larger architectures but the main character in the story is always the protein. Complicated architectures thus made are often called supramolecules.[3] Examples of supramolecules fabricated by self-assembly of proteins include cell organelles such as mitochondria and chromosomes. Looking through all this sequential process of biological fabrication, we notice that the most important step lies in the first part, where the genetic information is materialized into form of the protein.

Let us see now what is the outcome of all this biological manufacturing. Three examples of machines made by nature are shown in Fig. 2. First, special bacterium that lives in the Dead Sea in Israel fabricates a

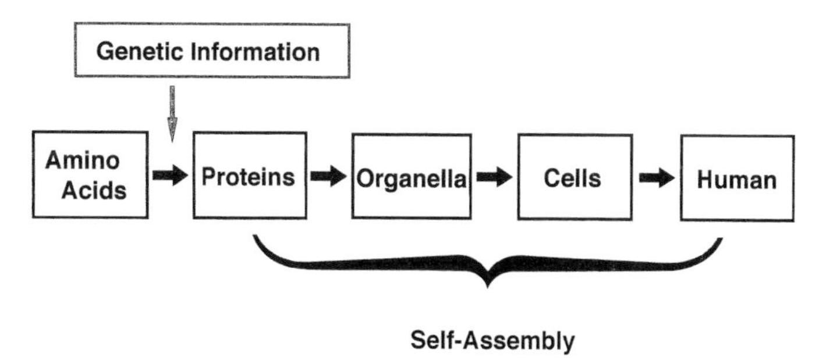

Fig. 1 A schematic model to depict the biological way of manufacturing complex forms. Genetic information is transferred exclusively to protein molecules in the form of specific structures. The subsequent manufacturing proceeds automatically by the proteins power of molecular recognition and the resultant self-assembly of constituent protein units.

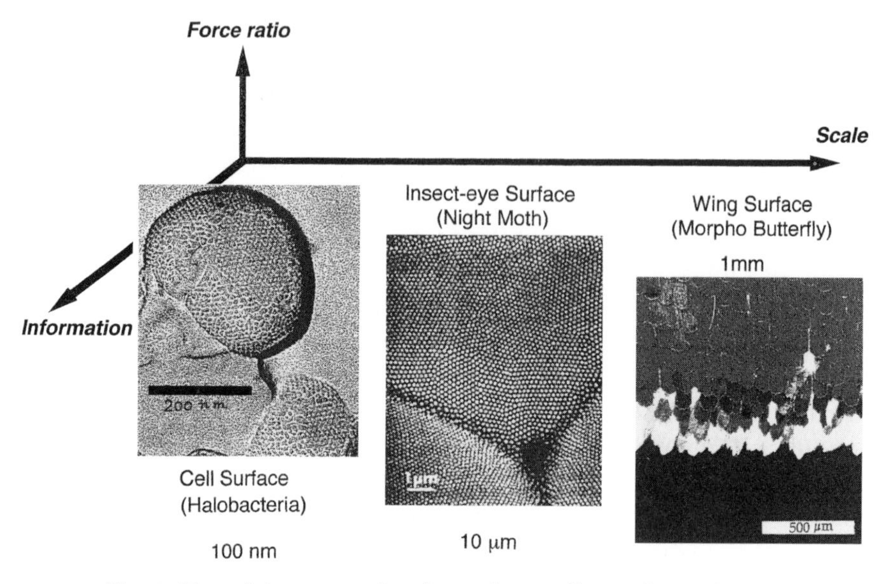

Fig. 2 Plot of three complex forms (naturally made machines) in the information–force–scale space. Cell surface: two-dimensional crystalline array of bacteriorhodopsin molecules. Insect-eye surface: two-dimensional array of 150 nm nipples (taken from ref. 20). Wing surface: two-dimensional arrays of wing scales of morph *sulkowskyi*.

two-dimensional protein crystal on its surface. This protein, bacteriorhodopsin, which converts light energy to chemical energy, can be transferred to an artificial cell surface (liposome). As shown on the left-hand side of Fig. 2, it also exibits two-dimensional crystalline arrays under appropriate solution conditions. Two other examples of two-dimensional assembly of identical units are particle arrays in the compound eye of a moth[4] and wing scales of a butterfly. Each of the arrayed particles on the moth's eye has a size comparable with the artificial cell. The size of the scales is almost equal to that of a single component in the compound eye.

These orderly two-dimensional patterns remind us of the micron-scale two-dimensional patterns seen in the highly successful microelectronics used in computers. Actually, tools, devices and machines made by humans often have a similar complexity to those observed in living organisms. The smallest extreme is the microchip of integrated circuits and the largest extreme is the space station which is completely human-made. In both cases, we can observe some regular pattern of repeating units on their surfaces irrespective of their size difference. On the microchip surface the repeating units correspond to the memory array and the repeating tiles on the surface of the space station correspond to the solar batteries. Let us think then what would happen when we put together two dissimilar but complex machines, nature-made and human-made. Isn't it very wonderful if the protein two-dimensional crystals as observed on the cell surface could be used as a type of functional elements in the human-made machines mentioned here?

Self-assembly of particles and proteins into two-dimensional arrays

We wondered if we could control the orientation of a protein, increase the degree of integration and create new functions without damaging its intrinsic functions. Inspired by the biological way of manufacturing, I began a project (Fig. 3) together with colleague researchers. This project which was called the 'Protein Array Project' has lasted for 5 years from 1990 to 1995. As proteins are key materials in the nature's manufacturing, we have also stuck to protein molecules to make complex architectures. As you see from the flow chart at the top of Fig. 3, the manufacturing process is very similar to nature's way. We tried to interrupt the genetic

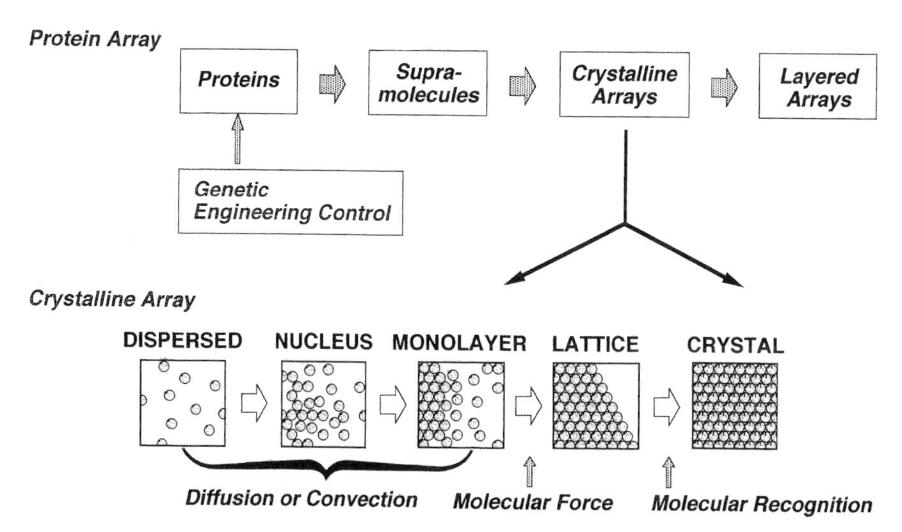

Fig. 3 Mission and goal of the Protein Array Project (1990–95). The Protein array project tried to apply genetic information control (molecular recognition) to arrange molecules in the form of an array suitable for industrial materials.[5] The goal of the project was to find the engineering for fabricating two-dimensional crystalline arrays of protein molecules as shown here.

process and take over its information to create proteins with an appropriate ability for molecular recognition by adopting genetic engineering. Then those protein units are assembled into a larger architecture step by step. Among the whole series of steps, we have exclusively focused on the process of forming a crystalline array in its two-dimentional form attached to a surface. Before finding any specific applications, we decided to develop a technique enabling us to make two-dimensional protein arrays at will. Remember that cells are invisible and a protein is one-hundredth of the size of a cell. They are not something you can pick up and sort with a pair of tweezers. We needed a new fabrication technique which works in the same way as nature. (see the bottom trace in Fig. 3).[5]

First, we start from a protein dissolved in water. Small nuclei are created and then collected to obtain a monolayer particle film. The protein molecules in this film are then packed together tightly to obtain a two-dimensional lattice. Finally, the orientation of the protein molecules is adjusted to obtain a crystal. In the early stage, the intermolecular force and mass flow are used as the means of assembly. In the last stage, a uniform orientation is achieved by specific interactions between protein molecules, that is, molecular recognition.

Incidentally, when we take a careful look at the whole process of this protein assembly, we can recognize that the main problem is how the protein can be collected in this two-dimensional fashion when they are dissolved and spatially dispersed in water. A traditional answer to this problem is the use of adsorption of protein molecules from an aqueous solution to the surface of a solid or a liquid.

Let us move on to a second illustration. If I put some Japanese coins in a glass plate and try to assemble them using gravity just as before, because the brim of the plate is round, the coins will not make a nice hexagonal lattice. Is it possible to make them self-assemble into a two-dimensional crystal without the help of gravity? It seems impossible, but it can be done by using water. There is a reason why I chose Japanese coins. One yen coins, each worth about half of a penny, are made of aluminum to save cost. Because of this, they are very light and float on water. Strange as it may seem, the aluminum coins floating on water attract each other. They come into contact one after another and eventually form a hexagonal lattice. This experiment tells us two things: that is, liberation from gravity, and the birth of a macroscopic attraction due to the surface tension of water. This is a typical instance of self-assembly. The water surface is now giving us a working place widely utilized to spread molecules on.

As shown in Fig. 4, a monolayer film of lipid molecules can spread on a water surface in a monolayer arrangement.[6] Protein molecules in

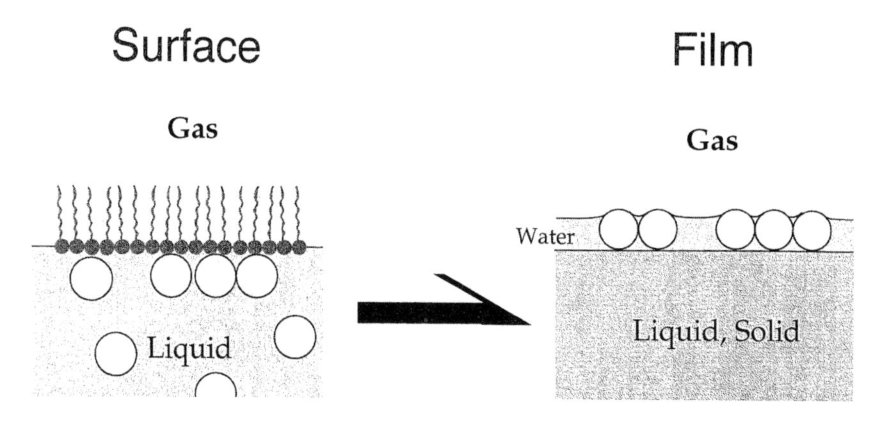

Fig. 4 Two methods employed in making two-dimensional crystalline arrays of protein molecules. Left: traditional method using lipid monolayer as the adsorption substrate for molecules.[7] Right: liquid film method innovated in the Protein Array Project.[8] (adapted from fig. 1 in K. Nagayama *et al.*, *Jpn. J. Appl. Phys.*, 1995, **34**, 3947.)

aqueous solution adsorb on the lipid monolayer as shown in the left side of Fig. 4.[7] However, it can be difficult to obtain a sufficiently high concentration on the surface as protein molecules also tend to dissolve to the bulky aqueous phase. Instead, as shown on the right-hand side, we considered using a completely two-dimensional box, namely a liquid film, to hold the protein. Our method of using liquid films is highly versatile and can be applied to various fine particles of varying sizes such as proteins, viruses, and colloids.[8] There are two points I want to make here. One is that the particles have to be able to move in this kind of liquid film. In this sense, adsorption onto the substrate plane is not beneficial but harmful. The other point is that there has to be an adequate attractive assembling or packing force between the molecules or particles.

As I mentioned in the beginning, the size difference manifests itself as the difference in the dominant force that governs the phenomenon. Table 1 shows different types of particles used in our experiments, and also their size and the assembling forces. Notice that there is a millionfold size difference between these samples. Depending on the sizes of samples, forces utilized to assemble matters are varied as shown in the most right column on the table. In the first and second experiments, marbles and coins have been used to demonstrate two-dimensional assembly by gravity and by the surface tension of water.

Table 1. Materials used in experiments on self-assembly

Materials	Scale	Forces
1. Marble	~1 cm = 10^{-2} m	gravity
2. Coin (aluminum)	~1 cm = 10^{-2} m	gravity and surface tension
3. Polystyrene particle	~1 μm = 10^{-6} m	flow (hydrodynamic force) surface tension
4. Proteins ferritin bacteriorhodopsin	~10 nm = 10^{-8} m	protein interaction or surface tension

Next, let us examine the assembly of polystyrene particles with a size of 1 μm, which is invisible and one-thousandth of the size of marbles and coins which I have shown so far. In this small world, gravity is almost negligible. Instead, the intermolecular force, particularly the adsorption between surfaces, becomes important.

The colloid particles are usually dispersed in water as a suspension. In our daily life we often observe that when suspensions of particles are spilled on to a flat place, for example a kitchen plate, and are dried, they leave white powder behind. Then a close up of the drying process of the colloid suspension provides some instructive phenomena about the particle assembling. The suspension of polystyrene spheres looks white because the 1 μm particles reflect light randomly. Putting it on a glass plate, we observe that the suspension of polystyrene particles spreads out and under a microscope we can see minute motions of the particles. This is nothing other than the famous Brownian movement. This random motion suddenly becomes directional when the container of water which holds the particles becomes thin. A beautiful unidirectional movement begins to develop under the microscope and the particles assemble in an ordered fashion. Both the Brownian motion and the unidirectional motion of particles look as if they occur without any force applied from outside. The key to understanding self-assembly lies in these two types of motion: they are the origin of the fabrication of objects in the small world, free from the dominance of gravity. The driving force of Brownian movenent is the collisions due to the thermal energy of water molecules. What, then, drives the remarkable phenomenon of the particles crystalline assembly?

The left side of Fig. 5 explains the mechanism behind it. First, we need a suspension in which the particles can be well dispersed and a substrate on which the suspension can spread to form a stable thin film. In the beginning, Brownian motion occurs in this adequately thick liquid film. However, once the thickness of the liquid film becomes

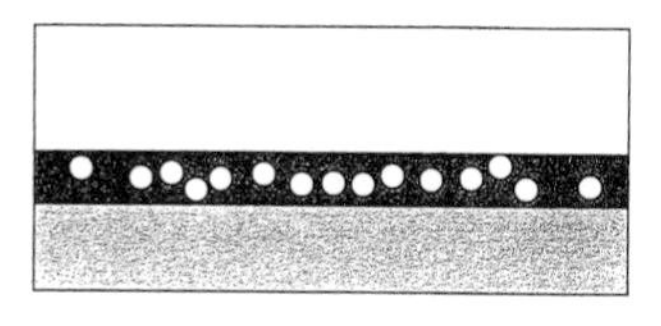

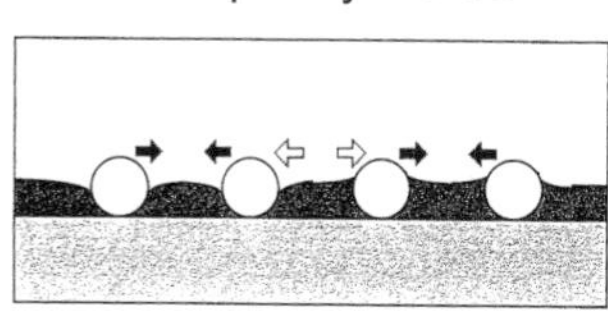

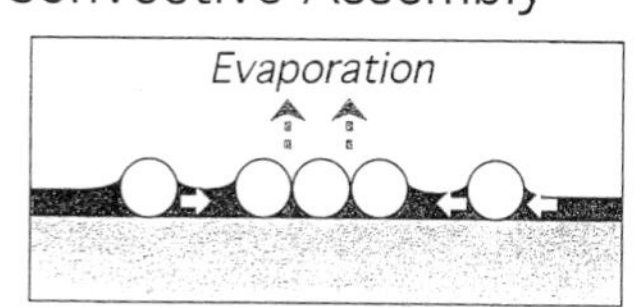

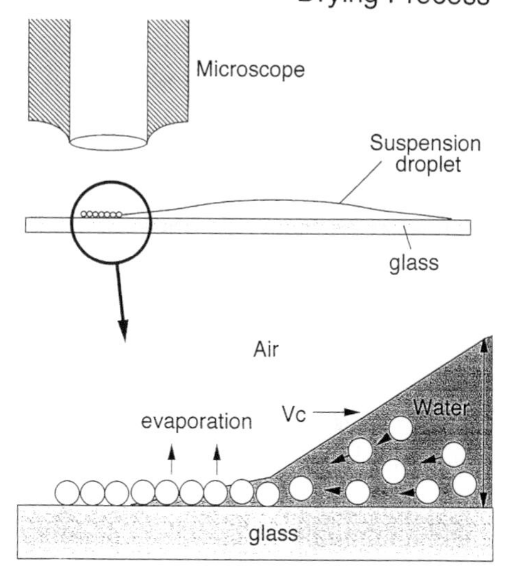

Fig. 5 Schematics illustrating the mechanism of crystalline array formation of colloidal particles which are initially dispersed in water.[10] Left: three requirements for the technology (refer to the text). Right: illustration of drying process of particle suspension.

comparable with the size of the particles, a completely new phenomenon arises. One is an attractive force originating from the surface tension between particles which stick out from the surface. This is the equivalent of the attractive force exhibited by the aluminum coins, which we saw earlier. This attractive force, which has been named 'lateral capillary force',[9] is quite intriguing. We discovered that in the two-dimensional world it has a similar distance dependence to the Coulomb force within a limited distance. Now the Coulomb force is the universal electric force on which, in fact, all intermolecular forces in our universe are based. Just as there are positive and negative electric charges and attractive and repulsive forces in Coulomb force, this lateral capillary force also has attractive and repulsive forms.[9] In assembly, it packs particles into a hexagonal array.

The second phenomenon has something to do with the flow of particles we have just seen. It looked as if only the particles were flowing, but in fact they were carried by the water current caused by the evaporation of water in this area. This is because when suspension dries by evaporation, water from the outer area flows toward the dry area to wet it. The situation is more clearly seen in the schematics in the right-hand side of Fig. 5. The phenomenon in which particles are carried and assembled by the water current is named 'connective assembly'.[10] This is a much faster process than Brownian motion, and one feels that industrial applications must be possible. A monolayer particle film is obtained by matching the movement of this meniscus to the rate at which particles are carried to the crystal boundary. If the movement of the meniscus is halved, twice as many particles are carried in and a bilayer particle film forms.

However, in the previous experiment, the velocity of the meniscus of the suspension was not explicitly controlled but left rather as it went. And the consequence is a non-uniform particle film with an alternating repeat of monolayer, and multi-layers. It is a difficult problem to design controlled fabrication to produce high quality particle films with a uniform monolayer.

Remember that this kind of small world was free of gravity. Let us take full advantage of this. A gravity-free environment means that the concepts of vertical and horizontal disappear. Therefore, horizontal experiments can be made vertical as well. The small tools with a vertical geometry shown in Fig. 6, where a glass plate was pulled out of suspension, make the engineering control very tractable.[11] With a vertical arrangement things turn out to be dramatically simplified and the particle film forms on the glass in a controlled manner. Conversion from horizontal thinking to vertical thinking, or vice versa, gives birth to a completely new idea: we Japanese have 150 years of experience of translating horizontal English written from left to right to Japanese, which is originally written vertically from right to left.

The experiment in Fig. 6 shows the controlled fabrication of crystalline films on a glass plate. A horizontal microscope is used to enlarge the scene on the screen. As shown on this screen, the particles are flowing upwards, apparently defying gravity. When the glass substrate is pulled more rapidly, an imperfect monoparticle film with voids is obtained, while well adjusted pulling velocity produces a perfect monoparticle film as you have seen by now.

The inset picture on the right-hand side of Fig. 6 reminds us of bubbles in lager beer, which are continuously born in the glass and then move upward; they look very similar but mechanism behind them is quite different. In our case solid particles heavier than water are carried upward by the water current stimulated by evaporation whereas the bubbles float to the surface by bouyancy.

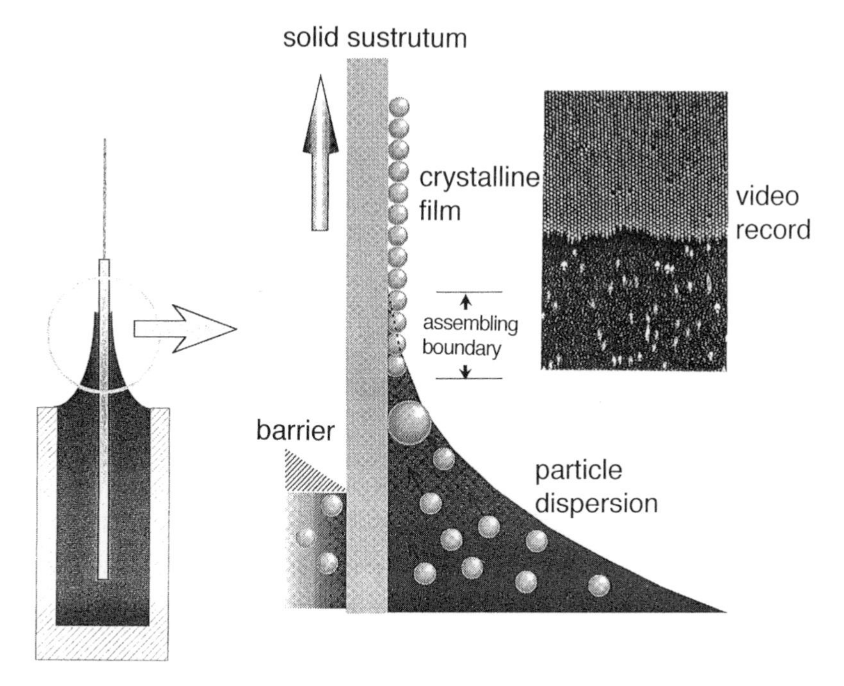

Fig. 6 Schematics illustrating the same process as in Fig. 5 but in a vertical set-up. Conversion from horizontal to vertical fabrication improves the control of the process for industrial purposes [11]

Let us now return to the two-dimensional assembly of proteins. Proteins are ultrafine particles about one-hundredth as big as the colloid particles we have just seen. Therefore, we encounter different problems from those we have met thus far. The greatest difficulty is that it is almost impossible to make a water film as thin as protein molecules. There are two reasons for this. First, such a thin layer of water would immediately evaporate or grow into a thicker film. Secondly, attempts to make a thin water film tend to end up with it rupturing into water droplets: recall that the condensation that we see on the window pane in winter is not a water film but water droplets. Now we can understand how difficult it is to make a very thin water film with a thickness similar to that of a protein molecule.

Table 2 shows the minimum thickness of a stable water film made on flat surfaces of various solids and liquids. In the case of polystyrene spheres, we used a water film on a glass surface. The lower limit of the thickness here is one-tenth of 1 μm, that is 100 nm. Below this limit a stable film cannot be made because the water film turns into water

Table 2. Stability of wetting films on various solid and liquid substrate surfaces

Substrate	Minimum thickness (d_{min})	Reference
Glass	~100 nm	10,12
Silicon	~100 nm	13
Mercury	~10 nm	14
Soap film[a]	~2 nm	15
Water on water[b]	0 nm	16

[a] Free film in the air.
[b] A protein solution spreading on a glucose solution.

droplets. The situation is different when we leave solid surfaces and turn to liquid surfaces. We tested three substrates, namely, a mercury surface, a water surface, and a soap film. (a water film as thin as 10 nm can be formed without rupture): experiments with those surfaces will be next shown.

We chose two proteins which had a size big enough to allow electron microscopic observation; one is an iron storage protein called ferritin, whose X-ray crystal structure is known.[17] As it is 13 nm across and contains a lot of iron, it is easy to observe with an electron microscope and is used as a standard sample; it also has a very regular shape with a cubic symmetry, just like dice. Is it possible to control the orientation of this protein to obtain various two-dimensional crystals? This was the central issue of our Protein Array Project. Another example consists of protein supramolecules made of bacteriorhodopsin. From the purple membrane of halobacterium, which lives in the Dead Sea and sustains a two-dimensional surface crystal structure as mentioned above, we can fabricate, a ball-shaped structure of protein supramolecule by using a surfactant. It has a diameter of almost 40 nm, which is extremely large for a protein.[15]

Next we describe how a protein suspension can be spread over a substrate consisting of an aqueous solution. To explain how it is possible to spread an aqueous solution on another aqueous solution without severe mutal mixing, a schematic is shown in Fig. 7. As water is much easier to handle than mercury, the apparatus is also very cheap and simple. To fabricate protein two-dimensional crystals on water we use a Teflon trough. The mercury apparatus costing £50 000 has turned into a £1 Teflon container: this kind of drastic cost reduction is a proof of our scientific success! Of course the mercury apparatus was not wasted, as it was necessary to make the initial breakthrough. However, at this stage of our research, it has become a historic relic. We have used videos to

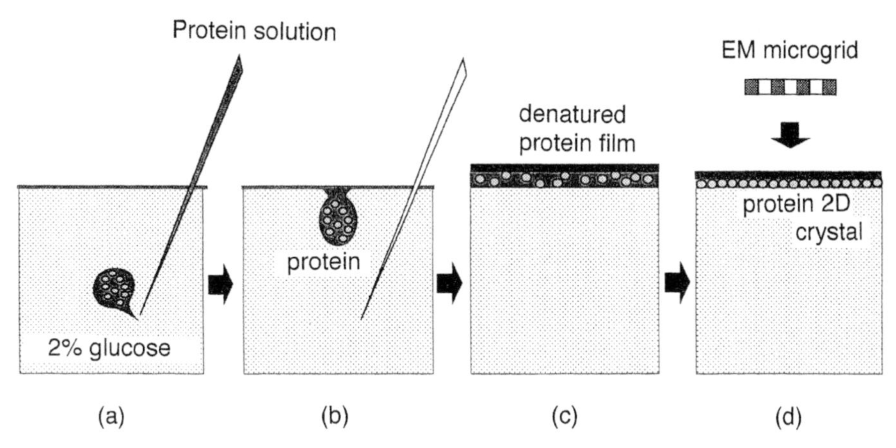

Fig. 7 Schematic illustration of the experimental procedure for spreading a protein suspension on a water surface to make two-dimensional crystals. EM: electron microscope.

show how ferritin molecules align on a glucose solution to give a two-dimensional crystal.

I would like to explain how the orientation is controlled in ferritin two-dimensional crystals. The examples of self-assembly we have seen so far did not explicitly use information or molecular recognition, which is actually the heart of the remarkable power of protein that I have been explaining. The ferritin took three different crystal structures, which was achieved by modifying the protein surface through genetic engineering, i.e. by altering ferritin's intermolecular interaction, or molecular recognition. Detailed observation of the crystal structure reveals that the 84th amino acid residue, its aspartate and the 86th amino acid residue, its glutamine simultaneously bind one cadmium ion.[17] Two identical bonding sites hold the ion from both sides as shown in Fig. 8 (Bottom left). The hexagonal two-dimensional crystal we have seen corresponds to a (111) cross-section of this three-dimensional crystal (top left in Fig. 8). This is because the molecular orientation in this cross-section allows the maximum number of cadmium ion interactions (top right in Fig. 8). In other words, the crystal is stabilized even by reducing its dimension if it is cut in the two-dimensional plane holding many ion binding bridges between adjacent protein molecules.

What would happen if we got rid of these cadmium ion binding hands by altering the binding site? We made a mutant by changing the 84th and 86th amino acids to a neutral amino acid (bottom right in Fig. 8). You may then expect an amorphous two-dimensional array due to the lack of binding sites, which was what we had expected initially.

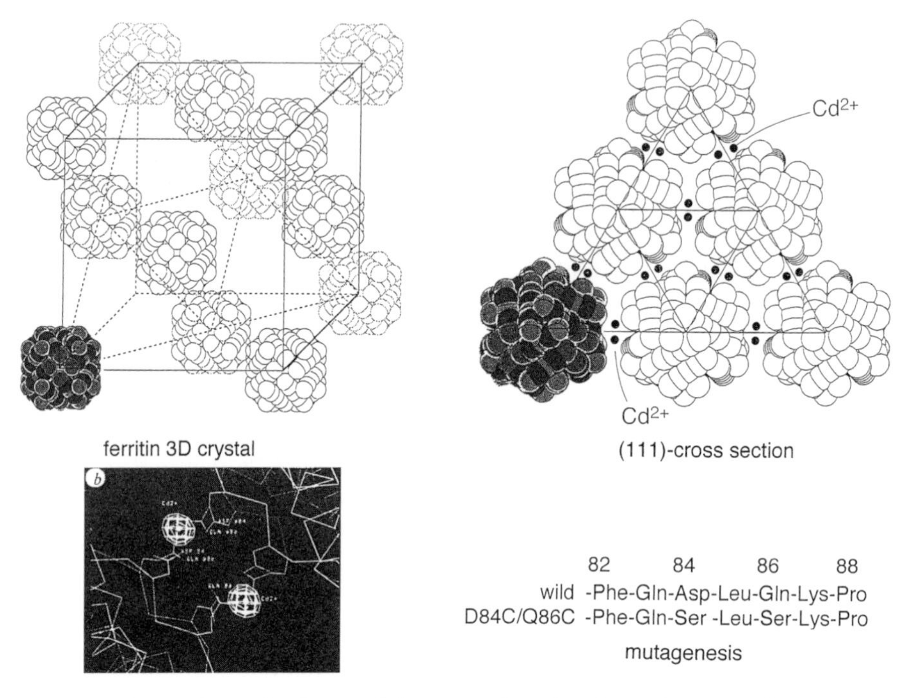

Fig. 8 Control of crystal forms and molecular orientation by mutagenesis.[18]
Top left: a schematic of a three-dimensional crystal of ferritin.[17] Top right: a schematic of the hexagonal two-dimensional crystal of ferritin, which corresponds also to the (111) cross-section of the crystal. Bottom left: a close-up view of the cadmium ion binding site; two-ions bridging adjacent ferritin molecules.[17] Bottom right: a mutagenesis to kill the cadmium ion binding site.

However, this expectation was not fulfilled; instead, a new oblique two-dimensional crystal was obtained (Fig. 9).[18] In this figure, the left-hand side is hexagonal and the right-hand side is oblique. At first glance they look alike, but the diffraction images shown above them reveal that one is a regular hexagon and the other a distorted tetragon. Why did such a crystal conversion occur in the mutant without the cadmium ion bonding? We believe that it was due to a significant change in the electrostatic potential on the protein surface.[18] Our interpretation is based on the fact that ferritin is made of 24 subunits. We changed two negatively charged side chains in each subunit to neutral ones, resulting in 48 units of charge difference per ferritin molecule. This change significantly altered the molecular recognition and brought about a new bonding between adjacent molecules, which was advantageous for oblique crystals.

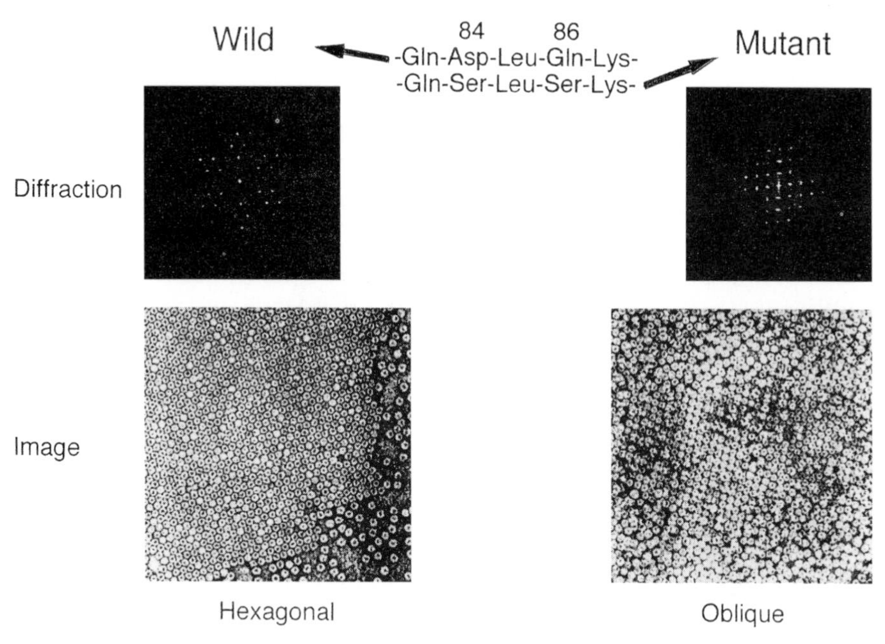

Fig. 9 Electron micrographs of two-dimensional ferritin crystals.[18] Top: numerically calculated diffraction images of two-dimensional crystals made of wild and mutant ferritin. Bottom: real images of hexagonal and oblique two-dimensional crystals of ferritin.

The protein mutation technique has made great progress since the introduction of genetic engineering, but what I have described is the first research project which used this technique to control protein assembly. The assembly of ferritin molecules is controlled together with their molecular orientation in crystals. The dice form of ferritin molecules brings about three typical orientations,[8] namely, one facing towards the supporting surface with the threefold symmetry axis perpendicular to the surface plane, and the other two with different symmetry axes (four- and twofold) normal to the crystal plane. This is what really we have expected for such a protein with cubic symmetry, and so we think that our major goal was mostly achieved.

Two technologies: nature's and ours

There is not much new under the sun except for scientific breakthroughs. Of course, nature has not changed since the olden days! Let us take another look at Fig. 2. Small-scale nature has a very refined configuration. One example I showed was protein crystals on the cell surface; the surface of the compound eyes of a night moth are covered with very tiny particles, and the surfaces of butterflies in the daylight are covered with 100 μm fragments called wing scales. These three examples cover a scale difference of 1000 times, but all of them are formed by means of self-assembly based on the intermolecular force. Furthermore, they are found at each level of biological hierarchical structures. This is the root of the generation principle of the profound multilayer structure of living things.

In Fig. 10 we see a morph butterfly, called *sulkowski*, one of the most beautiful creatures on Earth. Where does this beautiful gleam of the wings come from? This can be answered by further magnifying the wing scales of this butterfly. As shown in Fig. 10b (upper trace), the whole surface is covered with regular stripes placed at 1 μm intervals. Such a regular stripe structure, called a 'diffraction grating' in physical terms,[21] adds the same gleam as an opal to the morph butterfly wing, which does not have a colour of its own. Can we reproduce the beauty of this morph butterfly? We found that a crystallized monolayer film of fine particles we made produced a similar beauty (Fig. 10b, lower trace). Its intense gleam is generated by a 1 μm thick film consisting of only one layer of polystyrene spheres. Condensed aggregates of fine particles, whether a

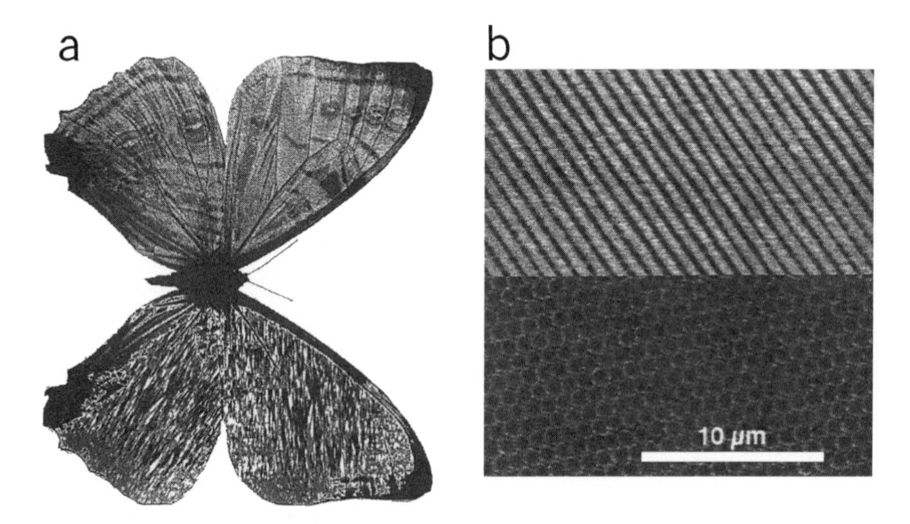

Fig. 10 Morpho colouring by the natural diffraction grating and artificial particle grating.[19] (a) morph *sulkowskyi* with natural wing (upper) and an artificial wing (lower). (b) Close-up view by scanning electron microscope of two surfaces of the natural (upper) and the artificial (lower) wing.

powder or a suspension, usually look white. However, arranging them in an orderly manner produces a new function. In this case, the grating on the surface of the butterfly is reproduced by an array of particles to obtain the beautiful gleam. Here's an application generated by imitating a structure in nature.[19]

Can I say that our research journey to find a route from nature's manufacturing to human technology comes to a happy end? The answer is yes and no. Now I understand that it is very difficult to truly reproduce nature's profound structure and its resultant beauty. In Fig. 11 the grating on the wing scales of the morph butterfly is further magnified to reveal each stripe covered with even finer pleats.[20] This fine structure endows the butterfly with a deep beauty found only in nature. When the morph butterfly is slightly rotated its colour remains light blue. Indeed, such a constancy in the colouring is quite natural when we recognize that nature demands individuality, i.e. that particular species, here morph butterfly *sulkowshi*, are identified by their own colours. In contrast when the imitation crystalline film is rotated, the colours change continuously. This phenomenon of changing colour, which is often

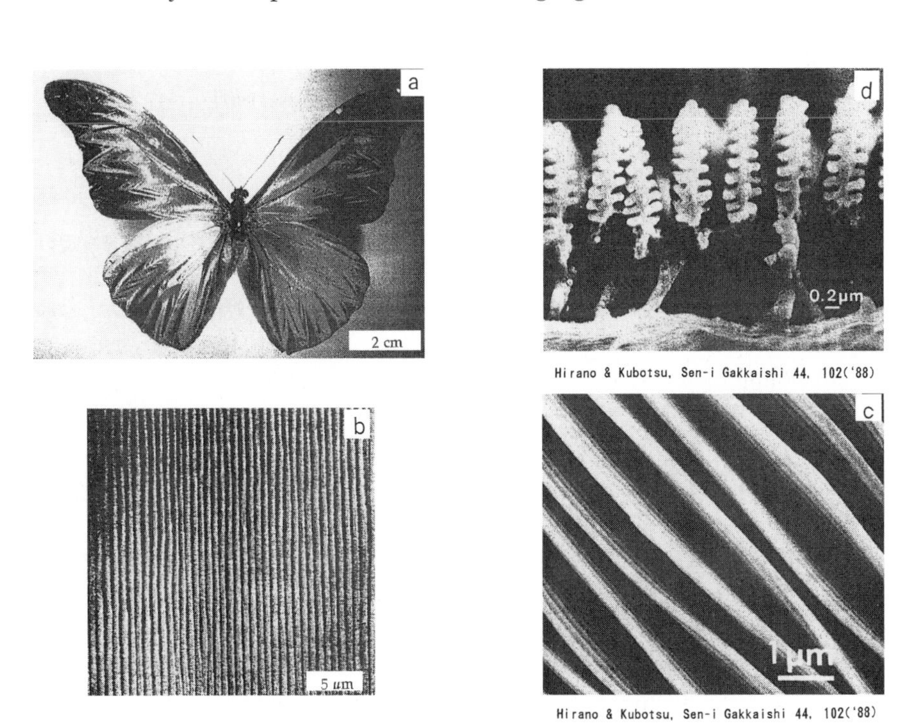

Fig. 11 Surface views of a butterfly wing on different scales. (a) morph *rhetenor*; (b) close-up of a wing scale; (c) close-up of stripes recognized in the close-up b;[20] (d) close-up of c.[20]

called 'play of colour', is physically grounded in simple diffraction grating theory;[21] You can understand it easily by analogy with the colour of the rainbow. So we have to ask why the morph butterfly does not change its colour, even though its gleam and beauty must arise from the same kind of grating. This mystery is still unsolved, although I believe it arises from the deepness of the natural multilayered structure shown in Fig. 11. So far, our clumsy human technology has not reached a level where we can make such beautifully complex hierarchical structures. Our best efforts require considerable energy to produce even clumsy artefacts with far less functionality. In the twenty-first century, one might hope that science and technology will develop some of the more subtle manufacturing techniques used in nature.

In closing, we may cite two great teachers from the past. The founder of the Friday Evening Discourses, Michael Faraday, said 'It is the great beauty of our science that advancement in it opens the doors to further and more abundant knowledge, overflowing with beauty and utility.' Going further back to the ancient Greek era, Aristotle said in *Ethics* 'If one way be better than another, that, you may be sure, is Nature's way.'[22] There really is much to be learned from Mother Nature, and, in particular, from small-scale nature.

Acknowledgements

Most of the material presented here has been taken from work performed under an ERATO program, Nagayama Protein Array Project. The author is grateful to my colleague researchers and the financial supporter, Research Development Corporation of Japan (now reformed to Japan Science and Technology Corporation).

References

1. Gulliver was created by the famous novelist, Jonathan Swift, but also the name of an educational product recently invented by A. Wada which can extend our physical sense to the larger and smaller universe with an aid of logarithmic scale of the space and time. Refer to an article, A. Wada, *Nature* (26 Jan.), 1995, xxxix.
2. This was first discovered with X-ray crystallography by M. Perutz and J. Kendrew supervised by Sir Lawrence Bragg, the Director of the Royal Institution in 1950s.
3. A large molecular architecture assembled through non-covalent bonding such as hydrogen bonds. Refer to J-M. Lehn, *Angew. Chem.*, 1990, **29**, 1304.

4. C.G. Bernhard and W.H. Miller, *Acta Physiol. Scand.*, 1962, **56**, 385.
5. K. Nagayama, *Nanobiology*, 1992, **1**, 25.
6. Lipid monolayer film: I. Langmuir, *J. Am. Chem. Soc.*, 1917, **39**, 1848. Protein monolayer film (denatured): I. Langmuir and V.J. Shaefer, *J. Am. Chem. Soc.*, 1938, **60**, 1351.
7. Protein monolayer film (native): P. Fromherz, *Nature*, 1971, **231**, 267.
8. K. Nagayama, *Supramol. Chem.*, 1996, **3**, 111.
9. P.A. Kralchevesky, V.N. Paunov, I.B. Ivanov, and K. Nagayama, *J. Colloid Interf. Sci.*, **151**, 79. P.A. Kralchevsky and Nagayama, *Langmuir*, 1994, **10**, 23.
10. N.D. Denkov, O.D. Velv, P.A. Kralchevsky, I.B. Ivanov, Y. Yoshimura, and K. Nagayama, *Langmuir*, 1992, 8, 3183. K. Nagayama, Colloid Surf. 1996, **A109**, 363.
11. A.S. Dimitrov and K. Nagayama, *Langmuir*, 1996, **12**, 1303.
12. A.S. Dimitrov and K. Nagayama, *Chem. Phys. Lett.* 1995, **243**, 462.
13. E. Adachi and K. Nagayama, *Langmuir*, 1996, **12**, 1836.
14. H. Yoshimura, S. Endo, M. Matsumoto, and K. Nagayama, *Ultramicroscopy*, 1990, **32**, 265. A.S. Dimitrov, M. Yamaki and K. Nagayama, *Langmuir*, 1995, **11**, 2682.
15. N.D. Denkov, H. Yoshimura and K. Nagayama, *Ultramicroscopy*, 1996, **65**, 147.
16. H. Yoshimura, T. Scheybani, T. Baumeister, and K. Nagayama, *Langmuir*, 1994, **10**, 3290.
17. D.M. Lawson, P.J. Artymiuk, S.J. Yewdall, J.M.A. Smith, J.C. Livingstone, A. Treffy, A. Luzzago, S. Levi, P. Arosio, G. Cesareni, C.D. Thomas, W.V. Shaw, and D.M. Harrison, *Nature*, 1991, **248**, 541.
18. S. Takeda, H. Yoshimura, S. Endo, T. Takahashi, and K. Nagayama, *Protein*, 1995, **23**, 548.
19. K. Nagayama and A.S. Dimitrov, *Film formation in waterborne coatings* (T. Provder, M.A. Winnik and M.W. Urban, ed.), ACS Series, **648**, 1996, 468.
20. T. Hirano and A. Kubotsu, *SEN-1 GAKKAISHI* (in Japanese) 1988, **44**, 102.
21. S.G. Lipson, H. Lipson and D.S. Tannhauser, § 9.2 In *Optical physics*, Cambridge, 1995.
22. Aristotle, *Nicomanachean Ethics.*

KUNIAKI NAGAYAMA

Born 1945 in Gunma Prefecture, Japan, gained a BSc in Physics and a PhD in Biophysics at the University of Tokyo, joined the Department of Physics of Tokyo University as a research associate in 1973. He spent his postdoctoral period at ETH-Zürich and enjoyed the development and application of two-dimensional NMR to protein systems. In 1983 he moved to JEOL Ltd as director of the newly founded Biometrology Laboratory. In 1990 he was appointed director of a national research project, ERATO Protein Array Project, by the Research Development

Corporation of Japan. In 1993 he returned to the University of Tokyo as Professor of Physics in the College of Arts and Sciences. From 1995 he has also held an additional post as Professor in the National Institute for Physiological Sciences. He is a member of the Japanese scientific societies of physics, chemistry and biophysics, and also a member of the International Reviewing Committee of the Human Frontier Science Programme, Strasbourg since 1993.

Metals combined—or, chaos vanquished

ROBERT CAHN

Just before the Russian revolution, in 1916, a group of chemists in Petrograd were examining alloys of copper and gold. These two metals can be mixed in the solid state, to form a single solid solution in any desired proportion; that is how low-carat gold for cheap jewellery is made. Kurnakov and his colleagues had been studying these alloys for years and gradually they had reached the conclusion that the alloys had anomalous properties at certain simple atomic ratios of gold to copper which, in their words, 'indicated a more complicated nature of the copper–gold alloys than was thought to be the case until now'. Some of their measurements, as reported[1] to the Institute of Metals in London in March 1916 (just before harsh events made such international visits impracticable) are shown in Fig. 1. In the slow-cooled alloys, the electrical resistance has sharp minima at atomic ratios of 1:1 and 3:1. Their conclusion was that 'the copper-gold system is an interesting example of the formation of definite compounds at the decomposition of the continuous isomorphous mixture of two simple elements belonging to the same group of the periodic system'. But, curiously, these compounds were only formed by slow cooling from high temperature, not if the alloys were cooled rapidly. The idea of comparing these two states was an intuition of genius on the part of Kurnakov, but he plainly had no idea what was going on here.

Four years previously, in 1912, Lawrence Bragg and his father William (each later director of this Institution) had discovered the use of X-ray diffraction as a tool for establishing the structures of crystals. In those days, new experimental techniques spread more sluggishly than they do in these days of fax and e-mail. It was not until 1925 that two Swedish physicists, Johansson and Linde, used X-rays to make sense of Kurnakov's mysterious findings. They found that the slow-cooled alloys

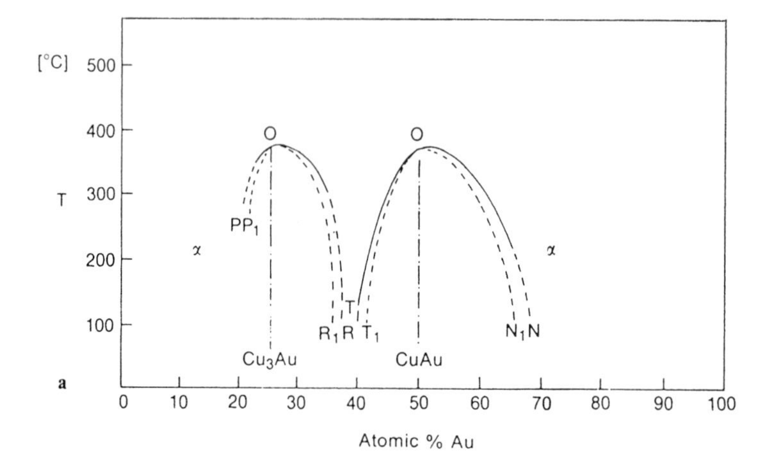
[°C]
T
500
400
300
200
100
O
O
PP1
α
α
R1 R T1
T
N1 N
Cu3Au
CuAu
a
0 10 20 30 40 50 60 70 80 90 100
Atomic % Au

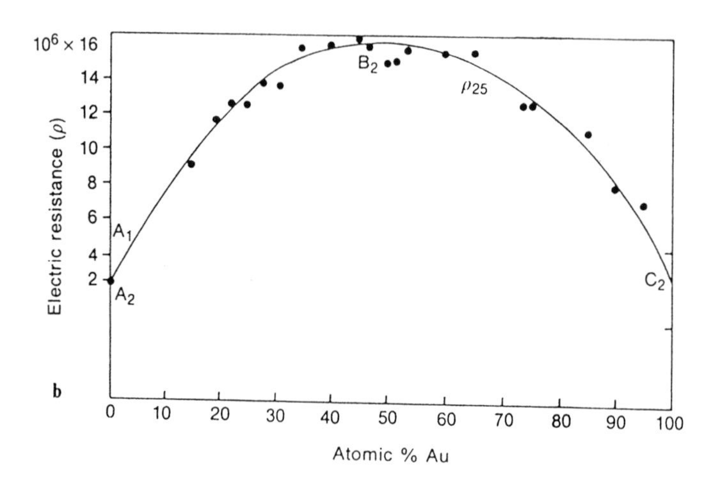
10⁶ × 16
14
12
10
8
6
4
2
Electric resistance (ρ)
A1
A2
B2
ρ25
C2
b
0 10 20 30 40 50 60 70 80 90 100
Atomic % Au

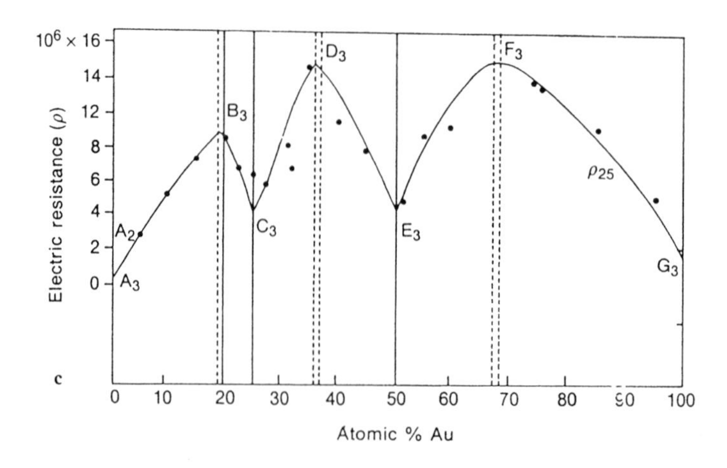
10⁶ × 16
14
12
8
6
4
2
0
Electric resistance (ρ)
A2
A3
B3
C3
D3
E3
F3
ρ25
G3
c
0 10 20 30 40 50 60 70 80 90 100
Atomic % Au

Fig. 1 (a) The 'thermic diagram' of the copper-gold system in the solid state; (b) the electrical resistance of copper-gold alloys water-quenched from 800°C; (c) the electrical resistance of copper-gold alloys annealed at 670°C and slowly cooled.[1]

of composition Cu_3Au or CuAu had *ordered* crystal structures as shown in Fig. 2, whereas the quenched alloys had a random distribution of copper and gold atoms; they were *disordered.* When the alloy was quenched in water, the ordered structures had no time to form, because the two types of atoms each require to make several jumps to finish up in the right places. Johansson and Linde had discovered *atomic order.* Once they had done so, the explanation for the Russians' curious findings was quite obvious: an orderly array of two different kinds of atoms which scatter drifting electrons with different efficacy is bound to conduct electricity much better than a random array. The obvious explanation is often the hardest to imagine beforehand.

Most intermetallic compounds with simple atomic ratios (1:1, 3:1, 4:1, even 13:2) are ordered, many of them all the way up to their melting temperatures; these last are said to be *strongly ordered.* It is the disordered state which is the exception among such compounds. Most of

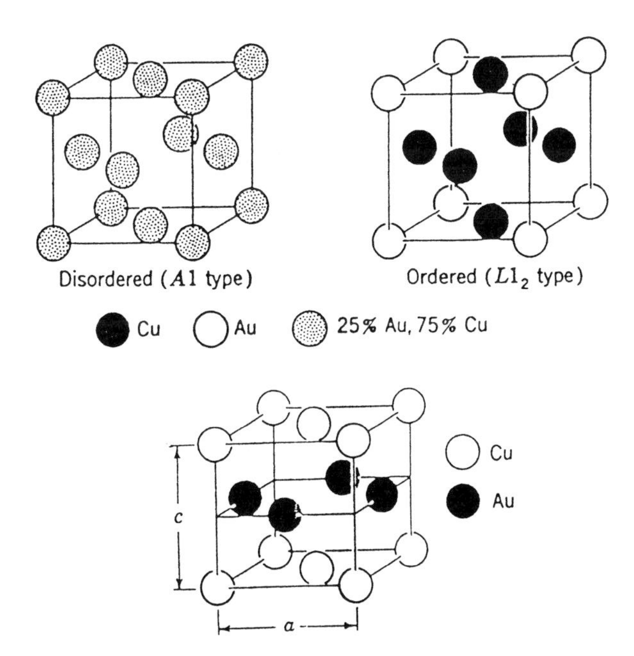

Fig. 2 The crystal structures of disordered and ordered Cu_3Au ($L1_2$) and ordered CuAu ($L1_0$).

the ordered structures, also called *superlattices*, are simple like those in Fig. 2; they all have crystallographic symbols to distinguish them, like those listed in the caption to that figure. Another very simple superlattice, of considerable practical importance, is the B2, which consist simply of one kind of atom at cube corners and another kind at cube centres.

The ordering is simply an expression of the fact that the strength of bonds between unlike neighbours is greater than the average of the strength of the bonds between the two kinds of like neighbours, and the strength of ordering is a measure of that difference. If the ordering is not too strong, such compounds disorder on heating because the universal urge towards disorder, or high entropy, overtrumps the tendency to achieve the strongest bonds possible.

Like most scientific explanations, this merely pushes the questions further back. Why are unlike bonds stronger? This takes us deep into the electronic structure of alloys, and there are no simple models on offer. Another important contributor is the difference in atomic diameters: large and small atoms bed down in a crystal most easily when they are disposed in an orderly manner. This simple consideration plays a major part in making sense of the tendency to order. The magnitude of the ordering energy (in effect, the difference in energy between the ordered and disordered states) can be estimated from experimental measurements of the heats of formation of compounds. The higher the ordering energy, the higher the temperature at which order disappears, and if it is high enough, the compound will melt without first disordering—it is a *permanent intermetallic*. This last is a very common situation.

Simulation

Nowadays, when scientists wish to clarify a complex situation, they simulate it in a computer. Computer simulation has become a third branch of scientific endeavour, midway between theory and experiment. But before electronic computers were available, people used physical simulation. One can demonstrate crystalline close packing, as in a pure metal, with a handful of equal-sized steel balls confined between two Perspex sheets and shaken down under gravity. A better model, because it better simulates the force-versus-separation characteristics of pairs of atoms, is a two-dimensional array of equal-sized bubbles floating on a bath of soap solution, as first tried out by Sir Lawrence Bragg half a century ago. But neither of these models can simulate ordering behaviour in a compound: if one mixes two populations of balls or bubbles of different radii, they form a wholly disordered (i.e. glassy) array. No crystal of any kind forms.

One kind of simulation has emerged, unexpectedly, from the world of precious stones. Precious opal, with its splendid play of diffraction colours, mostly comes from Australia, and a physicist there, John Sanders, in the early 1960s[2] began to examine such opals by electron microscopy, which can resolve much finer detail than the light microscope can. He found that precious opal (as distinct from the much commoner, and valueless, potch opal) consists of regular, crystal-like arrays of silica spheres of equal size, with a repeat distance of the same order as the wavelengths of visible light (i.e. about a thousand times greater than the repeat distance of true crystals). Such arrays were probably formed by the slow sedimentation of the spheres from suspension in water. A lively account of his earlier findings was published in the *Scientific American*.[3] Later, Sanders discovered that a few opals contain an array of silica spheres of two distinct sizes, and some of the 'crystals' thus formed have hitherto unfamiliar arrangements[4] (Fig. 3); this discovery led Sanders and one of his collaborators to examine the theory of the 'crystal' structures which can form for mixtures, in different proportions, of two populations of spheres of different size ratios—both the proportions and the size ratios being adjustable variables.[5] This last study then had considerable effect on studies of synthetic 'crystals' made of small particles of sub-micrometre size, a growing field of research in its own right.

Such synthetic crystals are called 'colloidal', a term which is reserved for particles of submicrometre size. In the extensive studies of such

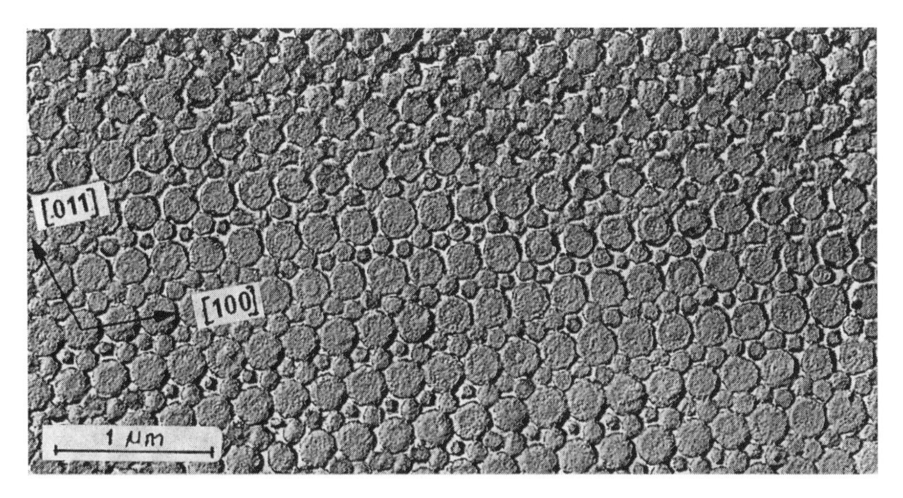

Fig. 3 An example of 'crystal' structure in a precious opal containing silica spheres of two different sizes.[4]

artificial colloidal 'crystals', small coated polymeric particles are most commonly used and, as with opal, they are allowed to settle from a suspension in a liquid. An early account of these studies came from a French laboratory.[6] Much thought went into an analysis of what attracted such colloidal spheres to each other; it is now accepted that the coated spheres acquire an electric charge, but they float in a liquid carrying opposite charges, and this can lead to very weak attractive forces instead of the expected repulsion. Because the forces are so weak, crystals 'freeze' slowly after the array has been 'melted' by shear, and so these model structures have been used recently to simulate, and examine at leisure, the freezing process.[7]

Superlattice formation in mixtures of such colloidal particles of two different sizes, influenced by Sanders' observations, has become a popular field of study. One such study[8] showed that, for a fixed size ratio, when the concentrations (in liquid suspension) of the two kinds of colloidal particle are systematically varied and plotted, each combination leads to a different kind of crystal structure, several of them hitherto unfamiliar. It is a curious feature of these materials that different 'crystal' structures arise as the proportion of solid spheres to suspension liquid is changed. This entirely new kind of crystallography has just been reviewed for the first time.[9] Figure 4 shows one such synthetic superlattice.

Because the nature of the interparticle force is quite different from the force between metal atoms in intermetallic compounds, the colloidal simulations have not cast as much light as had been hoped on the structure of such compounds, but the primordial importance of atom size has emerged very clearly. This story shows that physical simulations can have unexpected sources as well as highly unexpected outcomes. Incidentally, the attempts to interpret these varied observations has forced investigators to use computers to simulate the simulations! Computer simulations have now become preferable to physical simulations.

Heusler alloys

In 1904, a discovery was announced in Germany which astonished both metallurgists and physicists. Friedrich Heusler, the proprietor of a long-established factory for copper alloys in Dillenburg, near the university town of Marburg, published in an obscure Marburg periodical an account of a new copper–manganese–aluminium alloy, which was ferromagnetic when suitably heat-treated. This chanced to be discovered because the researcher tapped the sample with a magnetized hammer

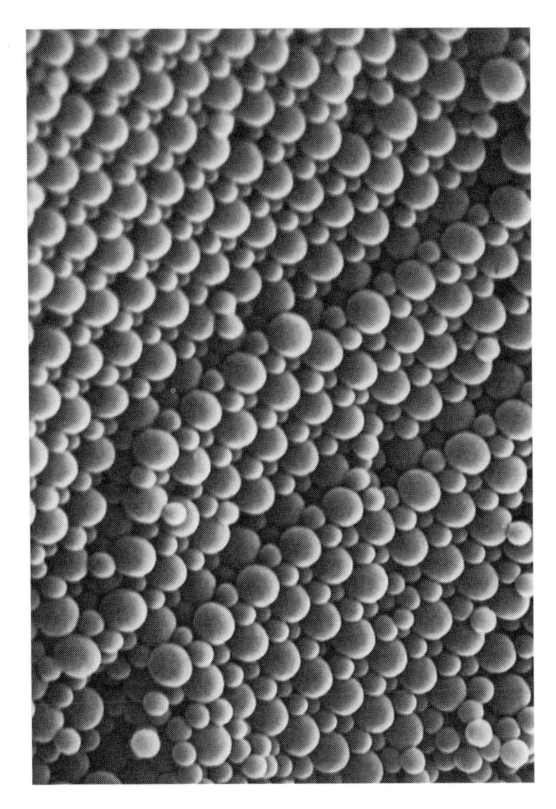

Fig. 4 An AB_2 colloidal 'crystal', deposited from a supension of particles of radius ratio 0.58; volume fractions $\phi_A + \phi_B = 0.536$. The particle radii are 0.36 and 0.21 μm.[8,9] (Photograph courtesy of Prof. R.H. Ottewill.)

kept to pick up dropped nails from the floor. The source of the surprise was that none of the constituent elements shows ferromagnetism. How was this possible?

Friedrich Heusler had already made a name for himself in 1889 by developing the 'Manganin' alloys (a tradename, still in use); these alloys, based on copper containing dissolved manganese and some minor additions, had been developed to provide an electrical resistivity almost invariant with change of temperature near ambient. This was crucial for electrical measurements requiring stable, reliable standards. His son has recently described the genesis of the manganins.[10] Friedrich Heusler then began to try out various changes in composition (on a wholly empirical basis), and this included additions of aluminium, then a new and unfamiliar metal. He published the observation of ferromagnetism in such alloys as a means of attracting attention to his factory and laboratory.

The alloys (with slight variations in composition) came to be known as 'Heusler alloys', and still are called so today.

For a long time, the behaviour of the Heusler alloys was a complete mystery. Then, in 1934, Friedrich's son Otto Heusler (who has just died, well into his nineties) published a study, by X-ray diffraction, of the alloys in their ferromagnetic form and found that they contained an ordered phase of composition Cu_2MnAl.[11] Just a few weeks earlier, the same phase and crystal structure were reported by two of Lawrence Bragg's young crystallographers in Manchester, Bradley and Rodgers.[12] As happens so often, the time was ripe and different scientists had the same idea. This crucial crystal structure is shown in Fig. 5. The studies also showed that the alloy in quenched form did not contain the ordered phase, which was formed by annealing at a few hundred degrees Celsius. Scores of subsequent studies refined the analysis of the changes resulting from various heat treatments.

Although Otto Heusler did point out at the beginning of his paper that Heisenberg had in 1928 'interpreted ferromagnetism in quantum-mechanical terms', the ferromagnetism of the Heusler alloys was still not understood. It was only a few years later that understanding at last came. Individual atoms of transition metals act like minute individual magnets because of their particular electronic arrangements, *and that includes manganese.* Neighbouring atoms interact by an 'exchange' force, which Heisenberg had first proposed. He also suggested that atomic neighbours needed to be far enough apart to interact 'positively', i.e. for the neighbouring atomic magnetic moments to line up parallel, and that is a precondition of ferromagnetic behaviour. What Heisenberg did not recognize was that if the neighbours come closer together, the exchange interaction changes sign, and the neighbouring magnetic

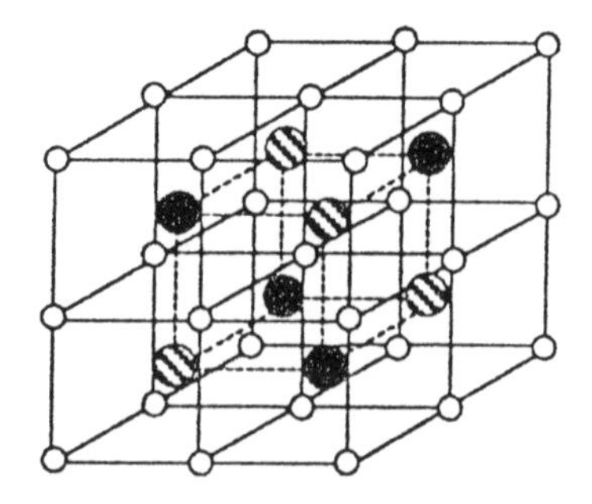

Fig. 5 The $L1_2$ crystal structure of the ordered phase, Cu_2MnAl, in a Cu-Mn-Al Heusler alloy. The open circles represent copper atoms.

moments line up *anti*parallel. The crystal then becomes *anti*ferromagnetic; half the atomic magnets line up one way, the other half line up oppositely, and there is no resultant microscopically detectable magnetism. (That is the situation in pure elementary manganese, and that is why manganese atoms had been erroneously supposed to be intrinsically non-magnetic.) Now, in the ordered structure shown in Fig. 5, the manganese atoms (black circles) are *all* separated by a distance of $a\sqrt{2}$ ($1.414a$), where a is the length of the small cube edge. When the atoms are in a random 'body-centred cubic' distribution, without a superlattice, many pairs of manganese atoms are closer together, at a distance of only $a\sqrt{3}/2$ ($0.866a$). (This is half the body diagonal of one of the small cubes.) The manganese atoms are the only intrinsically magnetic atoms in the alloy, and when they are separated by $0.866a$, they are aligned antiparallel and the crystal is antiferromagnetic, giving no external sign of magnetic behaviour. When the separation, in the ordered crystal, goes up to $1.414a$, the magnetic moments all line up parallel and the alloy becomes ferromagnetic.

The Heusler alloys were the first alloys in which the establishment of atomic order was shown to bring about externally detectable magnetic behaviour; now a number of other such alloys are known, and the Heusler alloys are not used practically. Nevertheless, these alloys have exerted a fascination over successive generations of metallurgists and physicists and they are still receiving much attention in research laboratories all over the world.

Applications of intermetallic compounds

As we have just seen, the Heusler alloys excited much interest among scientists but they proved not to have properties good enough to allow them to be used by industry. In recent years, however, following an unprecedented burst of metallurgical research, a number of intermetallic compounds have found increasing practical use as strong load-bearing materials for use at high temperatures.

The most important intermetallic compound in practical terms is trinickel aluminide, Ni_3Al, which has the same crystal structure as Cu_3Au (Fig. 2). This alloy has one quite remarkable characteristic, shown in Fig. 6, where its flow stress (i.e. the stress needed to generate irreversible, plastic deformation) is plotted. Unlike most other alloys, it becomes stronger as it becomes hotter, up to a limiting temperature of about 600°C. The explanation of this phenomenon, known as the *anomalous yield stress* syndrome, has for many years generated fierce debate

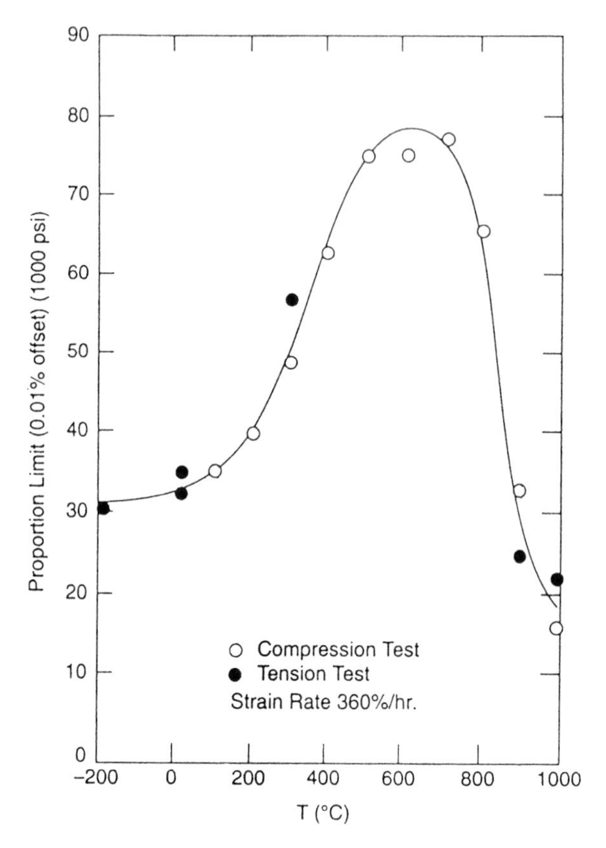

Fig. 6 The flow stress of polycrystalline Ni_3Al as a function of temperature. (After P.A. Flinn, 1960.)

among theorists and the fine detail of the interpretation is still under dispute.[13] There is no need to go into this debate here: what is important is the implication of this syndrome for applications.

The most important use of trinickel aluminide is as a constituent of *superalloys*, the family of alloys employed for the discs and turbine blades of aeronautic jet engines. These consist of nickel, aluminium, titanium, and a number of other minor constituent elements. Figure 7 shows a typical microstructure, magnified about 30 000 ×. The white cuboids consist of ordered $Ni_3(Al,Ti)$—trinickel aluminide with some of the aluminium replaced by titanium—dispersed in a *dis*ordered matrix (black) of nickel containing various dissolved elements. It is a crucial feature of these alloys that the crystal lattices of the two structures are strictly parallel and that their mesh sizes (cube edges) are virtually identical. This has the result that the crystal dislocations (defects which

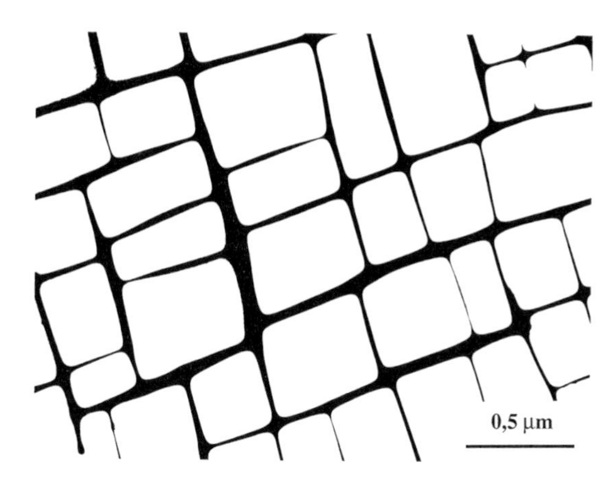

Fig. 7 Micrograph, made with an electron microscope, of the fine structure of a superalloy. The picture shows part of one crystal grain. (Courtesy Dr. T. Khan, Paris.)

'carry' plastic deformation when they move through the crystal) are largely confined to the narrow disordered 'corridors'; the highly perfect interface with the ordered cuboids acts as a one-way valve for dislocations. The results is that the resistance to slow plastic deformation under load at high temperature (known as creep) is greatly enhanced, and this is the most important property for a superalloy. Like pure Ni_3Al, the complex alloy becomes stronger as the temperature is raised. Superalloys have gradually been improved over the years by subtle changes to minor alloying constituents and heat treatment before use, until now the best of them can serve at absolute temperatures as high as 80% of the melting temperature, which represents a remarkable metallurgical achievement.

A superalloy is a mix of ordered and disordered 'phases'. During the past decade, interest has built up in using alloys consisting entirely, or very largely, of ordered crystal structures for load-bearing uses at high temperatures. The simple Ni_3Al has been improved by adding small amounts of boron (needed to prevent the alloy falling apart at the boundaries between adjacent crystal grains), chromium (to prevent oxidation damage in air) and zirconium (for further hardening), to generate a family of *advanced aluminides*, and alloy improvement has gone hand in hand with improved methods of casting, shaping, and welding the alloys. Much of this has been achieved by a team led by C.T. Liu at Oak Ridge, one of the large American national laboratories, in Tennessee, and these aluminides are now manufactured commercially in America. Although they are too dense and not quite strong enough when hot for use in jet engines, they have found use as forging dies, furnace 'furniture', radiant tubes, and many other applications which have recently been reviewed comprehensively.[14] The same article also discusses the treatment and uses of two other intermetallic compounds, the iron aluminides Fe_3Al and FeAl, which are less strong than Ni_3Al but have exceptional resistance to oxidation and also to attack from sulphur-containing gas products resulting from combustion of fuels.

The other intermetallic compound in which great hopes are invested is titanium aluminide, TiAl. This compound has received the lion's share of research attention in the past few years, and is now approaching the stage when it is used to replace superalloys in some components of jet engines. It is about 50% less dense than a superalloy, and as a major source of stress in the rotor of a jet engine is centrifugal force in whirling blades, the low density is a major consideration, quite apart from the lower overall mass of an engine that exploits this compound. TiAl suffers from very limited plastic deformability at room temperature and this means that it is highly sensitive to shock loads; this problem is gradually being minimized by very detailed control of the microstructure of TiAl which also contains a small fraction of a different phase, of composition Ti_3Al. TiAl has also found some terrestrial applications, the most important of which is for exhaust valves and turbocharger rotors in automobile engines: this has required a long and costly development programme by Nissan, a Japanese motor manufacturer, to improve the wear resistance and brazing technology.

If and when TiAl becomes a regular, major component of jet engines (and this will happen within the next few years if at all), as well as a standard component of mass-produced car engines, a long and elaborate programme of metallurgical research will have reached its final justification. A recent, concise review will give the reader a good impression of this programme as a whole.[15]

Shape-memory alloys

My last example of a family of intermetallic compounds in practical use encompasses alloys which remember their original shape after this has been changed by force majeure and return to it when they are heated. This is the family of *shape-memory alloys*, and the archetype is *nitinol*, an alloy close to the composition NiTi, originally discovered in 1963.[16] For instance, a spiral spring made of nitinol can be 'trained' by suitable heat treatment, to remember its original extended condition. It can then be squeezed into compressed conformation, and on subsequent heating, it will spring back automatically to the extended form, without the application of any stress. Indeed, if an external force tries to prevent the springback, the heated spring will insist on extending against the external constraint if the latter is not too powerful. It is ironic that the total brittleness of some intermetallics, which is a drawback in conventional structural applications, is an essential precondition of the behaviour discussed here.

The key to this behaviour is seen in Fig. 8. Here we see how a simple (metal) crystal can respond to an external force, by either slipping across a crystal plane or else by a *shear* process, called 'twinning', in which the sheared crystal is a mirror image of the unsheared structure. Either of these processes will leave the original crystal structure unaltered. If, however, the crystal is an intermetallic compound such as nitinol that is brittle, it cannot slip or twin. Instead of simply fracturing, as (say) a piece of brittle ceramic would do, it can undergo a *martensitic (structural) transformation.* In simplified form, this is a process like twinning (right-hand diagram in Fig. 8), but with a larger shear—a *martensitic shear* (named after a renowned nineteenth-century German steel metallurgist named Martens). A martensitic shear implies that the angle through which the crystal lattice rotates as a result of shearing is larger than that shown in the figure, the symmetry is lost, and as a consequence, the crystal structure is changed to a different one, which is stable thermodynamically at room temperature. The martensite characteristically takes the form of thin plates.

When the nitinol spring is compressed after training, the alloy changes its shape not by slip or twinning but by a martensitic shear. As stated, the new crystal structure is stable at room temperature, but when the alloy is heated to, say, 100°C (for instance, by plunging into boiling water), the original crystal structure becomes stabler than the martensitically generated structure and the crystal shears back to that original

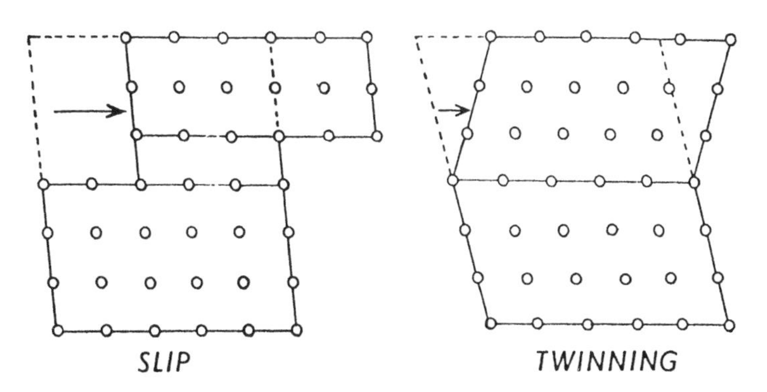

Fig. 8 Plastic deformation of a crystal by slip or twinning.

structure. The greater stability (at high temperature) of the original structure implies that the return shear can persist in the face of an external force trying to prevent it. That is the essence of the shape-memory effect. It is an essential condition of the process that the alloy is completely brittle, i.e. incapable of deformation by either of the processes shown in Fig. 8. Any hint of slip and the alloy will lose its shape memory.

The exact temperature at which the shape memory begins to operate, i.e. the shear-generated phase begins to unshear, can be modified over a range of several hundred degrees by a very slight adjustment of the composition of the intermetallic compound. Again, by change of composition the alloy can be modified to show a *two-way shape memory*; this means that it will change its shape in one sense when heated through the critical temperature and back to its original shape when cooled again.

Early research on this phenomenon was done with large single crystals of nitinol and, later, of other compounds such as a copper aluminide with additions of zinc or nickel (which is used today for the less critical applications), but nowadays, ordinary (and much cheaper) polycrystalline alloys are used instead. In any one crystal grain of such a polycrystal, the martensitic shear can take place on any of 24 crystallographically equivalent crystal planes, so that any form of externally imposed strain can readily be accommodated.

Shape-memory alloys are now manufactured commercially by several manufacturers and have found a number of major applications, which have recently been reviewed.[17] The most widespread applications involve *actuators*, that is, spiral or leaf springs which open or close a valve, commonly for some automotive or safety purpose. As one example among many, Fig. 9 shows a commercial pneumatic shut-off valve that will instantly divert a high-pressure air supply from processing equipment when a critical temperature (normally set at 57°C) is reached in the environment. A related device ensures that a domestic shower is disconnected if, through a malfunction, the water temperature suddenly rises to a point where there would be a risk of scalding. Two-way springs are used, for instance, in actuators which reduce the hydraulic pressure acting in an automatic gear-shift until the hydraulic fluid reaches a sufficiently high temperature for its viscosity to be low enough for smooth functioning. Other devices again are used to couple tubes end-to-end in places where access for spanners is difficult: a tubular coupling made of one-way memory alloy is slipped over both adjacent tubes and then warmed (e.g. with a stream of warm air) so that it shrinks and holds the two tubes very firmly. Related devices are used for multi-electrode electrical connectors that are easy to slip into place

Fig. 9 A heat-activated pneumatic shut-off valve operated by a shape-memory spring.

when cooled to subzero temperature and are then warmed to ensure firm electrical connections.

A different kind of shape-memory alloy exploits the phenomenon of *superelasticity*. This can be a nitinol of slightly anomalous Ni/Ti ratio or a Cu–Zn alloy. Here, the alloy remains throughout in its martensitic structure; to begin with, the martensite plates are restricted to just a few of the 24 symmetrically equivalent planes in each crystal grain. If now an external force is applied, these plates disappear and are replaced by new martensite plates on other crystal planes. This gives rise to a *fully reversible* change of shape, which is much larger than the shape change resulting from true elastic behaviour. Whereas a true elastic strain will be limited to 0.1% or less, a superelastic strain reaches several per cent. This happens without any change of temperature being required. Superelasticity is currently being exploited to make (very expensive) spectacle frames which adjust easily to the loads involved in frequent putting on and taking off the nose! More unexpectedly, drills for dentists working on root canals are now made of nitinol. Root canals are often curved, and a normal steel drill forced to bend as it rotates can break in 'low-cycle fatigue' after only a few hundred rotations. If the drill has superelastic characteristics it can put up with unlimited rotations while forcibly bent.

Like all the best materials, intermetallic compounds both pose fascinating scientific conundrums and have increasing uses in the world of industry.

Acknowledgements

I am grateful to Professor Konrad Heusler for information regarding his grandfather Friedrich Heusler and father Otto Heusler, to Professor Warlimont for making available samples of a Heusler alloy and to Dr. L. McDonald Schetky for providing samples and information relating to shape-memory alloys.

References

1. N. Kurnakov, S. Zemczuzny, and M. Zasedatelev, *Journal of the Institute of Metals* 1916, **15**, 305.
2. J.V. Sanders, *Nature* 1964, **204**, 1151.
3. P.J. Darragh, A.J. Gaskin, and J.V. Sanders, *Scientific American* 1976, April, 84.
4. J.V. Sanders, *Philosophical Magazine A* 1980, **42**, 705.
5. M.J. Murray and J.V. Sanders, *Philosophical Magazine A* 1980, **42**, 721.
6. P. Pieranski, *Contemporary Physics* 1983, **24**, 25.
7. D.G. Grier and C.A. Murray, *Journal of Chemical Physics* 1994, **100**, 9088.
8. P. Bartlett, R.H. Ottewill and P.N. Pusey, *Physical Review Letters* 1992, **68**, 3801.
9. P. Bartlett and W. Van Megen, Physics of hard-sphere colloidal suspensions. In: *Granular matter* (edited by Anita Mehta), Springer-Verlag, Heidelberg, 1993, p. 195.
10. O. Heusler, *Zeitschrift für Metallkunde* 1989, **80**, 757.
11. O. Heusler, *Annalen der Physik* 1934 (Series 5) **19**, 155.
12. A.J. Bradley and J.W. Rodgers, *Proceedings of the Royal Society (London)* 1934, **144A**, 340.
13. D.P. Pope, Mechanical properties of intermetallic compounds. In: *Physical metallurgy* (edited by R.W. Cahn and P. Haasen), North Holland, Amsterdam, 1996, p. 2075.
14. S.C. Deevi and V.K. Sikka, *Intermetallics* 1996, **4**, 357.
15. E.P. George, M. Yamaguchi, K.S. Kumar, and C.T. Liu, *Annual Reviews of Materials Science* 1994, **24**, 409.
16. W.J. Buehler, J.V. Gilfrich and R.C. Wiley, *Journal of Applied Physics* 1963, **34**, 1467.
17. L. McDonald Schetky, Shape-memory applications. In: *Intermetallic compounds: principles and practice* (edited by J.H. Westbrook and R.L. Fleischer), Wiley, Chichester, 1995, Vol. 2, p. 529.

ROBERT W. CAHN

Born 1924 in Germany, educated at Cambridge University (BA Natural Sciences, 1945; PhD Physics, 1950; ScD, 1963). Research at Harwell Laboratory, followed by periods in Birmingham University, Sussex University (where he organized Britain's first course in materials science), and Paris University. His research has been in physical metallurgy. Now attached to his old department at Cambridge. He has edited four editions of a standard text, *Physical metallurgy*, as well as numerous other monographs and encyclopedias, and has founded three scientific journals, one of which (*Intermetallics*) he is still editing. He has contributed some 100 pieces of popularization in materials science to the columns of *Nature*. He is a Fellow of the Royal Society.

On the air

MIKE GARRETT

The air in which we live and which we breath appears to be an almost limitless resource. It surrounds us all of the time and yet we cannot see it or touch it or even smell it although it can be felt when the wind blows. The planet earth is often referred to as a biosphere, although in reality a bio-shell would be more accurate, for if one were to reduce the world to the size of an orange then the thickness of the shell in which life can be naturally sustained would be less than a millimetre and the atmosphere is more than half of this.

Consequently, the air can be seen to be a limited resource and recently we have become aware of the impact human kind is making upon it and are concerned that this will lead to changes. In some ways this is an unreasoned viewpoint because the air is a constantly changing entity. It changes with height, temperature, and weather conditions and its composition is also continually changing so that the air that we breath today is not the same as that which we breathed yesterday or will breath tomorrow and much of this is a natural process. In the past the concentrations of oxygen and carbon dioxide have varied and are likely to have been much higher at times than they are today. For example, plants demonstrate increased growth when subjected to increased concentrations of carbon dioxide which tapers off when concentrations are about three times today's value, could this indicate that plants originally evolved when carbon dioxide concentrations were higher? Pollutants are often regarded as substances contaminating the atmosphere and threatening life by this definition perhaps the greatest pollutant of all time was the generation of oxygen, which, as a by-product of photosynthesis, wiped out a high proportion of obligate anaerobic life.

The atmosphere then has been experiencing change throughout its existence and will continue to do so into the future. The animals and plants in this world evolve to accommodate the changes and perhaps this is good for us. In many ways the atmosphere and even nature can be regarded as benignly hostile.

The air contains a mixture of many gases and gaseous compounds, some in high concentration and others much lower, the main gaseous constituents are shown in Table 1 in ascending order of magnitude. This shows gas concentrations down to 0.05 p.p.m and totals some 15 gases. The remaining gases and compounds fall far below this figure in concentration, but are very much higher in number. An example of the significance of apparently small concentrations is the scenting ability of dogs, such as the bloodhound. These animals have been accepted as evidence of identification in the courts of the USA and have been able to track the trail of a lost child in a metropolitan environment after 4 weeks. Let us assume that a lungfull of air for a bloodhound is about 1 litre, this contains about 27×10^{22} (270 000 000 000 000 000 000 000) molecules even if the concentration of scent was down to 1 part per quadrillion this would still mean that each breath would contain 27 million of those molecules, the bloodhound nose will be much more sensitive than this. Another way to imagine the magnitude of these numbers is that if all the molecules in a litre of gas were placed side by side they would reach the sun and back 25 times.

Having established the enormity of the numbers involved in the physical nature of gases it is also worth observing that the molecules are

Table 1. The atmosphere. Weight = 5 000 000 000 000 000 tons

Constituent	Formula	Molecular weight	Concentration in air		Boiling point
			%	v.p.m	°C
Hydrocarbons	–	–		<1	–
Hydrogen	H_2	2		<1	–253
Carbon monoxide	CO	28		<1	–192
Ozone	O_3	48		0–0.05	–112
Xenon	Xe	131		0.086	–108
Nitrous oxide	N_2O	44		0.5	–88.5
Sulphur dioxide	SO_2	64		0.1–1	–10
Krypton	Kr	84		1.139	–153
Helium	He	4		5.24	–269
Neon	Ne	20		18.21	–246
Carbon dioxide	CO_2	44	0.03		sublimes –79
Argon	Ar	40	0.932		–186
Water	H_2O	18	1.2		100
Oxygen	O_2	32	20.95		–183
Nitrogen	N_2	28	78.08		–196

moving and bumping into each other very rapidly indeed achieving speeds equatable with that of Concorde, although of course in perfect silence.

Applications

Let us now examine the major gases in the air and some of their industrial uses. The first group will be those gases formally known as the inert gases and currently termed the noble gases, helium, neon, argon, krypton, xenon, and radon. These gases, discovered by Lord Ramsey and Cavandish, have always been closely associated with the Royal Institution and mostly achieve only a small concentration in the atmosphere, the exception being argon. They are the products of radioactive decay deep within the earth's crust and slowly seep up from it into the atmosphere itself. Helium atoms have a high enough velocity ultimately to escape from the earth's gravitational field and be lost into space, the other gases are retained but in the case of krypton and xenon the low concentrations are probably due to adsorption of the gases on to the rocks in the ground. Although these gases have been regarded as inert they do have a very wide variety of uses.

Helium

Helium is the lowest density gas with the exception of hydrogen and because it does not react with oxygen provides a safe lifting gas for balloons and airships avoiding the risk of fire which devastated the airship industry in the 1920s.

Today helium is still used in airships, or more correctly dirigibles, and in some specialist balloons but one of the largest uses is in party balloons! Those which are made of latex deflate fairly rapidly as the gas diffuses through the rubber, those made with aluminized plastic, however, last for weeks because the gas is held at atmospheric pressure inside a low permeability material, which can be shaped and decorated in many ways.

Industrial uses of helium include its use as a shielding gas in arc-welding to prevent the oxidation of the molten metal, as a component of gas lasers, as a breathing gas in deep sea diving where it is mixed with a very small concentration of oxygen, and even in medicine where mixed with a higher concentration of oxygen it is used for patients who have lung obstructions, the low density of the helium making the gas easier to breath. Helium also plays its part in the space programme where it is used to pressurize the liquid oxygen tanks of the space shuttle prior to take-off. It is interesting to note that because helium has such a low mass any release into the atmosphere will eventually escape into outer-space. The gas in the shuttle of course gets there more quickly.

Liquid helium which boils at 4.2°K is effectively the coldest refrigerant known and apart from the many research uses it is also used to cool the superconducting magnet coils which are used in magnet resonance imaging machines.

Neon

Many uses of neon are associated with its red glow in a discharge tube and it forms the basis of the almost universal neon signs which decorate the cities at night; however, a more mundane use perhaps equally life-saving is the simple mains testing screwdriver, which warns the user of the presence of mains voltages. Neon is also used in lightning arrestors and as a refrigerant boils at 27°K, which is a useful temperature for refrigerators designed to cool high temperature superconductors.

Argon

Argon is most often associated with the electric light bulb of which it is the major component of the filling gas, but it also has a vital role in the steel industry where its inert properties make it ideal as an agitating and blanketing gas. Argon is also the gas used in the fluorescent lighting tube but one of the more unusual developments is its use in stunning chickens in a very humane way. This is becoming a preferred alternative to the previous method of electrocution.

Krypton

Krypton is the lighthouse gas being used in the enormously powerful bulbs which signal hazards around our coastline and krypton lasers are also proving useful in surgery especially eye surgery. A more recent development is the discovery that krypton can act as an enzyme or enzyme enhancer, and in this role several patents have been claimed, including one which assists in the conversion of corn syrup to fructose.

Xenon

Xenon is the heaviest stable noble gas and again one of its main uses is in lighting, where its use enables even small cameras to be fitted with a repeating flash gun which is a far cry from the magnesium-filled bulbs of only a few years ago. It also forms one of the components in the high-pressure sodium lamps which are now seen in our streets, providing a better light quality and higher efficiency than the old deep yellow units.

An unusual property of xenon is that it is also a narcotic gas with slightly stronger anaesthetic properties than the commonly used nitrous oxide. The gas appears to allow a quicker recovery from anaesthesia, fewer side-effects and a reduced hazard to operating theatre staff, although its cost has mitigated against its use until now. Recently, however, with the increasing scale of industrial gas production, more xenon is becoming available and prices are reducing, thus making it a viable alternative for operating procedures of more 1 or 2 hours.

Radon

This is the last gas in the noble gas series but it is unstable, having a half-life of about 4 days. It is present in the air only in a very small concentration but can achieve health affecting concentrations in some houses located on granite rocks. Earlier this century radon was used medically, both as a gas and dissolved in water, to treat various diseases from eczema to asthma. However, today its use is limited to radiation treatment because of the dangers from the residues of its radioactive decay.

Oxygen and nitrogen

The major components of the air are the gases nitrogen and oxygen and these remain the gases with the largest industrial use. Currently, approximately 420 000 tonnes of oxygen and 720 000 tonnes of nitrogen are separated each day from the atmosphere. The uses of these gases fall into two broad categories according to their properties. Oxygen is the life-giving gas which supports combustion and nitrogen, the relatively inert blanketing gas. Although of course nitrogen also forms the key gas in the production of ammonia. Some of the uses for these gases are given below.

Oxygen

Liquid oxygen is a pale sky blue colour and boils at −183°C. In large industrial plants the oxygen gas is separated from the air by cryogenic distillation and is often supplied at ambient temperatures through a pipeline to an adjacent site. Some of the gas is, however, produced as liquid and transported in vacuum-insulated road tankers and sometimes rail tankers to the customers where it is transferred to another vacuum insulated storage vessel. The reason for this 'bulk' distribution is the volume ratio of gas to liquid, about 850:1 at atmospheric pressure. This is nearly three times as efficient as storage in high-pressure cylinders for the same mass of gas without the need for thick metal casings to contain the pressure. For smaller users cylinders of compressed gas remain the most convenient method of supply and over 50 million are in use world-wide. Another method of producing oxygen directly in quantities ranging from a few litres a minute to hundreds of tonnes per day is by the use of pressure swing adsorption. This process makes use of molecular sieve chemicals such as zeolites that have an open crystalline structure giving immense surface areas and electric charge distributions, which attract gas molecules.[4]

The proportions of a gas mixture attached to these sites are essentially a function of temperature and pressure and the variation of these ratios with pressure changes can be used to separate gas using a cyclic method.

Another method of separating oxygen and nitrogen from the air is by the use of semipermeable membranes but because oxygen has a higher permeability than nitrogen the method is most often used as a nitrogen producer. One form of membrane, however (ionic membranes), is able to produce oxygen at very high purities by the process of ionic diffusion. The membrane materials such as zirconia operate at elevated temperatures, typically 300°C and have a relatively low yield. Their main use is as oxygen detectors in analysis but units large enough to provide breathing oxygen for fighter pilots are also possible and these use the high temperature, high pressure, combusting gases in the jet engine as their oxygen source.

Oxygen was one of the earliest industrial gases produced, being used (among other things) to produce the high temperature flame needed for theatrical lime lights. Many of its uses still range around its oxidizing properties especially cutting, welding, blast furnace enrichment, scarfing, and even Apollo, but oxygen is also the life-supporting gas used in hospitals and other emergency situations to aid respiration and sustain metabolism. This is not limited to humans or even other mammals, many bacteria are reliant on oxygen for optimum growth and these are utilized in many waste-water treatment plants in which industrial oxygen addition is becoming increasingly significant for optimum operation. Other uses include addition to polluted rivers to sustain fish life until the pollution has passed as well as directly in fish farming itself.

In many pollution instances the oxygen balance is very delicate and provision of relatively small amounts of oxygen can lead to absolutely spectacular results when nature takes charge again and re-establishes normal conditions. Such an example was a grossly polluted lake in South Africa which when cleaned up turned bright pink with feeding flamingos.

Nitrogen

This is the major residual gas when oxygen is separated from the air and was for many years regarded as a waste gas. It began to be supplied in liquid form, boiling at –196°C, in the late 1950s using a similar distribution and storage system to that of liquid oxygen. Today the ratio of nitrogen and oxygen produced in a developed nation is about 1.1:1 showing that this has now become the most widely used industrial gas of all. The individual properties of extreme cold and inertness are widely used in industry in the preparation and preservation of food, drugs, medical supplies, and fire prevention as well as rubber de-flashing and shrinkfitting of gear wheels, but nitrogen also finds a major use in the manufacture of plastics and fertilizers.

A more recent use has been the combination of nitrogen, CO_2 and oxygen in concentrations widely differing from air for the preservation of fresh produce both in storage and in transit (Fig. 1). In these applications the gas can be produced by the use of a semipermeable membrane or an adsorption system very similar to that described above but utilizing a carbon-based molecular sieve (Fig. 2). Much higher concentrations of nitrogen have also been used in the preservation of artefacts and relics such as Egyptian mummies, furnishings, etc. Liquid nitrogen is also used as the cryogen for the preservation of bull semen, tissue samples both animal and plant, and even to store human corpses, although in the latter case, the chances of successful restoration to life provide cold comfort. More importantly, many seeds and other small organisms are

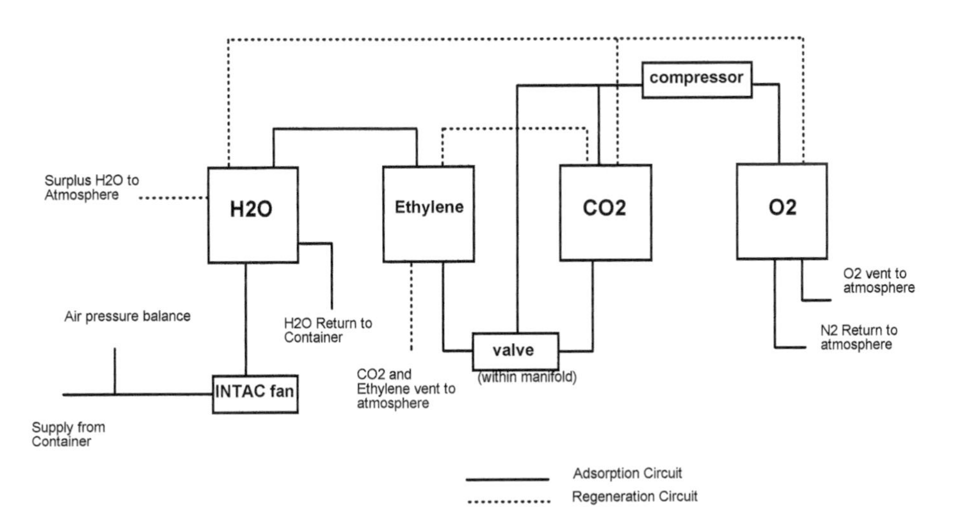

Fig. 1 INTAC schematic.

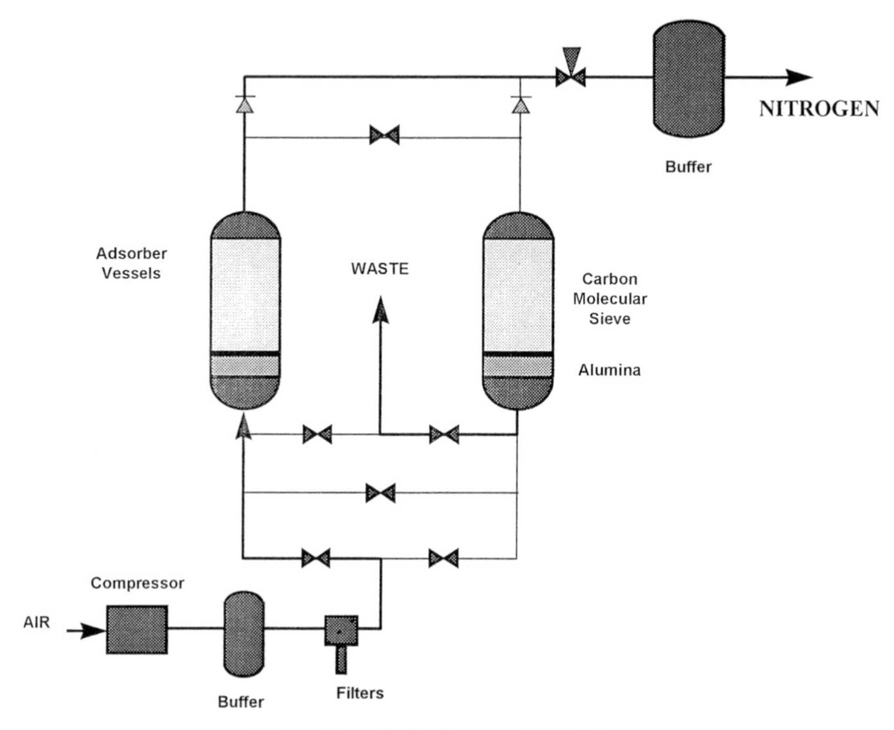

Fig. 2 Two-bed nitrogen PSA plant.

kept in liquid nitrogen called suppositories to maintain a wide gene database as a precaution against excessive monoculture.

This brief description of the uses of the air gases summarizes a vast and growing technology essential to modern industry. There are very few products which do not benefit from the use of industrial gases at some stage in their production or use including the human race with the medical usefulness of concentrated oxygen. The efficiencies of separation of the gases from the air are also improving, although the energies achieved in mammalian lungs still remain a target.

MIKE GARRETT

Born 1936, has had a long and varied career with BOC, initially at the R&D Centre working on novel forms of gas liquefaction and separation, and later the development of gas lubricated bearings for helium expansion turbines which formed the basis of the first large hydrogen bubble chamber at Harwell. He then became involved in superconductivity and cryostats for deep low temperature work and, in the early '70s, oxygen manufacture by Pressure Swing Adsorption. In 1983 he became Head of BOC's Centre of Excellence for the separation of oxygen by non-cryogenic means, developing the INTAC system for the transportation of produce in controlled atmospheres. Currently he is BOC Gases European Director for Innovation. He has been awarded over 150 patents.

Insects and chemical signals: a volatile situation

JOHN A. PICKETT

Introduction

Most animals, and many plants, use volatile chemical signals to regulate their behaviour. These signals have diverse roles and one chemical substance can convey different messages under different circumstances. In human beings, such signals are detected primarily by the olfactory epithelium within the nasal organ, and although often underestimated against our other highly developed senses, human olfaction can be extremely sensitive, particularly where compounds containing heteroatoms such as sulphur and nitrogen are concerned. We can detect some sulphur-containing compounds at well below 1 p.p.b. and despite having rather a bad press, they can have extremely positive effects when present in trace amounts. For example, the crisp and fresh aroma of blackcurrant, *Ribes nigrum* (Saxifragaceae), fruit and foliage derives from a particular chemical structural type. During the industrial processing of blackcurrant fruit, this aroma is usually lost but can be regained from other sources, for example the mercaptoketone from the buchu plant, a member of the Rutaceae in the *Barosma* genus. In a different context, the same kind of chemistry can have a less appealing role, giving rise to the smell of stale or sunstruck (sunlight-damaged) beer and the strong scent characteristic of tom-cats. Indeed, even the cat's perception of this type of chemical structure can cause some confusion and they may become excited by the foliage of *R. nigrum* or the flowering currant, *R. sanguineum*. Thus, from our own experiences, we can see that minute amounts of volatile organic chemicals can have positive or negative connotations, depending on the context in which they are perceived.

Compared with human beings, insects and other invertebrates can be dramatically affected by the chemistry of their environment. Although many plants have evolved to defend themselves by producing toxic chemicals, it is also common to find insects exploiting such chemistry for their own defence. An excellent example is the cinnabar moth, *Tyria jacobaeae* (Arctiidae), the larvae of which obtain highly carcinogenic and toxic pyrrolizidine alkaloids from plants such as ragwort, *Senecio jacobaeae* (Asteraceae). The toxins are sequestered, probably facilitated by conversion to the less toxic and more soluble *N*-oxides, for final storage as the free base and compartmentalized away from the susceptible physiology of the insect. Both adults and larvae of this moth advertise their poisonous nature by bright aposematic colouration. A great deal of work[1] has also been done on monarch butterflies in the Danaeidae. As well as sequestering the pyrrolizidine alkaloids, these butterflies take steroidal cardenalides from plants in the *Asclepias* genus (Asclepiadaceae), which act as heart poisons in vertebrates. However, another member of this family, the tree nymph or Japanese newspaper butterfly, *Idea leuconoe*, sequesters a totally innocuous compound, methyl 4-hydroxybenzoate,[2] and will even take the synthetic crystalline chemical from the bottle; in this case, the compound is probably used by the male as a signal agent in the mating process. It is on the use of non-toxic volatile chemicals as signals controlling insect behaviour that this chapter concentrates. The general term for such compounds is 'semiochemical', from the Greek *semeion* (*σημειόν*), meaning *sign* (the more generally accepted meaning in Greek is *point*), therefore *sign-chemical*. Pheromones represent a particular type of semiochemical employed exclusively between members of the same species.

Perception and identification

The insect's 'nose' principally comprises the antennae. In the case of the well known moth antenna, the surface is covered with hair-like projections, or sensilla. Many of these have a mechanosensory role, but some are olfactory sensilla having a porous cuticle through which volatile semiochemicals can pass. The semiochemical is then captured by a relatively low molecular weight binding protein and passes to the receptors, which are proteins embedded in the dendritic extension of the olfactory nerve cells normally situated within the body of the antenna. Via the axons of these cells, the signals pass to the central nervous system for processing and eventual motor neuron stimulation. The wonderful structure of such antennae can be seen on an otherwise rather insignificant insect, *Theresimima* (= *Ino*) *ampellophaga* (Zygaenidae), the vine-bud moth (Fig. 1). Their aerial-like appearance

Fig. 1 Male vine-bud moth.

has often led people, quite wrongly, to suggest an electromagnetic mechanism for olfactory perception.

To study olfactory signals in moths, or indeed any insect, the first step is to demonstrate unequivocally a behavioural response. This can be done by means of an olfactometer or wind tunnel in which the insect is introduced to semiochemical sources and, by elimination of other cues such as vision and contact effects, a behavioural response can be monitored numerically and analysed statistically. Samples of volatile semiochemicals are usually obtained under natural conditions, if possible, either from the live insect or from an intact plant, often by absorbing the material on to a porous polymer and re-eluting with a solvent. The samples can be analysed by gas chromatography (GC), using high-resolution quartz capillary columns and a stationary phase of an inert polymer with various levels of polarity. By linking such chromatography directly to mass spectrometry (MS), it is now possible to obtain spectra in sub-nanogram (10^{-9} g) quantities. Insects, however, can detect much smaller amounts and may respond to minor components in complex mixtures. Simply to identify the most abundant compounds would not give a complete picture of the signals involved and, as was exemplified in the case of human olfaction, trace components can be of key importance. Tremendous efforts have therefore been made to record directly from the insect antenna. By placing electrodes across the antenna, a standing potential is obtained which is perturbed by the polarization of olfactory nerve cells ensuing when the cells are activated. This gives the so-called electroantennograph (EAG) and is a common technique used in many laboratories, particularly with moths (Lepidoptera). These preparations can be routinely linked to high resolution GC (Fig. 2) to give a measure of the relative activity and importance of individual components eluting after chromatographic separation. At Rothamsted, we have pioneered the technique of recording from olfactory sensilla, or even individual nerve cells within these organs, using electrolytically sharpened tungsten electrodes. The olfactory role for these cells is established by stimulating the preparation with a sample containing the semiochemical of interest, prior to linking with the GC for the full GC-coupled single sensillum or single cell recording (SCR). The output comprises a trace of all of the compounds present in the sample and a simultaneous recording of the responses of the olfactory cell. An example is shown for the raspberry beetle, *Byturus tomentosus* (Byturidae), which aggregates in spring on flowers of hawthorn, *Crataegus monogyna* (Rosaceae) (Fig. 3). Tentative identifications by GC-MS are then confirmed by use of synthetic compounds with structures authenticated by high-field nuclear magnetic resonance spectroscopy (NMR). After the activity of the identified compounds has been

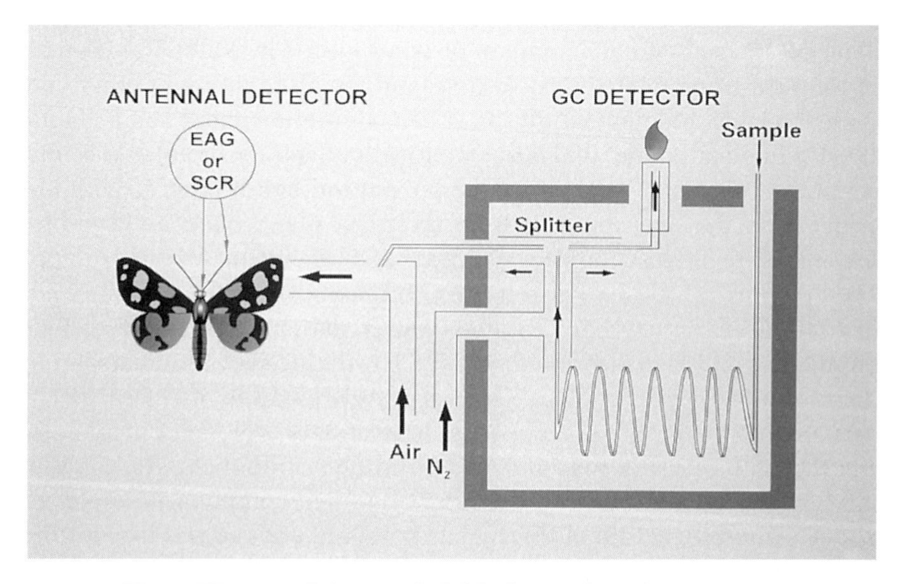

Fig. 2 Diagram of the coupled GC-electrophysiology system.

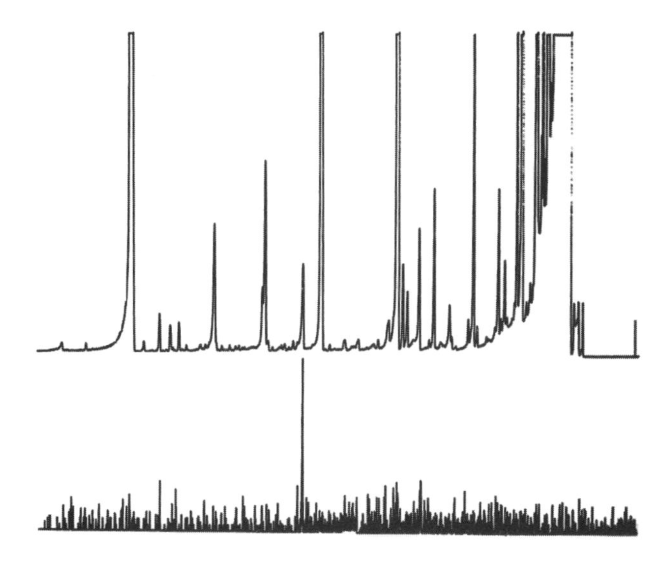

Fig. 3 Coupled GC-SCR. Upper trace: GC of volatiles from hawthorn flowers; lower trace: response of raspberry beetle olfactory cell to one minor component (C.M. Woodcock, unpublished data).

confirmed by electrophysiological studies, behavioural assays then ensue and, finally, the chemicals will be taken into the field for assessment in pest management strategies. The semiochemicals are either made available commercially, and already a large number of insect pheromones have been developed by this approach, or exploited in, for example, intercropping techniques where the plants themselves produce the appropriate range of semiochemicals.

Sex and aggregation pheromones

The best known examples of pheromones are those which bring the sexes together for mating or aggregate insects in appropriate egg-laying or overwintering sites. Many of the sex attractant pheromones for moths have been identified, but some intriguing phenomena still remain to be investigated, as we have recently found in a collaboration with the eminent British entomologist, Sir Cyril Clarke, F.R.S. He introduced us to a fact, observed by himself and his colleagues, that some moths within the family Arctiidae (tiger or footman moths) will indulge in cross-species mating, for example the cinnabar moth, *T. jacobaeae*, and the scarlet tiger moth, *Callimorpha dominula*. Sir Cyril suspected that this may be caused by the two species, normally separated geographically and temporally, having the same sex pheromone component. By analysing the volatile effluvia from the females using the coupled GC-EAG system, we were able to show that, indeed, a single compound produced by female *T. jacobaeae* and *C. dominula* stimulated the antennae of males from both species (Fig. 4). Chemical analysis revealed that the compound is the highly sophisticated (3*Z*,6*Z*,9*S*,10*R*)-9,10-epoxyheneicosa-3,6-diene, having two double bonds with specific geometry and an epoxy group present in a particular optical isomeric form.[3]

Of course, most of our efforts are not directed at such beautiful moths as the two mentioned here, and often involve very small and seemingly uninteresting insects which are, none the less, important pests. In much of world agriculture, a major problem is caused by sucking insect pests, which include the aphids (Aphididae), a large family of homopterous insects. A relatively limited subfamily, the Aphidiinae, have evolved to exploit the herbaceous plants that proliferated in the Cretaceous period, providing a much better source of food than their original tree hosts. During the spring and summer, the aphids reproduce asexually, or parthenogenetically, on their herbaceous (secondary) hosts. However, they must return to their tree (primary) hosts in the autumn in order to mate and produce eggs to survive the winter, particularly in temperate climates where the herbaceous plants will die off. It was clearly of great interest to identify the pheromones enabling male aphids to find the sexual females (oviparae) on the primary host in the autumn. The

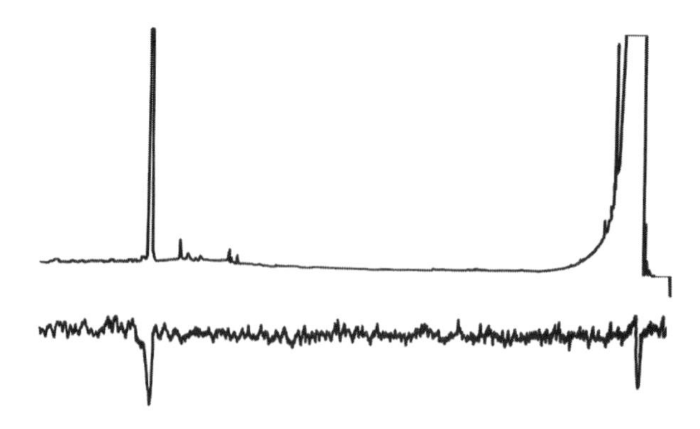

Fig. 4 Coupled GC-EAG. Upper trace: GC of volatiles from female scarlet tiger moth; lower trace: antennal response of male cinnabar moth.

oviparae are laid by winged sexual female precursors, the gynoparae, but are themselves flightless, or apterous, attracting the males by a sex pheromone released from porous plaques on their hind tibiae: the females raise their abdomens and hind legs from the leaf surface to allow evaporation of the pheromone into the air (Figs 5 and 6). As with the moths described previously, the volatile effluvia produced by 'calling' females could be collected and then, by placing tungsten electrodes into olfactory organs (the secondary rhinaria) on the antenna of the male, it was possible to locate the cells responding to the female sex pheromone. Using the coupled GC-SCR technique, originally with the vetch aphid, *Megoura viciae*,[4] the peaks for the sex pheromone components were located and subsequently identified by GC-MS, synthesis and X-ray crystallography. For many aphids,[5–7] the sex pheromones include structures I and II (see Fig. 7), which are isoprenoids in the cyclopentanoid class. For the damson-hop aphid, *Phorodon humuli*, the sex pheromone comprises the two diastereoisomeric structures III.[8] Indeed, this was the first aphid with which we demonstrated the potential of the sex pheromone to attract males in the field, a possibility which had been denied by many aphidologists prior to our identification of the compounds involved. Recent studies with collaborators from Imperial College,

Fig. 5 Sexual female vetch aphid releasing pheromone to attract males.

Fig. 6 Vetch aphids mating.

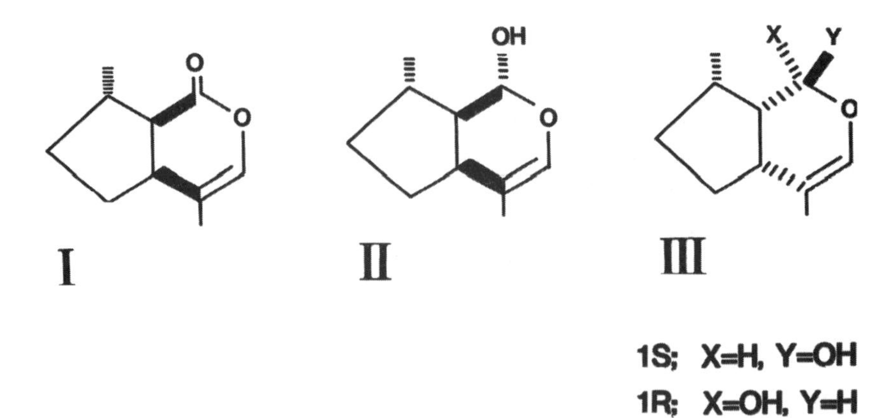

Fig. 7 Aphid sex pheromone components.

Silwood Park, have shown that these very small male aphids are able to fly upwind towards a pheromone source against wind speeds of approaching 2 m/s. This demonstration was achieved using sophisticated computer tracking techniques, combined with modern sensitive anemometry. Working with colleagues at HRI East Malling,[8] we were able to trap large numbers of male aphids as they migrated from the senescing hop plants towards the sexual females on damson trees (*Prunus* spp.), which act as the primary host for this insect. At this stage, the question might arise, 'Why not destroy damson trees in the vicinity of hop gardens, thereby avoiding the overwintering of this important and highly pesticide-resistant pest?' Unfortunately, such destruction has, where attempted, resulted in loss of wind-breaks and subsequent instability of the hop garden concerned.

The kind of chemistry involved in aphid sex pheromone biosynthesis is quite common in the insect world. Structurally related compounds are used for defence, for example by leaf beetles (Chrysomelidae) and ants in the *Iridomyrmex* genus (Formicidae), which produce chrysomelidial and iridodial, respectively. As the isoprenoid pathway responsible for the biosynthesis of these compounds probably developed before the separation of the animal and plant kingdoms, some plants are also able to produce them. Indeed, cyclopentanoids are produced by members of the Lamiaceae (= Labiatae), particularly in the *Nepeta* genus, and by *Actinidia polygama* (Actinidiaceae), a close relative of the kiwi fruit, *A. chinensis*. The identification of these compounds as aphid sex pheromone components thus explained the phenomenon, observed in Japan, of aggregation by an aphid predator, the lacewing *Chrysopa septempunctata*, on *A. polygama*,[9] presumably mistaking the plant for a colony of its prey. Aphids, however, do not normally confuse their own pheromones with plants producing the cyclopentanoids, possibly because the plants contain inappropriate isomers. Alternatively, the aphids may detect that the cyclopentanoids are from a plant by their association with other compounds inhibiting the normal attractancy of the pheromone components.

The cyclopentanoids can also be used in a completely different and rather surprising context, as excitants or attractants for marshalling the large cats (Felidae) within zoo enclosures, e.g. for cleaning purposes. So again we see a tremendous parsimony in biosynthesis where one structural type is used by insects, as sex pheromones and in defence, and in completely different taxa, from plants to highly evolved mammals.

Alarm and deterrent pheromones

Many insects, when attacked by predators or parasitoids (termed as such because, unlike parasites, the activities of a parasitoid result in the death of the host), warn other members of their colony of danger by release of alarm pheromones. Thus, ants release a pheromone from the mandibular glands,[10] which recruits other workers and soldier ants to attack the aggressor; a substrate contaminated with such an alarm pheromone may be ejected from the colony. Aphids also use alarm pheromones, although very few respond aggressively, with the exception of the sugar cane woolly aphid, *Ceratovacuna lanigera*, which, when attacked, rears up and fights to defend itself. The aphid alarm pheromone is released from sticky droplets produced by small tubes (cornicles) protruding from the upper surface of the rear of the abdomen. This secretion should not be confused with honeydew, which is a means of eliminating excess sucrose acquired from the phloem of the plant during the normal feeding process. The alarm pheromone for many aphid species is a simple sesquiterpene hydrocarbon, (*E*)-β-farnesene; its usual role is to cause aphids in the colony to disperse from the area.

Another form of deterrent pheromone may be produced on oviposition. The function of this is to reduce competition on host plants and to avoid possible cannibalism of new hatchlings by older larvae. Thus, when the cabbage seed weevil, *Ceutorhynchus assimilis* (Curculionidae), has laid its egg within a pod of oilseed rape, *Brassica napus* (Brassicaceae), it marks the site, and indeed the whole pod, with an oviposition deterrent pheromone produced from its abdomen. Later arrivals seeking oviposition sites detect the pheromone and move to other parts of the plant.

Host and non-host interactions

Many of the olfactory cues that insects use to find their host plants involve chemicals found commonly throughout the plant kingdom. It was originally thought that so-called generalist receptor cells were involved in the perception of such compounds. We have found, however, that this is not necessarily the case and that highly specific cells may be employed. For example, in the peach-potato aphid, *Myzus persicae*, the green leaf volatile (*E*)-2-hexenal is detected by a highly specialized receptor cell, and closely related compounds give little or no response even at orders of magnitude higher stimulus concentrations. We also observed that some insects have cells which frequently occur in pairs and respond to different compounds, often from unrelated biochemical pathways.[11] We believe that such cell pairings may allow the insect to detect relative differences in amounts of these common compounds, thereby giving another layer of discrimination between plant species. Additionally, there are species-specific compounds such as the isothiocyanates produced by catabolism of glucosinolates, typical of members of the Brassicaceae (= Cruciferae) and a few other related families, and where we already have excellent examples of highly specific olfactory cells in insects.[12]

During these investigations, we noted the presence of olfactory cells which did not respond to host plant chemicals or, indeed, to other sources of semiochemicals eliciting positive behavioural responses. Some of these cells have now been found to respond to specific components from plants that do not provide suitable hosts. Thus, the black bean aphid, *Aphis fabae*, although unable to feed on members of the Brassicaceae has, within its olfactory make-up, cells specific for the isothiocyanates, and indeed it has been shown that such compounds are repellent to this insect or act by inhibiting the normal attractancy of the host plant.[13,14] Furthermore, many of the herbivorous insects so far investigated are able to detect compounds typically released by damaged plants and this may allow avoidance of heavily colonized plants. It is possible that plant induction signals may in future be identified by 'plugging into' such systems using the insect's own olfactory mechanisms. Although the potential of this approach has not yet been fully realized, non-host plant chemicals are already proving valuable in the field as strategically identified repellents and slow-release formulations of plant stress compounds, applied to cereal crops, can cause up to 50% reduction in aphid colonization.[15] Such semiochemicals, being produced by plants, are now the target of molecular biological approaches for their exploitation. Thus, causing a plant to produce, from its vegetative parts, a compound typical of stress and thereby induced defence, can result in a lower infestation of pests that are repelled by such chemistry. Insects attracted by it, of course, would still find themselves on an inappropriate host if colonization was attempted.

Predators, parasitoids, and other beneficial insects

Predators and parasitoids, mentioned previously, employ an interesting range of semiochemicals for prey and host location. They may utilize the chemistry that the host employs in its own mating or aggregation behaviour and an exciting example of this is provided by the aphid sex pheromones. As was found in Japan with *C. septempunctata*, we have recently shown, in collaboration with Korean colleagues, that the lacewing *C. cognata* is attracted in the field to sources of cyclopentanoids. In the United Kingdom, again working with scientists at Imperial College, Silwood Park, we have found that a range of aphid parasitoids are attracted into traps releasing aphid sex pheromones. This response appears to be innate rather than learnt and occurs not only in the autumn, when the aphids themselves are using these semiochemicals, but throughout the year. There is considerable potential for incorporating this aspect of behaviour into practical aphid control methods. There are large populations of such beneficial insects in arable agricultural ecosystems but, unfortunately for the farmer, parasitoids seldom come into the crop early enough to reduce aphid numbers below economic thresholds. By employing the innate response of aphid parasitoids to the sex pheromones, it is proving possible to attract them into the crop earlier in the season and also to conserve populations on specially sown headlands and set-aside areas.

Parasitoids can also make use of 'SOS' semiochemicals, or synomones, released from plants damaged by herbivorous insects. Here, the response is largely learnt, with the parasitic organism using cues which have previously given some reward in terms of host location. The induction of synomone production in plants is extremely sophisticated. For two species of legume-feeding aphid, *A. fabae* and the pea aphid, *Acyrthosiphon pisum*, although compounds common to both aphid species can be detected in damaged beans, there are certain compounds produced by the plant that are specifically generated when one or other of the species is feeding, and this also relates to parasitoids specializing on different aphid species, which colonize the same plant taxa.

The parasitoids referred to so far are members of the insect Order Hymenoptera, which also includes the ants and pollinating insects such as honey bees and bumble bees (Apidae). A great deal of work has been done on honey bees at Rothamsted, and indeed one of the first pheromones to be characterized was the honey bee (*Apis mellifera*) queen pheromone.[16] Honey bees produce a wide range of pheromones and also interact with plant compounds in very subtle ways. Recently, our electrophysiological techniques have been brought to bear on this issue and have shown interesting results which, we hope, will relate back to our exploitation of other Hymenoptera, particularly the aphid parasitoids. Here, in collaboration with colleagues at the INRA/CNRS laboratory at Bures-sur-Yvette in France, we have combined GC and electrophysiological studies with a behavioural bioassay involving a conditioned proboscis extension response (CPE). The bees are held in a flow of air and conditioned, by means of a sucrose reward, to extend the proboscis on application of various semiochemical cues. When the bees are conditioned to a synthetic mixture, coupled GC-EAG demonstrates that the antennae can detect all the compounds present. However, the simultaneously recorded CPE reveals that the bees only respond behaviourally to certain components in the mixture. If the conditioning stimulus comprises volatiles from a flower, for example oilseed rape, coupled

GC-CPE again shows that the bees are only responding to a few key compounds in this complicated mixture. Furthermore, it was noted that the EAG response in the bee is increased by a sensitization taking place during the conditioning process. These studies are proving of immense value in understanding olfaction in these highly evolved insects and will also be useful for assessing whether new crop plants obtained by recombinant DNA techniques could have any deleterious effects on beneficial insects.

Practical considerations

The wider understanding of insect chemical ecology arising from these studies has great scientific value in elucidating natural processes and the various types of semiochemicals have considerable potential in pest management as an addition to, or indeed replacement for, broad-spectrum pesticides acting by toxic mechanisms. The commercial penetration is, to date, extremely small, but there is great pressure from the public and from food retailers for more of these approaches to be used in agriculture and in protecting against insect pests involved in disease transmission. The new techniques of molecular biology offer novel ways of exploiting semiochemicals[17] and as the world's population is rising rapidly, such methods will be essential, particularly in resource-poor areas where expensive pesticides simply cannot be afforded. Indeed, it is already apparent that, in understanding the mechanisms by which insects locate host plants and avoid non-hosts, new ways of preventing crop losses can be developed by intercropping with strategically selected plants. In East Africa, we have a programme with ICIPE (Nairobi) to reduce stem borer damage in subsistence cropping of sorghum and maize, funded by the Gatsby Charitable Foundation as part of its strategy of financing collaborative projects between international research centres in Africa and advanced research centres in the UK. Here, plant species that repel stem borers, and also cause increased parasitism among those insects that do infest the crop, are being used in combination with trap crops which encourage oviposition and can themselves be used as cattle forage.[18]

Conclusions

As mentioned at the outset, there are many interactions within insect chemical ecology in addition to those involving volatile semiochemicals.

As well as the toxicants already referred to, there are chemosensory interactions whereby the insect can detect nutrients, and also the so-called contact antifeedants, which are themselves important semiochemicals. Here, we have concentrated on insects and chemical signals in which a volatile situation is manifest.

Acknowledgements

IACR receives grant-aided support from the Biotechnology and Biological Sciences Research Council of the United Kingdom. This work was in part supported by the United Kingdom Ministry of Agriculture, Fisheries, and Food.

References

1. Rothschild, M., and Reichstein, T. (1976). *Nova Acta Leopoldina*, Supp. **7**, 507–50.
2. Nishida, R., Kim, C-S., Kawai, K., and Fukami, H. (1990). *Chem Express*, **5**, 497–500.
3. Clarke, C.A., Cronin, A., Francke, W., Philipp, P., Pickett, J.A., Wadhams, L.J., and Woodcock, C.M. (1996). *Experientia*, **52**, 636–8.
4. Dawson, G.W., Griffiths, D.C., Janes, N.F., Mudd, A., Pickett, J.A., Wadhams, L.J., and Woodcock, C.M. (1987). *Nature*, **325**, 614–16.
5. Dawson, G.W., Janes, N.F., Mudd, A., Pickett, J.A., Slawin, A.M.Z., Wadhams, L.J., and Williams, D.J. (1989). *Pure Appl. Chem.*, **61**, 555–8.
6. Pickett, J.A., Wadhams, L.J., Woodcock, C.M., and Hardie, J. (1992). *Annu. Rev. Entomol.*, **37**, 67–90.
7. Dawson, G.W., Pickett, J.A., and Smiley, D.W.M. (1996). *Bioorg. Med. Chem.*, **4**, 351–61.
8. Campbell, C.A.M., Pettersson, J., Pickett, J.A., Wadhams, L.J., and Woodcock, C.M. (1993). *J. Chem. Ecol.*, **19**, 1569–76.
9. Hyeon, S.B., Isoe, S., and Sakan, T. (1968). *Tetrahedron Lett.*, 5325–6.
10. Blum, M.S., Padovani, F., and Amanté, E. (1968). *Comp. Biochem. Physiol.*, **26**, 291–9.
11. Blight, M.M., Pickett, J.A., Wadhams, L.J., and Woodcock, C.M. (1995). *J. Chem. Ecol.*, **21**, 1649–64.
12. Blight, M.M., Pickett, J.A., Wadhams, L.J., and Woodcock, C.M. (1989). *Aspects Appl. Biol.*, **23**, 329–34.
13. Nottingham, S.F., Hardie, J., Dawson, G.W., Hick, A.J., Pickett, J.A., Wadhams, L.J., and Woodcock, C.M. (1991). *J. Chem. Ecol.*, **17**, 1231–42.
14. Hardie, J., Isaacs, R., Pickett, J.A., Wadhams, L.J., and Woodcock, C.M. (1994). *J. Chem. Ecol.*, **20**, 2847–55.
15. Pettersson, J., Pickett, J.A., Pye, B.J., Quiroz, A., Smart, L.E., Wadhams, L.J., and Woodcock, C.M. (1994). *J. Chem. Ecol.*, **20**, 2565–74.

16. Butler, C.G., Callow, R.K., and Johnston, N.C. (1961). *Proc. R. Soc. B*, **155**, 417–32.
17. Hick, A.J., Pickett, J.A., Smiley, D.W.M., Wadhams, L.J., and Woodcock, C.M. (1997). *Phytochemical diversity: a source of new industrial products*, pp. 220–36. The Royal Society of Chemistry, Cambridge.
18. Khan, Z.R., Ampong-Nyarko, K., Chiliswa, P., Hassanali, A., Kimani, S., Lwande, W., Overholt, W.A., Pickett, J.A., Smart, L.E., Wadhams, L.J. and Woodcock, C.M. (1997). *Nature*, **388**, 631–2.

JOHN PICKETT

Born 1945, received an honours degree in chemistry from the University of Surrey in 1967 and then went on to complete a PhD in organic chemistry. Head of the Biological and Ecological Chemistry Department at IACR-Rothamsted since 1984, he previously held the post of Principal Scientific Officer in the Department of Insecticides and Fungicides at Rothamsted Experimental Station. He is a long-standing Fellow of the Royal Society of Chemistry, a Fellow of the Royal Entomological Society, and a Member of the American Chemical Society. He has published widely and has been invited to lecture all over the world, but still works practically in the laboratory, particularly in semiochemical identification by mass spectrometry. He was awarded the Rank Prize for Nutrition and Crop Husbandry in 1995 and was elected Fellow of the Royal Society in 1996.

Tower Bridge—enduring monument or Victorian folly?

ROY F.V. AYLOTT

Introduction

One's first impressions of Tower Bridge generally are of its scale, age, and strange design. It is large and of unusual design, but it is not old in comparison with the Tower of London, with which is it inextricably linked. Tower Bridge is only 102 years old, whereas the origins of the tower date from the eleventh century.

Perhaps the most surprising thing about Tower Bridge is its silence in operation, particularly as it is an iron and steel structure clad with stone. But its smooth, silent operation derives from an amazing hydraulic system. Tower Bridge is unique, not only because it is the only bridge of its kind in the world, or because it displays the ingenuity and engineering skills of the Victorians at their zenith, but also because it holds a particular place in the hearts and minds of all Londoners.

It stands as a symbol of many things, particularly of our heritage; the power of Britain and its former Empire in the nineteenth century; the strength and reliability of steam and hydraulic engineering; the grandeur and ornate folly of Victorian Gothic architecture; the importance of the City of London; and much more.

Building feats such as the Britannia Bridge and the Forth Rail Bridge brought unprecedented fame and fortune to their builders. Now, however, we see them only as forerunners to Tower Bridge, arguably the greatest Victorian engineering achievement of them all. The need for, and the eventual building of, Tower Bridge were highly controversial issues, and had it not been for the ingenious design concept finally adopted, it may never have been built.

Tower Bridge has in every way fulfilled the hopes of its designers and promoters. The design and construction of the bridge and its motive power are outstanding examples of Victorian craftsmanship, and the bridge has regularly carried loads far greater than were envisaged at the time of its construction, with the minimum obstruction to river traffic. The machinery for raising the bascules is as good today as it was in 1894 and has never failed in its task. It is fitting that a complete set of the original engines and pumps remains on permanent exhibition alongside their modern counterparts.

Today we are accustomed to the sight of Tower Bridge as arguably London's most famous and well loved landmark. Indeed, Tower Bridge is synonymous with London in the same way as the Eiffel Tower with Paris, or the Statue of Liberty with New York. There can be little doubt that Tower Bridge is now seen as an enduring monument. But this was not always the case. The controversy which preceded its construction raged for over 60 years, as there was considerable opposition to the building of a new bridge in the Pool of London downstream of London Bridge, over what was then the busiest river in the world.

In support of his proposals for a bascule bridge the City Architect, Horace Jones, did refer to the fact that this type of construction would lend itself to bold architectural treatment and make a worthy landmark, but I am sure even he could not have foreseen the impact it would have, and the affectionate regard in which it would come to be held, not only by Londoners, not only by the British people, but by countless millions world-wide to whom it symbolizes our great City.

Historical background

Britain experienced amazing growth during the reign of Queen Victoria in industry and democracy, as well as overseas. At home, the development of steam power and civil engineering precipitated an industrial revolution that changed the nation from being mainly agricultural to industrial. Inevitably, this industrial expansion was centred on cities and towns, which in turn experienced massive growth in population. This was not just due to migration from rural areas, but was equally due to an increase in the size of families brought about by greater year-round employment prospects and associated wealth. The population as a whole increased by some 50% during this period.

London's growth was primarily linked to the expansion of trade, particularly with the ever-expanding empire. The importance of London as a port resulted in the building of new docks to the east of the City because London Bridge prevented shipping passing upstream. Even John Rennie's new London Bridge which was completed in 1831 had only 29 feet (8.8 m) headway above high water and remained a barrier to shipping. The opening of these new docks to the east was inevitably followed by the building of homes for the dockworkers. But London's growth was not, however, limited to the east.

The introduction of the steam engine brought with it the construction of the railways radiating from central London to suburban towns such as Aylesbury, Bedford, and Croydon, and beyond. This resulted in the rapid development of the smaller towns and villages along the route of these railways, partly sponsored by the private railway companies. The key incentive offered to potential commuters was a better life-style in the 'country', with the ability to quickly get to one's place of business by train in no longer time than it previously took to travel across London by carriage. In some instances, workers' homes were built to rent with the benefit of a one penny train fare to central London.

During the nineteenth century the population of London grew rapidly, with 1 million people, 39% of London's population, living east of London Bridge; this being roughly equivalent to the combined populations of Birmingham, Leeds, Liverpool, and Manchester. Every day, thousands of carriages and pedestrians had to make lengthy detours to cross the river. The traffic during 24 hours over London Bridge, a bridge only 54 feet (16.5 m) wide, taking the average of 2 days' observations in August 1882, was 22 242 vehicles and 110 525 pedestrians—the number of people daily crossing London Bridge was equivalent to half the population of Edinburgh.

The westward expansion of London had brought with it a succession of bridges starting at Putney in 1729; albeit this was only initially a wooden viaduct. Nevertheless, in the next hundred years, no fewer than eight new road bridges were built, with another five, exclusive of railway bridges, being added by 1880. But the main problem lay downstream of London Bridge, where the public were poorly served.

The earliest proposal to build a new bridge by the Tower of London was in 1813, when Sir Samuel Browne and James Walker proposed the erection of a high level suspension bridge. This was followed in 1824 by a proposal to build the 'St Katherine's Bridge of Suspension'; the joint promoters sought to raise £392 000 by public subscription, and offered prospective shareholders a potential yield of 10% per annum, based upon income from tolls estimated at in excess of £100 per day; a clear indication of the demand for such a crossing. Despite this, the proposal did not proceed.

The first proposal intended to relieve the situation to go ahead, was Sir Marc Isambard Brunel's Thames Tunnel project between Wapping and Rotherhithe. Conceived in 1824, this scheme suffered many problems, and it was not until March 1843 that the tunnel finally opened to pedestrians only; the idea of accommodating vehicles having been abandoned due to severe difficulties and cost. Moreover, the foot of the tunnel was some 75 feet (22.95 m) below water level at full tide. Despite the early recognition of need, it was to be more than 60 years before a solution was found.

The main reason for the delay was the conflict of interest between the public need for a crossing and the commercial interest of shipping and wharves in the Pool of London. London was known as 'the Warehouse of the World' due to the vast rows of vessels moored in the river, often three abreast with barges alongside, so that 'one might walk from deck to deck for many a mile!'.[1] A survey by Horace Jones in 1828 confirmed that the number of ships passing upstream of St Katherine's Dock by the Tower during a 6-day period, with masts ranging between 40 feet (12.2 m) and 95 feet (29 m) in height, was 144.

Although it was generally accepted by the middle of the century that any new crossing should be just to the east of the Tower of London, at Little Tower Hill, the dilemma was how to build a suitable crossing without steep gradients which would inhibit horse drawn carriages, while maintaining access for vital shipping. In an attempt to alleviate the situation, a 7 feet (2.12 m) diameter cast-iron subway was constructed in 1871 by a joint stock company, linking Great Tower Hill on the north side to Pickle Herring Stairs on the south side of the River Thames. It was intended by the engineer, Peter Barlow, to operate using a rope-drawn car, but after several accidents, it reverted to pedestrians only.

Despite a toll of a halfpenny, nearly 1 million people a year used the subway to cross the river. However, the subway did nothing for road traffic, and the pressure for a 'total' solution remained. The subway eventually closed when Tower Bridge opened, and was sold to the London Hydraulic Company and much later, in the 1980s, acquired by Mercury Communications plc.

How to solve the problem?

In 1855, the Metropolitan Board of Works was set up, with one of its primary responsibilities being to improve roads and bridges. The Corporation of London conferred with the new authority on the question of removing the tolls from Southwark Bridge (1819), a privately owned bridge, hoping to increase usage and to relieve the ever mounting pressure on London Bridge. But, no common view was forthcoming.

However, in 1864, the corporation resumed direct negotiations with the Southwark Bridge Company, resulting in the company agreeing to remove the tolls if the Corporation would reimburse its losses. Southwark Bridge was officially opened as a toll-free bridge by the Lord Mayor, William Lawrence, on 8 November 1864. Subsequently, on 12 June 1868, the Corporation purchased the bridge for £200 000, funded from the Bridge House Estates.

Unfortunately, even though the number of vehicles crossing the bridge each day increased from 1014 (June 1863) to 4113 (March 1866), this was to have little effect on traffic using London Bridge, as the siting of the approach roads and the steep gradients were a serious deterrent to traffic. Thus it was, that the need for a new crossing east of London Bridge was continually urged upon the corporation and the Metropolitan Board of Works; it not being clear at that time who would eventually be responsible.

There was no such confusion on the part of local traders, who presented a number of petitions to the (Corporation's) Court of Common Council who, in 1874, instructed its Bridge House Estates Committee to consider and report on the best means of increasing the capacity of London Bridge. This was followed, on 28 January 1875, by a resolution of the Ward of Candlewick being presented to the Court of Common Council praying for additional facilities. Then, in June 1875, a deputation representing City Business Houses stressed to the Court of Common Council the urgent need to resolve the matter. In all, between 1874 and 1885, some 30 petitions and other presentations were received by the Court of Common Council.

The Bridge House Estates Committee was instructed to consider any possible means of relieving the growing congestion on London Bridge. In December 1875, the Committee, advised by Horace Jones and having considered various schemes, recommended the widening of London Bridge. However, this recommendation was not well received and, in January 1876, a third petition was presented by 'Merchants and Traders largely interested in the conveyance of goods in around the eastern part of the City of London' referring to the proposed widening of London Bridge and strongly urging the provision of a new bridge near, or to the east of, the Tower.[2]

This petition was immediately referred to the Bridge House Estates Committee with instructions to weigh the relative advantages and costs of a bridge over the river or a tunnel beneath it, and authorised the Committee to confer directly with Her Majesty's Government about its proposals. A week later, a Special Bridge or Subway Committee was established, and assumed responsibility for dealing with all references in relation to additional accommodation for traffic between the north and south sides of the River Thames east of London Bridge.

The alternative proposals

Over the next 8 years, the Corporation examined and rejected at least 19 alternative schemes in addition to proposals for widening London Bridge. These included: four low-level bridges by Mr F. Barnett, Mr J.P. Drake, Mr F.J. Palmer and Messrs Kinipple and Morris, with different forms and widths of swing opening for the passage of ships into the Upper Pool; a rolling bridge by (Sir) George Barclay Bruce, 'shuttling' constantly across the river on piers, leaving space for shipping at all times; three designs for high-level bridges by Sir Joseph Bazalgette, all served by a 300 feet (92 m) diameter spiral ramp on the south side; a similar high-level bridge by Mr T. Claxton Fidler; an alternative high-level bridge by Bazalgette with hydraulic hoists at each end, thereby eliminating the need for major property acquisition on both approaches; a similar high-level bridge by Mr Sidenham Duer; a steam 'ford' by Mr C.T. Guthrie, comprising submerged rails with a travelling platform above high water level; a submerged wrought iron archway cum tunnel on the river bed by Messrs Maynard and Cooke; a sub-riverine arcade with ramped approaches by John Keith; a three-span mean-level bridge by Henry Vignolles; in addition, Bazalgette also produced further designs for two low-level bridges, another high-level bridge, and a tunnel.

The comparative estimated cost of these schemes varied from £144 000 for Bruce's rolling bridge, to £2 980 000 for one of Bazalgette's low-level bridges, including approaches, which was some £1 000 000 more than the cost of his high-level designs. Most of the schemes were estimated to cost between £500 000 and £600 000. But as none of these schemes was deemed to be acceptable and therefore built, the validity of the various estimates was never tested. However, Sir Joseph Bazalgette, Engineer to the Metropolitan Board of Works, was one of the most eminent and experienced Victorian engineers and his estimates were more likely to have been realistic.

On the face of it, Bruce's rolling bridge concept, which would always leave space for ships to pass and did not require ramped approaches, looked extremely cheap. However, at best, its projected capacity was limited to 100 vehicles and 1400 passengers for each crossing. Although the river crossing would take only 3 minutes, the time needed for loading and unloading resulted in an overall crossing time of 10 minutes. It was

Fig. 1 Mr F.J. Palmer's proposal for a duplex bridge, 1877.

therefore thought to be inadequate for expected demand. In addition, the powerful ship owners and wharfingers objected on the grounds of safety and the inevitable interference with the movement of ships.

By early 1877, the Special Bridge or Subway Committee had concluded that, subject to sufficient funds being available, a bridge over or a tunnel under the river should be constructed east of London Bridge. The Committee recommended that they should be empowered to examine means of fulfilling this objective and invite further schemes. However, consideration was adjourned, pending further advice from Horace Jones as to the relative merits of high and low-level bridges and a tunnel under the river.

The Committee reported further on 3 May 1877, strongly recommending some form of low-level bridge linking Little Tower Hill with Horselydown Lane. In reaching its conclusion that a new bridge was preferred to widening London Bridge, the Committee also had the benefit of a detailed analysis made by Horace Jones of traffic movements across London Bridge as monitored by the Commissioner of Police, and comprehensive details of all shipping passing into the Upper Pool.

In March 1878, Sir Joseph Bazalgette reported to the Metropolitan Board of Works, proposing the construction of a high-level bridge with 65 feet (19.8 m) headway. This scheme was adopted by the board, and resulted in a Tower Bridge Bill being deposited in Parliament in the next session. The bill was successfully opposed both by the Corporation and by the Thames Conservancy Board as the restricted clearance would have prevented much of the shipping from entering the Upper Pool, and the proposed spiral ramp for vehicles and pedestrians on the southern approach was unacceptable.

During the consideration of Bazalgette's scheme, Horace Jones was called to comment on the design concept. Doubtless it was hoped that the two men would collaborate, but Jones reported on 16 October 1878 that: 'Sir Joseph Bazalgette did not propose or desire to pursue the subject jointly with me.'[3] Indeed, Bazalgette proceeded to prepare two alternative designs for high-level bridges and several other entirely different schemes. Jones found that Bazalgette was inclined to arrogance and '... generally ignored the evidence, views, wishes and interests of the wharfingers and shipowners...'.[4]

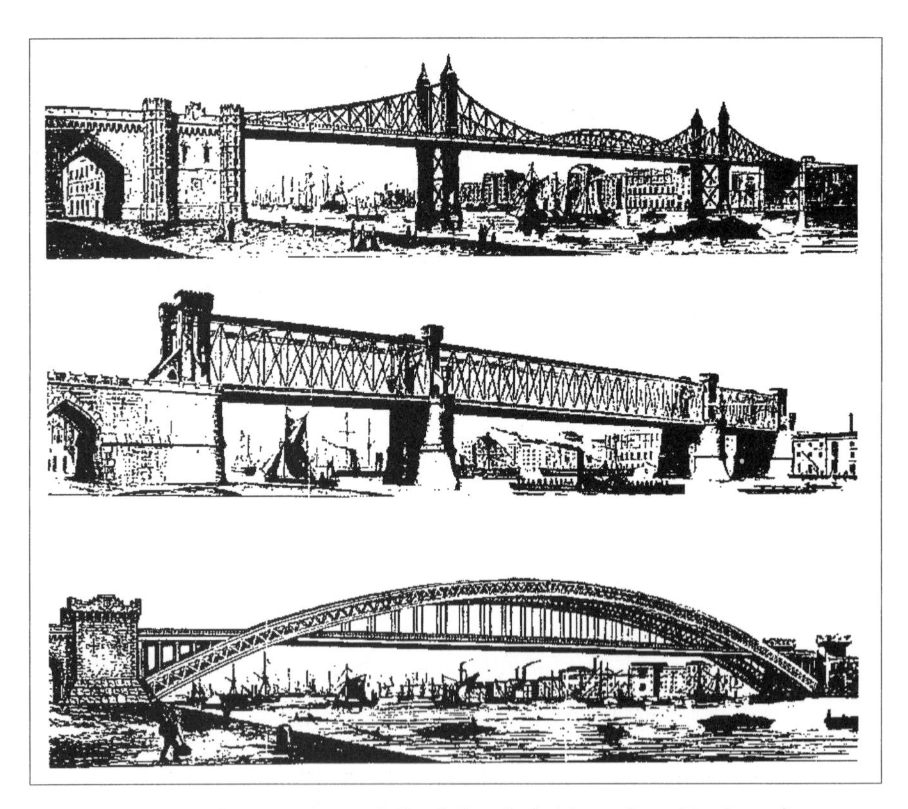

Fig. 2 Three proposed high-level bridges by Sir Joseph Bazalgette, 1878. The main drawbacks of this solution were the lack of headway for ships, and the steep gradients either end, which would impede horse-drawn vehicles.

Horace Jones then decided to produce his own design for a low-level bridge, having the same headway as London Bridge, namely 29 feet (8.8 m), with a centre span of some 300 feet (92 m) comprising two hinged platforms or bascules, to be raised by steam or hydraulic power. In his report to the Committee in October 1878, he recommended the construction of a low-level bridge at an estimated cost of £750 000, and outlined the several advantages of this type of construction as follows:

— easy gradients for vehicles and pedestrians;
— economy of construction, with little encroachment on either bank, thus necessitating very slight alterations of the adjoining streets and properties;
— a wider, unobstructed passage for shipping compared with a swing bridge;
— less interference with the tide-way, there being only two piers;
— beauty of form, the chief features of the bridge being capable of architectural treatment, it might be rendered the most picturesque bridge on the river; and
— ease and speed of operation using the machinery proposed.[5]

The Committee incorporated Jones's recommendation in a report to the Court of Common Council. But the report did not receive universal acclaim because of certain vested interests and the matter lapsed for several years. Notwithstanding continued public agitation for a new bridge, the Court of Common Council even dissolved the Special Bridge or Subway Committee on 24 January 1879 and transferred its outstanding references back to the Bridge House Estates Committee.

In October 1881, the Metropolitan Board of Works organized a conference to consider further the question of a new crossing. The board and its engineer, Bazalgette, were still strongly in favour of a high-level

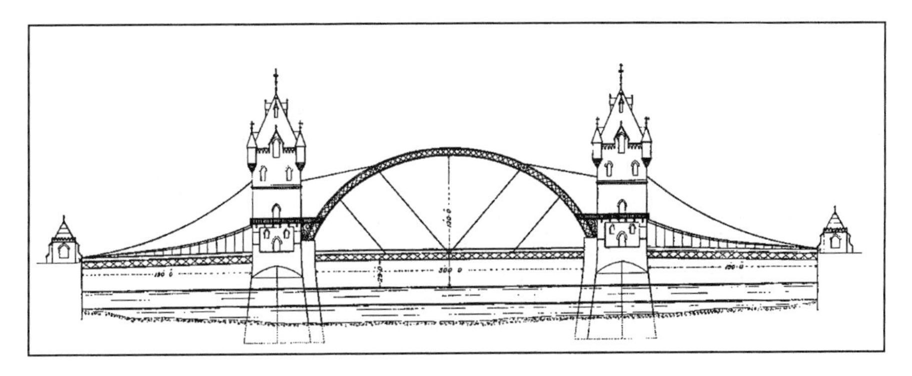

Fig. 3 Original bascule bridge proposal by Mr Horace Jones, 1878.

bridge, but they found little support for this view. Indeed, the Bridge House Estates Committee even suggested that: '... the need for any bridge or tunnel did not seem sufficiently proved to justify taking part in the promotion of a scheme ...';[6] a view that most observers found unacceptable, and led to accusations against members of the committee like Sir Henry Isaacs, a prominent wharfinger, of simply protecting their own interests.

The following May, a public meeting was held at the Mansion House to protest about the delay in providing a new river crossing. Despite this the Bridge House Estates Committee, having considered the various alternatives anew, recommended to the Court of Common Council that a bill should be presented to parliament for the establishment of a free ferry at the site of the proposed bridge. A deputation was even sent to Portsmouth, Gosport, and Southampton to view various floating bridges in use there. But, having consulted the Thames Conservancy Board, this idea was not pursued.

Frustrated with the lack of progress, in 1883, the London Chamber of Commerce organized an exhibition of maps, plans or models of bridges, tunnels, and other ideas for crossing the river below London Bridge. Eleven schemes were put forward and exhibited, but nothing conclusive transpired as none of the designs on view offered anything new. However, 1884 was finally to be the year of decision. Indeed, by March there were three bills before Parliament. First, there was The Metropolitan Board of Works (Thames Crossings) Bill; secondly, The Tower (Duplex) Bridge Bill, introduced by private promoters, Messrs Bell and Miller; and, thirdly, the Corporation's own Lower Thames Steam Ferries Bill.

These three bills were considered by a Select Committee of the House of Commons, who sat for 25 days before reporting on 4 July to the House that it was of the opinion that two crossings are required, and should be sanctioned by parliament; one, a low-level bridge at Little Tower Hill; the other a subway at or near Shadwell. The Select Committee went on to express the hope that the corporation might be induced to undertake the construction of the bridge and the Metropolitan Board of Works the subway at Shadwell.

On 24 July, the Bridge House Estates Committee reported that it agreed generally with the Select Committee's conclusions, and recommended to the Court of Common Council that a low-level bridge, with mechanical opening(s), be erected at Irongate Stairs, at the end of Little Tower Hill. The bridge would be funded out of money to be raised on the credit of the Bridge House Estates and the Committee should be empowered to obtain a design, together with an estimate of cost, with a view to a bill being deposited in Parliament later that year.

The final solution

Before the Corporation applied to Parliament to grant powers to construct the new bridge, a deputation of the Bridge House Estates Committee accompanied by Mr F.T. Reade, a civil engineer, due to the indisposition of Horace Jones, inspected a number of mechanical bridges in the north of England and on the Continent. This followed the comments of the Parliamentary Select Committee that: '... they thought the question of a swing bridge on the pivot principle of two openings, each about 100 feet (30.6 m), had been solved satisfactorily at Newcastle, where such a bridge has been constructed over the River Tyne, and has been proved capable of accommodating in a satisfactory manner both the shipping and the road traffic. They therefore unhesitatingly recommended this method of crossing the river at this site, as being the most effective for relieving the congested condition of London Bridge.'[7]

The deputation visited bridges at Newcastle, Edinburgh, Antwerp, and Rotterdam, and were much impressed by the bascule bridges seen in Rotterdam. As a result of the visit, Horace Jones was invited to review his earlier design for a bascule bridge in collaboration with John Wolfe Barry, who was appointed as engineer to the project.

Less than 3 months later, on 28 October 1884, the Bridge House Estates Committee, after considering three alternative schemes produced by Jones, was able to recommend the promotion of a bill to construct a bascule bridge at an estimated cost of £750 000. The Court of Common Council adopted the Committee's recommendations, and the City Remembrancer was instructed to promote the necessary bill in parliament.

The design finally chosen was basically similar to Jones's 1878 proposal, but modified in two important ways. First, the width between the main towers was narrowed from 300 feet (92 m) to 200 feet (61.2 m), which dramatically reduced the weight of each bascule. Secondly, the central arch was replaced by horizontal girders, which reduced the risk of tall ships striking the bridge if they were not in the centre of the river. The provision of the horizontal girders also permitted the incorporation of facilities for pedestrians to cross at the high level while the bridge was raised for ships.

The need to reconcile the requirements of land traffic with the very important interests of the trade of the Upper Pool had presented Jones and Barry with no small difficulty. That they succeeded was a testament to their ingenuity. This part of the river was always crowded with craft of various kinds, and this made the bascule concept unusually appropriate. Any opening bridge revolving horizontally would have stagnated a large area of the river frontage, whereas vertical opening bascules not

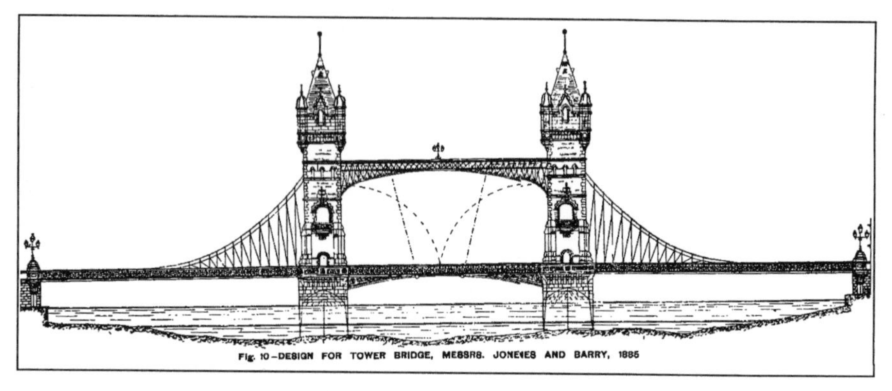

Fig. 4 The scheme submitted to parliament in 1884 by Jones and Barry.

only occupy a minimum of space but, at an early stage in the process of opening, also afford a clear passage for ships, which increases rapidly as the (opening) process continues.

Sea-going ships were generally moored head to stern in two or more parallel lines (tiers) in the Upper Pool on each side of the river, leaving a central channel of 200 feet (61.2 m) or so free for the passage of ships and smaller craft. On the landward side of the tiers, channels were maintained for the passage of barges and smaller craft to and from the wharves and warehouses. This configuration of ships and other vessels governed the leading dimensions of the proposed bridge, to the extent that, the central opening was to be 200 feet (61.2 m) wide, with the clear headway being 135 feet (41.3 m) when open and 29 feet (8.8 m) when closed. Each of the side spans were to be 270 feet (82.6 m) clear, with the width of the piers being 70 feet (21.4 m), which roughly equated to the width of two tiers of ships. More importantly, the proposed width of the piers was sufficient to accommodate the (bascule) counterbalance chambers.

Despite the critical attention to detail, the design proposed encountered strong opposition in Parliament from the wharfingers. Moreover, the Corporation's bill was also opposed by the Thames Conservancy Board, because of the perceived impact on shipping.

Parliamentary approval

The Corporation's Tower Bridge Bill was deposited in Parliament in the autumn of 1884. The bill was considered in succession by strong

Committees of both Houses of Parliament and certain concessions had to be made by the corporation to overcome a concerted opposition to the proposed bridge.

In his evidence to the House of Commons Committee, Barry made it known that he had consulted a number of eminent engineers and stated: '... therefore the bridge that is now before the Committee is not merely Mr Jones's and my proposal, but it is a proposal that before it took any tangible shape was endorsed by all these gentlemen.'[8] Asked if he would stake his reputation on 'Mr Jones's flaps or wings' always opening at the right moment and at the requisite speed, he replied: 'If I was not convinced in my own mind that this was a thoroughly workable scheme, I am quite sure I should not be sitting in this chair.'[9]

Commenting on the frequency and duration of bridge openings, Barry also stated that: '... from his observations, which had extended over five weeks, it appeared that the bridge on the average would have to be opened about 22 times a day to allow such vessels through as could not pass under it, and that the operation of passing a ship through would take about five minutes, and even in the worst case observed, when several vessels followed each other, the stoppage of the road traffic would have been but 20 minutes.'[10]

One of the most important witnesses to give evidence in support of the new bridge was Captain W.F. Luders of the port of Copenhagen. Harbour Master Luders was responsible for seven bascule bridges in Copenhagen, the most important of which had been completed in 1869 and was operated by hydraulic power. The time taken to open this bridge was about 1 minute, while the average number of vessels passing through the bridge every day was 20, although as many as 55 had passed in one day. The maximum daily road traffic over the bridge was about 500 vehicles and 3500 pedestrians. Luders also confirmed that he had not observed any difficulties arising from the speed of the current through the bridge.

The Thames Conservancy Board's overall objection to the bridge was rejected by the House of Commons Committee. However, the conservators were successful in inserting a clause in the bill obliging the corporation to maintain at all times during the construction of the bridge a clear waterway of 160 feet (49 m) in width, a requirement which was to cause much delay to the project, as will be explained later.

In the Upper House, the main concerns related to the question of whether the wharfingers should receive compensation for depreciation of property. After prolonged discussion, the House of Lords Select Committee decided that the corporation should in some way provide against the possibility of injury to the wharfingers businesses. A clause was inserted in the bill which provided that:

> If, at the expiration of four years after the opening of the Tower Bridge for traffic, the owner, lessee, or occupier of any of the wharves or quays between the bridge and London Bridge shall allege that such premises are depreciated in value by reason of danger or delay caused by the bridge to vessels destined for such premises, and shall give the necessary notices and claim compensation, then, if an agreement be not come to with the Corporation for the settlement of such claim, it shall be referred to arbitration to decide whether there has been depreciation, and the amount payable, if that should be found to be sustained. The aggregate amount of compensation in respect of all interests in any such premises shall not exceed two years' purchase of the assessable value in force on the 1st of January, 1886; and such aggregate amount shall be in full satisfaction of all compensation payable to all parties interested in such premises. No claim shall be sustainable unless made within seven years of the opening of the bridge for traffic. No compensation shall be given for any danger or delay caused during the construction of the bridge to vessels destined for such premises. No compensation shall be awarded for any quay or wharf on the south side of the Thames south of a line 200 feet back from the river front or its quay.[11]

In respect of a petition from the ferrymen plying between Irongate Stairs and Horselydown Stairs near the bridge, it was decided that the Corporation should pay compensation for damage done through interference with their rights; such compensation to be settled by agreement or by arbitration.

The Select Committee recommended that, in view of the great need for the bridge, the additional clauses should be accepted and the bill enacted. Royal Assent was received for the Corporation of London (Tower Bridge) Act 1885 on 14 August 1885. The sting in the tail was the requirement within the act for the bridge to be built within 4 years.

Nevertheless, congratulations were now in order, and the Chairman of the House of Commons Committee, Mr R.D.M. Littler, QC, observed:

> That the scheme will be an enormous public advantage I think can hardly be denied; that the design is ornamentally satisfactory, I think can hardly be denied. It is intended to be to a certain extent in harmony with the buildings of the Tower; it is intended to be an ornament to the river and it is intended to be provided without taxing anybody to the extent of a single halfpenny, which is a very large element in these days of large local and Imperial taxation. In point of fact, just as the City of London (Corporation) have made a present to the metropolis of its grandest park, that of Epping Forest, they are now prepared to make a present to the metropolis of another bridge, which,

while not forgetting London Bridge, will be its most important and most valuable bridge.[12]

How was it that the corporation could fund this major project without a levy on local ratepayers or support from the Central Exchequer? A number of references have been made to the Bridge House Estates; it would therefore seem appropriate at this point to give a brief history of this ancient charity.

The bridge house estates

The origins of the Bridge House Estates trust exercised by the Corporation of London extend back to the tolls and taxes collected in respect of the early wooden London bridges almost 1000 years ago. Originally, the bridge's revenues must have been obtained from charitable contributions and grants of tolls and taxes by the Saxon and early Norman Kings.

The Bridge House Estates trust as we know it was established in 1201 by King John, when he directed that the rents and profits from houses to be built on the bridge should be devoted to the future sustenance of London Bridge. This probably resulted from the frequent demands upon the Royal Exchequer to fund the maintenance or rebuilding of the earlier timber bridges at this important river crossing.

Whether or not the Romans built a bridge across the Thames at Augusta (London) is a question which has been much debated. The lack of any evidence or remains of piers or abutments in the area would seem to support the negative view, but there are references to a Roman bridge or crossing of some kind circa AD 43. However, it is recorded that in 993, Danish King Olaf sailed up river to Staines to rape and pillage without hindrance, whereas the following year his successor, King Sweyn, encountered a bridge and many of his men were drowned as, in their hasty rage, they took no heed of the bridge.

Subsequently, the Saxons built a 'bridge', most probably a series of movable pontoons, which would still enable vessels to travel upstream. The thirteenth century Icelandic writer, Snorre Sturlason, in his account of the Battle of Southwark (1014) between the Danes and Saxons, described the bridge over the river between the City and Southwark as '... so wide that if two carriages met they could pass each other ...'[13] The battle resulted in the bridge being destroyed by King Ethelred and his ally, King Olaf of Norway, to protect the City when the Danes occupied Southwark. King Olaf's part in this battle is preserved to this day by four London churches which bear the name St Olave.

A 'second' wooden bridge was destroyed during a violent storm and flood on 16 November 1091, which also destroyed more than 600 houses in the City. In 1136, in the reign of King Stephen, a fire consumed the southern part of the City from Aldgate (in the east) to St Paul's Cathedral (in the west) including London Bridge. The historian, Stow, informs us that the bridge was not only repaired but newly made in timber by Peter of Colechurch in 1163. Peter, who was Chaplain of St Mary Colechurch (which stood in Poultry until being destroyed in the Great Fire of London in 1666) was then commissioned by King Henry II and the citizens of London to build a more durable bridge in 1176.

The King and other benefactors throughout the country, including the Archbishop of Canterbury, contributed towards the cost of the new bridge which took 33 years to build. So well was the work done, that it remained a monument of engineering skill for over 620 years. Unfortunately, Peter did not live to see the completion of his great work (in 1209); he died in 1205 and was buried in a chapel dedicated to St Thomas à Beckett which was erected in the middle of the new bridge.

In 1212, 3 years after its completion, the houses on the bridge were destroyed by fire, the bridge also being badly damaged. The following year, King John contributed once again to the bridge's repair and also introduced a new levy 'God's pence' to be taken from foreign merchants.

King Henry III, John's successor, was no friend to the citizens of London and, on 20 May 1249, he ordered his Treasurer, his Chamberlain, and the Constable of the Tower, to seize the City of London, the County of Middlesex, and London Bridge, and to pay their revenues into the Royal Exchequer. In 1269, he granted the custody and revenues of the bridge to his consort, Queen Eleanor, for a term of 6 years. But after his death in 1272, a long dispute arose between the Queen and the citizens with respect to her claim to the disposal of the bridge and its revenues. King Edward I appointed a Commission of Justices to inquire into the custody of the revenues. The citizens complained that the Keepers of the Bridge appointed by the Queen had collected all issues of the rents and lands, but had '... expended but little in the amending or sustaining of the said bridge, whence danger may easily arise...'[14]

King Edward did his best to make amends for the injustice of his predecessor and, in 1280, authorized a further appeal for contributions stating that: '... the Bridge of London is in so ruinous condition that not only the sudden fall of the bridge, but also the destruction of innumerable people dwelling upon it, may suddenly be feared.'[15]

A more reliable source of revenue and management was thought to be necessary.

On 4 February 1282, the King empowered the Mayor of London, Henry Walles, to associate himself with two, or three, or more discreet citizens to take customs or toll from persons using the bridge. Everyone crossing 'the water of Thames' from either side was to pay a farthing, and every horseman one penny, and the charge for every pack carried on a horse across the bridge was one halfpenny. Other grants for the support of this great public thoroughfare were made by the King in a Charter dated 24 May 1282, together with a grant of three vacant portions of ground contiguous to the Wool Church, St Paul's Cathedral and St Michael's at the Corn Market, to be built on or let on lease by the Mayor and Commonality for the support of the bridge forever without hindrance or impediment.

Thus control over the bridge and its revenues was vested in the Corporation of London, and all matters of great moment were decided by the Common Council. An instance of this occurred in 1390, when the Wardens prayed allowance from the Common Council of £38.8s.7½d, which their collector of rents, William Leddrede, failed to account for; he was duly sent to prison and the sum was remitted by the Corporation on the understanding that it should not be taken as a precedent.

In 1592, the Common Council established a committee with full power to make, bargain and conclude all leases connected with the Bridge House Estate and other City property. This arrangement continued until 1818, when the Common Council resolved that the committee for letting City Lands and Bridge House Estates should be constituted as two separate committees with different personnel. The separate committees were once again merged in 1968, with day-to-day responsibility for the maintenance of the City's bridges being transferred to the Planning and Communications Committee, which is responsible for highway and traffic matters generally. The responsibility of the Bridge Masters (the former Wardens or Proctors) ceased for all practical purposes in 1855, but two persons are still elected annually by the Common Hall to what is now an honorary office.

Today, the Corporation owns all four road bridges which connect the City with Southwark. The care and maintenance of these bridges is borne, at no cost to the ratepayer or taxpayer, by the proceeds of the Bridge House Estates of which the corporation are trustees. Moreover, the Bridge House Estates have funded the building, upkeep, and replacement of London Bridge, Blackfriars Bridge, Southwark Bridge, and Tower Bridge as necessary over the centuries.

The Bridge House Estates consists of various types of property situated within the City, the London Boroughs of Southwark and Lewisham, the County of Essex, and other parts of London. Revenue is derived mainly from ground rents and long-term leases, but there are many weekly and monthly tenancies. There is also now a considerable investment portfolio which has accrued as a result of surpluses of income over expenditure. Nevertheless, during the nineteenth century, the Corporation expended no less than £2 489 000 from the Bridge House Estates on its four bridges, a vast sum by today's values.

It may be added that the Corporation possesses over 800 deeds of grant of lands and other documents relating to London Bridge, the grantees being variously 'the Mayor of Commonalty and Wardens of the Bridge' or, 'the Bretheren and Sisters of the Bridge' or, 'God and the Bridge'. The Corporation has always maintained careful accounts, which survive in an unbroken series from 1381 to the present day, within the office of the Comptroller and City Solicitor.

Building Tower Bridge

Horace Jones was appointed as City Architect and Surveyor in 1864, having previously designed many important public and commercial buildings. Among his many major works for the City Corporation were additions to Guildhall and the great markets of Billingsgate, Leadenhall, and Smithfield. Jones's reputation and integrity were legendary, and his special appointment to design Tower Bridge, in addition to his other duties, followed by his knighthood in August 1886, was hardly surprising. Unfortunately, he did not live to see the fulfilment of this, his biggest challenge; he died suddenly in May 1887, without even the piers and foundations completed.

In John Wolfe Barry, Jones found a most able and determined partner who saw the bridge through to its completion, albeit with many changes along the way. Barry's family were noted architects, his father Sir Charles having designed the Houses of Parliament, while his brothers designed the Royal Opera House and Dulwich College. His early experience was on railways, which led to him setting up his own practice in 1867. What is amazing is that while working on Tower Bridge, he was also responsible for the building of Barry Dock near Cardiff and assisted with government commissions on major Scottish and Irish public works and advised on aspects of the Suez Canal.

Although Barry assumed overall responsibility for Tower Bridge following Jones's untimely death, it was Jones's assistant, George Stevenson, who was to become the major architectural influence and be responsible for the major changes in the bridge's final appearance. Indeed, he is reputed to have said that he would not make a start before his choice of stone facing had been approved in preference to the red bricks originally proposed by Jones.

Tower Bridge is fundamentally an iron and steel bridge, with its towers being fully clad with stonework, which was particularly fashionable during the late Victorian period. This was of particular importance here, given the proximity of the Tower of London. Indeed, some of Jones's early sketches for the bridge, and in particular, the approaches, strongly replicate the turrets and fortifications of the tower as, at one time, there was a stipulation requiring suitable mountings for guns.

Nevertheless, when Barry considered the detailing in conjunction with Stevenson, he could see two possible ways of cladding the steelwork, either by using cast-iron panelling or by using stone. Either would work, but he believed that embedding the steel in masonry would be one of the best ways of preventing corrosion. He was afraid that: '... some purists will say that the lamp of truth has been sadly neglected in all this, and that the old architects would not have sanctioned such an arrangement ...',[16] but hoped that: '... we may forget that the towers have skeletons as much concealed as that of the human body, of which we do not think when we contemplate examples of manly or feminine beauty...'[17]

Stevenson's final designs and drawings are minor masterpieces; being meticulously executed in Indian ink and watercolour. Together with his assistant, W.T. Hanman, he enriched the facades of the stonework with elaborate window frames, balconies and arches at every opportunity. He also produced intricate patterns for the cast-iron balustrades and decorative panels. Every opportunity was taken to produce the ultimate Gothic monument: 'The most monumental example of extravagance in bridge construction in the World', as described in *The Guinness Book of Structures.*

Work eventually began on 22 April 1886, but was soon interrupted for the laying of a memorial stone by HRH Edward, the Prince of Wales, on behalf of Queen Victoria on 21 June 1886; the first day of the fiftieth year of her reign. In the end the bridge was to take 8 years, not 4, to complete, due to delays, strikes, and other difficulties, and two extensions of time in 1889 and 1893 had to be sought from parliament.

When work finally began in earnest, it was to start work on the two piers for the bridge. These are most complicated structures, quite unlike the piers of an ordinary bridge. The piers house the counter-poise and machinery to operate the bascules, as well as supporting the main towers which carry the suspension chains and the high level walkways. The contractor for this work was John Jackson of Westminster, and this was to be the most difficult of all the eight contracts which made up the whole project.

Fig. 5 Masonry arch at the roadway at the north abutment tower.

Fig. 6 Masonry arch over the roadway at the north main tower.

The work entailed teams of navvies excavating inside 12 permanent caissons sunk into the London clay below the river bed. The caissons, which together roughly formed the shape of the pier, were open bottomed 'boxes' of wrought iron, 19 feet (5.85 m) high, with sharp cutting edges to penetrate the ground. Initially, divers worked inside the caissons excavating the gravel and the upper part of the clay forming the river bed and loading the spoil into barges using cranes.

Additional temporary sections were added to the caissons as the cutting edges were driven deeper into the solid London clay below the river bed. The inside was then pumped out so that the navvies were then able to dig out the remaining clay in the dry. The caissons (permanent and temporary) finally measured 57 feet (17.4 m) high, with the bottom being 20 feet (6.1 m) below the river bed. The clay was then excavated a further 7 feet (2.12 m) deeper and outwards which both widened the foundation and linked up with the adjacent caissons to form a continuous foundation for the pier.

The whole depth of the permanent caissons and the spaces between them were then filled with concrete, upon which brickwork and masonry were built within the temporary caissons to a height of 4 feet (1.2 m) above high water, the total weight of each pier being some 70 000 tons (71 120 tonnes).

The 1885 Act had stipulated that 160 feet (49 m) of clear water had to be maintained at all times for shipping. This restriction seriously impeded the works and caused many delays and Jackson's contract was not finally completed until January 1890. The contract had also included the construction of the abutments, which were built within traditional cofferdams, and, according to Barry: '... though formidable in size and depth, presented no new features of construction such as have been explained with regard to the piers...'[18]

On completion of the piers and foundations, new contracts for the iron and steel superstructure and the masonry work, which had been awarded to Sir William Arrol and Company Limited and Messrs Perry & Company respectively in May 1889, became the dominant features of the works for the many spectators who daily lined London Bridge to observe events.

All the wrought steel which, according to James Tuit, Arrol's Site Engineer, amounted to some 11 000 tons (11 176 tonnes), was manufactured in or around Glasgow, although all fabrication etc. was carried out at Arrol's works. Wherever possible the steel was prefabricated before despatch, albeit that the maximum weight of each piece was limited to about 5 tons (5.08 tonnes) for ease of handling. Predominantly hydraulic riveting was employed both at the works and on-site to build the massive structural framework. The steel sections were shipped from the Clyde to the Thames and then transferred on to barges for the final delivery to site on to temporary bridges linking the abutments to the piers.

When Tower Bridge was built, the shock of the Tay Bridge disaster was still fresh in people's minds. As a result, the Board of Trade required that all bridges be designed to withstand wind pressure of 56 lbs/ft^2 (270 kg/m^2)—a very high pressure. This meant that the structure and the hydraulic power provided at Tower Bridge were both massive and lavish. Accordingly, the main towers each consist of four huge, octagonal steel columns, braced by three horizontal landing girders and intermediate diagonal wind bracing. Each of the main columns is 119 feet 6 inches (36.6 m) tall, and 5 feet 9$\frac{3}{4}$ inches (1.78 m) internally between opposite faces, i.e. large enough for someone to carry out inspections inside the columns.

On top of the octagonal columns sit a series of rollers, which allow the main chains supporting the fixed shore spans of the bridge to rest and move, so as to accommodate any changes in temperature or unequal distribution of traffic loading. With the bridge fully loaded, these rollers have to support and distribute up to 1000 tons (1016 tonnes); consequently, Barry designed a system of blocks and plates to ensure equal distribution of weight on to each of the two parallel sets of four rollers under all loading conditions. Barry was assisted in the design of the structures by Mr H.M. Brunel.

The arrangement on the abutment towers is similar, but here the shore span chains are linked to the land ties, which are connected to massive concrete anchorage blocks beneath the approach roads. The whole superstructure is in a sense four bridges in one. First, there is the low-level central span of 200 feet (61.2 m) between the main tower piers comprising the two opening leaves or bascules. Secondly, there are the two suspended shore spans, and finally, these are linked by the chains to the high-level central fixed span, which comprises two short cantilevered sections from the top of each of the main towers, connected by a 120 feet (36.7 m) long central suspended section. The central fixed span is the same width as the opening span, but 140 feet (42.8 m) above high water, rather than 135 feet (41.3 m) as prescribed by parliament.

It may be asked why the chains are so described, when they can be clearly observed to be steel girders which weigh approximately 1 ton/foot run (3.3 tonnes/m). Their shape and function is none the less that of a chain as in any suspension bridge, but stiffened to prevent deflection, by doubling the section and bracing across the chord with steel angles. In this way local deflections of the chains and distortions of the bridge deck are avoided.

The opening leaves of the low-level span had to be largely constructed in the raised position, in order that shipping was not obstructed. Indeed, most of the paving was also carried out by the Acme Flooring Company with the bascules raised. This comprised of creosoted blocks of Memel pine, fixed to one another with oak dowels, and laid on greenheart planking; the final stage of jointing with asphalt being undertaken later when the bascules were lowered for testing. The finished weight of each bascule including kentledge is some 1200 tons (1219 tonnes).

A most rigorous load test of the south bascules was conducted shortly before the bridge was opened, using steam rollers, traction engines, and

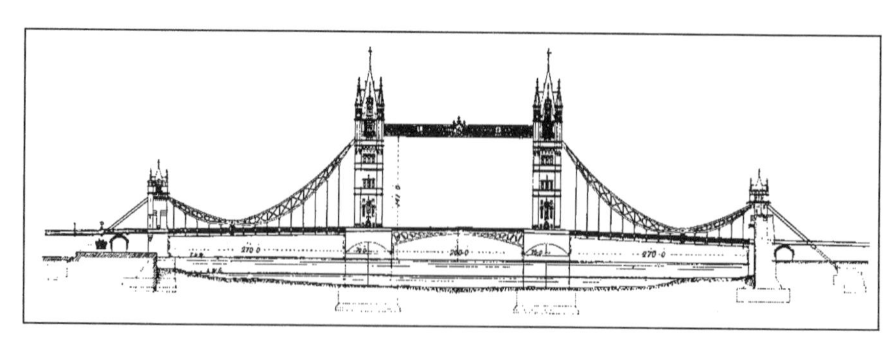

Fig. 7 The bridge as constructed; it differs from Fig. 4 mainly in the architectural detail and in the arrangement of the chains, which necessitated the addition of the abutment towers.

trollies laden with waste granite weighing some 150 tons (152 tonnes). George Cruttwell, Resident Engineer, in his account of the load test stated: 'The deflection at its extremity amounted to only $1\frac{3}{8}$ inches (35 mm), and no permanent deflection remained after the removal of the load.'[19]

To prevent corrosion, the main steel columns were wrapped in oiled canvas as the masonry was built around them, prior to which the columns were washed with neat cement as further protection against rust. Most surprisingly, given all the attention to architectural detail, the records show that the exposed steelwork received three finishing coats of paint of an approved quality, the last coat being bright chocolate brown in colour.

The granite facings came from the Eddystone quarry at De Lank in Cornwall, each stone being cut and dressed before shipping. The decorative Portland stone arches, parapets, finials and gargoyles were carved either on-site or *in situ* by Mabeys of Westminster. The ornamental cast-iron balustrades and decorative panels, weighing in total almost 1200 tons (1219 tonnes), were made by Messrs Fullerton, Hodgard, and Barclay of Paisley. Both these and the ornate lamp standards were designed by George Stevenson.

Despite intense competition from electricity, gas lighting was chosen and installed by W. Sugg & Company of Westminster. Sugg's were so proud of their success that the company took a full-page advertisement in *Building News* on 29 June 1894 to announce that 'The Tower Bridge is lighted entirely by gas, by means of upwards of 200 Sugg's patent high-power flat-flame gas lamps.'

George Cruttwell's records noted that on average 432 men were employed during the 8-year construction period. Reports on loss of life differ. Cruttwell records the total as 10, four in sinking the foundations, one on the approaches, and five on the superstructure. Whereas Barry, in a lecture given to the Institution of Civil Engineers, identified six deaths, at least one of which was the result of sudden illness or of a fit, not an accident. Whoever was correct, the figures were low for a project of this magnitude.

How the bridge is raised

'Nothing in their way at all equal to them is to be seen anywhere in the World, and to be near them in action is to witness one of the most imposing displays of hydraulic power that even Sir William Armstrong and his company have ever yet afforded.' So stated the *Architect and Contract Reporter* magazine on 6 July 1894, its reporter having witnessed the raising of the bascules. This operation is achieved in about $1\frac{1}{2}$ minutes in total silence by the bridge driver simply moving a series of levers, similar to those used in a railway signal box.

All the control gear for raising and lowering the bascules was located in identical control cabins sited on either side of the bridge. The control cabins were manned 24 hours a day so as to be ready for an approaching ship whether proceeding upstream or downstream. The 1885 Act required the bridge to be raised for ships on demand, and to have precedence over road traffic and pedestrians. This was achieved by a series of signals; semaphores by day and signal lamps by night. During foggy weather, a gong was used to give an audible signal.

The full operating sequence to raise the bascules commenced with the closure of the bridge to road traffic and pedestrians under the direction of the Head Watchman. Once this had been done, the bridge drivers released the four long, wedge-shaped nose bolts which link the two bascules together at the centre of the opening span. The pawls or tail locks at the rear end of the bascules were then released, so freeing the bascules to rotate on their massive 21 inch (625 mm) diameter steel pivots.

Interlocking gears made it impossible to carry out the operations in anything other than the correct sequence, or to commence any manoeuvre before the preceding manoeuvre was completed. Thus was safety ensured. The drivers then set in motion the hydraulic system which raises the bridge through a series of interconnected pistons and geared wheels which engage with and drive the toothed quadrants fixed on either side of the bridge to the rear end of the bascules.

The hydraulic power at Tower Bridge was generated from four coal burning, double-flued Lancashire boilers located in the arches beneath the southern road approach. Only two boilers were used at any one time, fired to supply steam at 85 lb/in^2 (5.98 kg/cm^2) to the two 360 HP steam pumping engines located in adjoining arches. One engine alone could provide more than enough power to operate the bridge.

The steam pumping engines fed water at a pressure of 700 lb/in^2 (49.25 kg/cm^2) to six accumulators, two large ones in the accumulator house alongside the engine room, and two smaller ones in each of the main river piers. Their function was to ensure that there was at all times a constant supply of water at the required pressure in the pipework of the hydraulic system for release, on demand, to raise the bascules. Barry pointed out that: '... this pressure is nothing unusual, but its magnitude will be appreciated when we remember that in the boiler of a locomotive engine the steam pressure is usually not more than one-fifth of the above amount.'[20]

Two pairs of bascule-drive engines were located in each main pier. These operated on the reciprocating cylinder principle, and communi-

Fig. 8 Sketch of one of the four Lancashire boilers.

Fig. 9 Sketch of one of the two main accumulators.

cated rotary motion, through a series of geared wheels, to the toothed quadrants on the back end of the bascules.

As previously stated, due to the requirements of the Board of Trade, the hydraulic power installed at the bridge was lavish. Indeed, the machinery provided was capable of delivering twice the requirement of

Fig. 10 Sketch of one of the two 360 HP steam pumping engines.

the Board of Trade of 56 lb/ft^2 (270 kg/m^2). Barry decided that: '... an ordinarily strong wind will not give a pressure [on the bascules] exceeding about 17lbs./sq ft (82 kgs/sq m), and therefore the hydraulic engines are arranged in pairs, one engine of each pair exerting a power equal to a 17lbs. wind, and the other equal to a 39lbs. wind, the two together being equal to the extravagant pressure of a wind of 56lbs.'[21]

The machinery was designed and manufactured at the Elswick plant of Sir William Armstrong, Mitchell & Co. Ltd and installed by John Gass under the supervision of Samuel Homfray. In all the years, up to the introduction of electricity as the primary source of power in 1976, the bridge never failed to open on demand; a testimony to the wonderful hydraulic system and to all concerned in its design, manufacture, installation, and subsequent maintenance.

The royal opening

'The Opening of the Tower Bridge on Saturday was a picturesque and stately ceremonial, perfectly performed under the most favourable conditions', reported *The Times* on Monday, 2 July of that great day, 30 June 1894, when the Prince and Princess of Wales officially opened Tower Bridge. After weeks of meticulous planning, the corporation celebrated the Royal occasion in a style befitting this, its most magnificent achievement and gift to the nation.

Hundreds of guests representing every part of the United Kingdom and the British Empire filled the tiers of seats on the bridge's approaches awaiting the arrival of their Royal Highnesses. At 12 noon, the Royal Cavalcade arrived at Little Tower Hill and drove across the bridge to the cheers of the awaiting throng, before returning across the bridge to allow the Royal party and other dignitaries to alight within a special pavilion designed by the City Surveyor, Andrew Murray.

Following a short, loyal address by the City Recorder, the Prince of Wales turned a silver chalice mounted on a pedestal and declared the bridge open, and HMS Landrail then led a flotilla of ships decked with flags through the bridge in procession. According to *The Times*: 'For a moment, the great crowd hushed in silence. Then in a deafening shout of applause, which soared, as only a British cheer can soar, above the thunder of the Tower guns, above the ringing notes of the trumpets, and above the wild din from the sirens and the whistles of the steamers, they gave vent to their admiration and delight at the marvel they had been privileged to see. They had indeed witnessed a spectacle not easily to be forgotten.'

The Times was quite profuse in its praise for the Corporation both for building such a remarkable bridge: 'one of the structural triumphs of this age of steel', and for the consummate skill with which the day's events had been carried out. *The Graphic* and the *Illustrated News* were similarly profuse in extolling the bridge's virtues and produced special commemorative supplements. However, not all the reports were so complementary.

The Builder went as far as to say: 'We decline to waste any plates in giving illustrations of the so-called architecture'; they described the towers as about '... as choice specimens of architectural gimcrack on a large scale as one could wish to see.' *Building News* was sorry that: '... the labours of engineer and architect were so soon to be wasted. A see-saw bridge was a mistake and repeated raising and lowering of the bascules would be found tedious, then unnecessary ...' Even a month later, on 27 July 1894, *Building News* reported that: 'Tower Bridge was of course a sham, as no real art can tolerate a casing.'

The bridge at work

What appears to have escaped comment is that the bridge did not begin to operate fully until 9 July 1894, under the management of the first Bridge Master, Bertie Cator. Nevertheless, in its first year of operation, the bridge was raised 6160 times and was crossed by 8000 vehicles and

60 000 pedestrians each day. On average, it took 6 minutes to open the bridge from receipt of a signal from the Master of an approaching vessel having a mast of 30 feet (9.2 m) or more. Not surprisingly, most of the vessels arrived or left the Upper Pool in the 2 hours before or after high water.

During the first year, the bridge opened on average 17 times a day, and in the second year, 18 times a day. From the Bridge Master's records, this figure appears to have been remarkably consistent across the years, averaging 15 times a day up to the 1950s. But progressively thereafter activity in the Upper Pool declined as trade moved first downstream and then finally ceased. In consequence, bridge lifts reduced to no more than once a day on average by the 1970s, many of which were 'maintenance' lifts to check operational equipment or train staff.

As a safeguard to minimize delays in operation in the early days, the Bridge Master maintained a tug to help clear river traffic out of the path of inward or outward bound ships, and he kept horses stabled beneath the southern approach to replace any animals pulling carts that might collapse or even drop dead and block the road. Both the tug and horses were rapidly phased out before the First World War, reflecting the reduction in sailing ships and the arrival of the motor vehicle.

Although hydraulic lifts, capable of carrying 25 people, could make 25 journeys each hour up to the high-level walkways when the bridge was raised, little use was made by pedestrians as the wait while ships passed through the bridge was relatively short. Most people who bothered to climb the 206 steps, or used the lifts, did so to enjoy the wonderful panoramic views over London. Owing to the lack of usage, and some say public safety, the walkways were closed to the public in 1910.

The bridge led a truly charmed life during the World Wars, emerging unscathed from the First, and sustaining a small amount of damage to the south tower and its southern shore span in the Second. In comparison, the Tower of London suffered 15 direct hits! It seems that Tower Bridge was of more use to German bombers as a landmark to guide them to the rest of the City.

The Bridge has, however, witnessed its fair share of incidents over the last 100 years. In 1912, for example, Frank McClean became the first man to fly between the bascules and the walkways in a Short biplane, much to the amazement of onlookers. Since that time six other flyers have repeated the feat.

In 1943, Mr W.F.C. Holden submitted a plan to provide 200 000 feet^2 (18 740 m^2) of 'liveable' floor space inside a glass-encased 'Crystal Tower Bridge'. But though the plan was given serious consideration, thankfully, it was never put into action.

Then in 1952, a bus full of passengers had to leap from one bascule to another, when one of the bascules began to rise while the bus was still driving along it, through a misunderstanding between the Head Watchman and the two bridge drivers. The positive outcome from this near miss was the installation of a public address system, a closed-circuit monitoring system, and new road and pedestrian gates, with just one bridge driver having responsibility for bridge lifts.

Amazingly, the original hydraulic system operated without fail until economics dictated that electricity replace steam as the primary power source in 1976, with oil replacing water in the hydraulic system at the same time. This was something that was first investigated in 1915 because of the difficulty of maintaining the labour force during the First World War. The question of conversion was raised again in the 1920s, the 1930s, and the 1950s, but on each occasion it was not pursued because of the continuing success of the original machinery. Nevertheless, the Corporation eventually bowed to the inevitable and a great tradition was lost forever. The number of people employed at the bridge reduced to 15, in comparison with the 80 who were originally employed when the bridge opened.

Even more remarkable than the success of Armstrong's hydraulic system, is the fact that the main pivot is still entirely original and its bearings have never once been stripped down, simply lubricated. The fact that the bridge has been raised some 400 000 times, and been traversed by some 750 million vehicles, is a further testament to our Victorian engineers, albeit in part due to the lavish approach previously referred to. Is it any wonder that, despite the weight and volume of modern motor vehicles, this unique bridge still fulfils its role for river and road traffic alike, albeit the heaviest vehicles have been prohibited from crossing the Bridge since 1979.

Between 1979 and 1982, Tower Bridge underwent a most drastic change. The bridge was given a £2.5 million facelift and opened as a tourist attraction for the first time. The walkways were reopened, the engine rooms given public access, and the cast-iron parapets and decorations were replaced with lightweight glass reinforced plastic reproductions. 30 June 1982 was quite an occasion, too, when the Lord Mayor, Sir Christopher Leaver, opened the new exhibition on the anniversary of the opening of the bridge in 1894, marked by a large procession of ships and a bevy of honoured guests. One of these, Miss Beatrice Quick, had been the first child to cross the Bridge 88 years before.

In May 1988, dramatic floodlighting was introduced to London's most famous landmark, switched on by the Right Hon. Margaret Thatcher, the Prime Minister. In October 1993, in preparation for its centenary, a stunning new exhibition inside the bridge entitled 'The Celebration Story' was opened to the public by the Lord Mayor, Sir Francis McWilliams. With its state-of-the-art animatronics, interactive computers, model reconstructions, and modern audio and linguistic techniques, Tower Bridge was once again in the forefront of modern technology.

The centenary of the bridge was another occasion for celebration, with HRH the Prince of Wales unveiling a commemorative plaque using one of the original bridge driver's levers. That evening thousands of Londoners witnessed one of the most exciting and spectacular firework displays ever, while the Corporation hosted a celebration dinner on board the Royal Yacht Britannia which was moored in the Upper Pool, in the shadow of the Bridge.

Into the twenty-first century

When postulating the future of his bridge, during a lecture to the Institution of Civil Engineers in 1893, Barry observed that: '... if the road traffic becomes of greater importance, and the sea-going river traffic grows less, I suppose the fate of the Bridge will be to become a fixed bridge. How soon this will happen no one can tell.'[22] He added that: 'Tower Bridge is no ordinary bridge, and in no ordinary position. The structure and its machinery are full of the most elaborate and complicated work of all kinds.'[23] He hoped that it: '... will be considered to be not unworthy of the Corporation of the greatest City of ancient or modern times...'[24]

While the question of making the bridge a fixed bridge has been suggested from time to time over the years, the issue has only been formally considered by the Corporation on one occasion, in September 1970, when the Court of Common Council agreed that Tower Bridge should continue as an opening bridge, rather than leaving the bascules permanently in the lowered position. At the same time, it was agreed that parliamentary powers should be sought for the early modernization of the machinery and improvements to the structure.

The following year, an impassioned letter to *The Times* deplored the City's decision to replace '... the remarkable mechanism, which is the most perfect example of integrated hydraulic power in the world.' The writer begged that the Bridge be '... designated a museum of industrial archaeology, and the machinery kept intact, and made available to the public ...' To a large extent this has been done, and the exhibitions within the bridge tell its story and preserve much of the magnificent machinery for visitors to view in awe at its beauty and scale.

However, far from planning to make the bridge itself a museum piece, or seeking to change its role by replacing Mr Jones's 'flaps or wings' with a fixed span, the Corporation and I are looking ahead to the celebration of its second centenary. While we must ensure that the bridge's integrity is not abused by heavy traffic, it is imperative that this symbol of London, 'this Enduring Monument', in the words of HRH Edward, Prince of Wales, when opening the bridge in 1894, remains a fully functioning bridge.

Tower Bridge is the most fitting tribute to the ingenuity and skill of those who conceived and built such a wondrous solution to the needs of Victorian London. This so-called 'folly' can still perform its unique role and will continue to do so well beyond the lifetime of my audience. You can be assured of this, because of the equally unique funding of all the Corporation's bridges over the river, namely the Bridge House Estates.

Selected bibliography

J.W. Barry, *The Tower Bridge*, Boot, Son & Carpenter, London, 1894.

J.E. Tuit, *The Tower Bridge*, *The Engineer*, 1894.

C. Welch, *History of the Tower Bridge*, Smith, Elder & Co., London, 1894.

G.E.W. Cruttwell, *Minutes of Proceedings, the Institution of Civil Engineers*, 1896, Vol. CXXVII.

S.G. Homfray, *Minutes of Proceedings, the Institution of Civil Engineers*, 1896, Vol. CXXVII.

Hansard and Minutes of the Parliamentary Select Committees relating to the Corporation of London (Tower Bridge) Bill Session 1884–85.

References

1. Nelson's Handbooks for Tourists, *The River Thames*, 1859; quoted in H. Godfrey, *Tower Bridge*, John Murray Ltd, London, 1988, p. 10.
2. Welch, p. 151.
3. Welch, p. 158.
4. Welch, p. 159.
5. Welch, p. 161.
6. Welch, p. 163.
7. Tuit, p. 38.
8. Hansard.
9. Hansard.
10. Hansard.
11. Corporation of London (Tower Bridge) Act 1885.
12. Hansard.
13. Welch, p. 28.
14. Welch, p. 30.

15. Welch, p. 31.
16. Barry, p. 48.
17. Barry, p. 48.
18. Barry, p. 35.
19. Cruttwell, p. 19.
20. Barry, p. 40.
21. Barry, pp. 43–4.
22. Barry, p. 63.
23. Barry, p. 64.
24. Barry, p. 64.

ROY F.V. AYLOTT

Born 1938, he received his initial training in the gas industry before entering local government in 1960. He has since held various professional positions with both urban and county authorities prior to joining the City of London Corporation in August 1982. He qualified as a chartered municipal engineer in 1963, and was elected as Municipal Engineer of the Year in 1991 in recognition of his contribution to road safety and the Government's review of street works legislation. He is an adviser to the Association of Metropolitan Authorities and a member of Council of the Institution of Civil Engineers.

One hundred years of the electron

SIR BRIAN PIPPARD

The hundred years that have elapsed since the discovery of the electron[1] have seen enormous advances in science, and it is not physics alone in which the electron has played a star role. Nor can the advances be attributed solely to the electron; the discovery of X-rays in 1895, radioactivity in 1896, Planck's quantum in 1900, and Einstein's special theory of relativity in 1905 have also been central to the forging of what is sometimes called modern physics. There is no way of treating any one of these ideas separately, but I shall do my best to focus on the electron. In fact, what purports to be a historical account of a century's progress will turn out to be concentrated on a few years in the twenties, when the problem of finding a replacement to Bohr's theory of the hydrogen atom, and dealing with many electrons running around in any but tidy closed orbits, was the principal spur to the invention of quantum mechanics; and it was quantum mechanics, above all, that precipitated a radical revision of the picture we form of the physical world. It was so intense a crisis that most physicists prefer not to think about it, but are happy to accept that in quantum mechanics we have the most versatile and powerful technique ever devised to give a precise account of the observed world of physics.

As I shall have no time to recount the vast range of successes for which the electron must take much of the credit, let me say at the outset some of the things I shall ignore. First, as in private duty bound (to quote the Bidding Prayer) let us note and immediately forget the sort of topic I spent a good part of my research career on—electrons in metals and especially superconductivity. If I once started on this I should get no further. For the opposite reason—sheer ignorance—I shall say almost nothing of the electron's key role in the chemical bond. Nor have I the time to do justice to the cathode ray oscilloscope, or the skill to celebrate the electron microscope and its powerful relations. But there is one technical application of the electron that must not be ignored, for electronics has been responsible for the most pervasive of social revolutions and, incidentally, for a host of innovations in scientific research. There is no need to expand on this—you have only to imagine what your life would be without modern electronic equipment—but a few minor historical points are perhaps worth making.

The thermionic valve and transistor

Electronics can be said to have begun[2] in 1904 with Ambrose Fleming's patent on what he called a valve—because its operation resembled that of a valve in a pump—and we now call a diode, but it owes little or nothing to J.J. Thomson's announcement 7 years before, if only because Fleming had started work even earlier, and was building on the Edison effect—the observation in 1883 that a hot wire can emit a current in one direction only. To be sure, the Edison effect was carefully studied by Owen Richardson[3] and correctly explained as due to electron emission, but Fleming's approach was little affected by new physical discoveries. Nor was the slightly later addition by Lee de Forest of a control grid, to constitute the first triode. This was the time when long-distance radio communication was developing rapidly, and Fleming's valve would have been an instant success had it not been for the discovery of the crystal rectifier which was cheaper, and seemed more convenient, than a valve for detecting signals.[4]

The Great War (1914–18) gave immense impetus to the development of radio communication, and it is in this period that we see the rise of circuit techniques, especially at the hands of Langmuir and Armstrong in the United States, and a group at Siemens in Germany. By the end of the war the essential concepts of valve amplifiers and oscillators had been established, and there was massive scope for the commercial production of valves and circuit components. Public broadcasting began in the early twenties, and it is at this time that we find the first radio amateurs with their simple wire aerials, tuned circuits, crystal and cat's whisker, and earphones. By the thirties many households had valve-operated receivers (costing, in real terms, at least 100 times as much as a portable transistor radio). The crystal and cat's whisker were all but forgotten, so that when Southworth,[5] of Bell Telephone Laboratories, decided to study wave-guide propagation at centimetric wavelengths, frequencies too high for thermionic diodes, he was forced to ransack junk shops for relics of amateur obsession. When the Second World War initiated the development of radar sets at centimetric wavelengths, the solid-state rectifier came into its own, and with it techniques for purifying germanium and silicon which had an immense influence on post-war electronics. By that time quantum mechanics had clarified many of the previous enigmas of electron motion in solids; in particular, the principles underlying the operation of semiconductors—including the meaning of electrons and holes—were rather well understood, as was the rectifying action at the junction of a metal and a semiconductor.

Immediately after the war Shockley and others at the Bell Telephone Laboratories began their attempt to realize a dream they had cherished since the pre-war days, to replace the amplifying valve with a solid-state device. Wartime researches on semiconductors, especially the production of ultra-pure materials, gave them the start they needed, and in 1948 they announced the first transistor.[6] This heralded the birth of a huge research effort, involving the physics of semiconductors and the technology of large-scale production. It was not until the sixties, however, that it was possible to deposit dense arrays of semiconductor circuits on silicon chips. The resulting explosion in computers and every form of electronic control must have exceeded the wildest expectations of the most ambitious Bell researchers.

Electronics in the laboratory

Let us turn back to the years after 1918, when thermionic valves became generally available (at a price). It was then that amplifiers made their appearance in academic research, even in the study of nerve conduction by Adrian in the early twenties.[7] The earliest physical application of an amplifier that I can find is a short paper by Barkhausen which reports two very different discoveries.[8] The more famous is the Barkhausen effect—a coil is wound round an iron bar so that any sudden change in magnetization will induce a voltage impulse which can be amplified and connected to an earphone. On bringing a magnet near the bar he heard a succession of clicks which indicated that sizeable regions (domains) were suffering a change, possibly a reversal, of magnetic moment. In his second experiment he connected the amplifier directly to a long wire aerial, and heard from time to time a note of descending frequency (which he describes well by the word *piou*). These whistlers were studied in great detail by Owen Storey[9] as a research student in Cambridge. They start as electric pulses from lightning strokes, and follow a closely guided path along a line of the Earth's magnetic field. Electrons high above Earth, moving in helical trajectories under the influence of this field, slow down and disperse the electric waves so that the lower frequencies take longer than the higher frequencies to travel between hemispheres. If the wave bounces to and fro, as sometimes happens, one hears a succession of whistlers, each more drawn-out than its predecessor. The experiment shows that the electron cloud extends well beyond the ionosphere, and indeed it allows the electron density to be estimated. This short paper of Barkhausen's was a worthy introduction of electronics into the physical laboratory.

Quantized electron orbits

By 1919, in the world of atomic theory, Bohr's ideas[10] about quantizing the electron orbit in a hydrogen atom had been extended, particularly by Sommerfeld[11] who showed how elliptical orbits could be quantized. The classical Keplerian orbit, a closed ellipse with the nucleus at one focus, gave the same energy levels as Bohr's circular orbits. When modified for the relativistic change of mass, most significant when the electron came closer to the nucleus and therefore moved faster, the orbits ceased to be closed and formed a rosette pattern. Sommerfeld was able to extend the rules for quantization to cover this case, and found that the energy levels developed a fine structure; orbits that without relativistic effects were degenerate now had slightly different energies according to the eccentricity of the ellipses. As a result he could explain the fine structure of the spectral lines which had been observed shortly before.

Some years later Uhlenbeck and Goudsmit[12] suggested that the unexpectedly complicated splitting of spectral lines by a magnetic field—the anomalous Zeeman effect, so-called to distinguish it from what Zeeman and Lorentz expected in 1897 but which is hardly ever observed—could be understood if the electron possessed an intrinsic spin and an associated magnetic moment. The spin and moment could be orientated only parallel or antiparallel to the magnetic field. Its presence caused the energy levels of hydrogen to be rearranged, but the details predicted by Sommerfeld were unchanged. Thus spectroscopy, in the absence of a magnetic field, gave no further evidence for spin, but the explanation of the Zeeman effect was convincing. The magnetic moment of the electron is responsible for the overwhelming majority of examples of paramagnetism and ferromagnetism in solids, and the flipping of a spin from one orientation to its opposite by an oscillating field is observed as spin resonance, a fascinating phenomenon of great analytic power in the study of condensed matter.[13]

Sommerfeld's quantization of elliptical orbits was one of the ideas that encouraged Bohr to build up the Periodic Table of elements by progressively occupying different quantum levels.[14] His initial successes encouraged Stoner[15] and others to make improvements, and it was Pauli's Exclusion Principle[16] that tidied the whole matter up—no two electrons could have the same set of quantum numbers. Even before the invention of electron spin Pauli had suggested that there should be an extra quantum number that could take only two values, and spin fitted the bill neatly. All the same, the conception was an arbitrary extension of the single-electron atom, as is clear from a diagram (Fig. 1) prepared by (or for) Bohr to show the electronic configuration in radium. The simple elliptical orbits assume each electron to be acted upon by an inverse-square force from the nucleus; the mutual interaction of the electrons, which would wreck the closed-orbit pattern, and eliminate the possibility of quantizing according to the Bohr–Sommerfeld rules, is ignored. Leaving aside a complex atom like radium, one must note that attempts in the early twenties to apply Bohr rules to even the two-electron atom helium made no progress. For all its spectacular success, and undoubted historical importance as a heuristic model, the Bohr scheme was no springboard for a more general attack on quantum

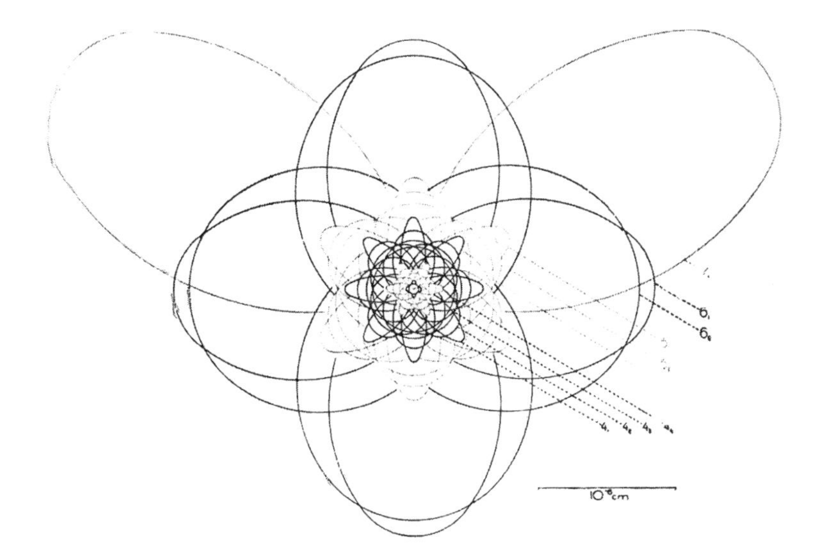

Fig. 1 Diagram of the conjectured Bohr–Sommerfeld orbits of the 88 electrons in the radium atom, as used by Bohr to illustrate his lectures on the Periodic Table. The originals have disappeared.[17]

problems. The first breakthrough was Heisenberg's matrix mechanics which owed nothing to de Broglie's rather primitive association of waves with particles.[18] It was Schrödinger's[19] more general formulation of de Broglie's idea that provided a technique that gave the same results as Heisenberg's but was mathematically more manageable, and certainly more immediately acceptable to physicists brought up on classical problems.

Direct experimental evidence to support the wavelike character of electrons came very soon after Schrödinger's first publications; Davisson and Germer[20] revealed well-defined peaks in the scattering of electrons from a nickel crystal, and G.P. Thomson[21] obtained diffraction rings when electrons were passed through thin polycrystalline films. The phenomena were exactly parallel to X-ray diffraction, and the wavelength λ of the electron wave could be measured, to confirm de Broglie's proposal $\lambda = h/p$, p being the momentum of the electron. De Broglie showed that the requirement for an electron wave to fit tidily round a circular orbit led immediately to Bohr's quantum condition, but he did not take the argument further. Schrödinger wrote down the three-dimensional differential equation for a wave, with due allowance for variations of momentum, as for example in a hydrogen atom where the momentum is higher near the nucleus, and the wavelength correspondingly less. The solution of the wave equation is similar in principle to the determination of resonant acoustic modes in a hollow vessel; corresponding to each 'resonance' there is a characteristic energy, and Bohr's quantum theory of the hydrogen atom follows in detail. Each possible stationary state of the atom is described by a pattern of nodes and antinodes of the wave-function Ψ, and it was not long before spinning models were made which, when photographed with a long exposure, gave visual pictures of these modes.

The new quantum mechanics was seized with enthusiasm by young mathematical physicists, and each application was so successful that the army of followers grew rapidly. There was no obstacle in principle to the analysis of many-electron atoms, although the practical obstacles to complete solution were insuperable. Intelligent approximation, however, allowed Hartree to develop theories of atomic structure by allowing each electron wave to be governed by the field of the nucleus and the average of the other electrons.[22] The lack of periodic classical orbits that had stymied Bohr-like models was now no trouble. Even before this, Heitler and London[23] had solved Schrödinger's equation, again approximately, for two electrons in the vicinity of two protons, and had established the possibility of quantitative explanation of the covalent chemical bond.

The interpretation of quantum mechanics

But all was not well for those who sought the underlying meaning of quantum mechanics, and were not content simply with a technique which gave the right answers. Some, like Schrödinger, hoped that the electron could be considered as nothing more than a wave-packet, but it soon became clear this was no answer. A wave-packet can be disintegrated by scattering, but the electron remains a single entity; the wave may be broken into distinct parts, the electron appears at one place only. It was soon accepted that the wave intensity at any point determines the probability of observing an electron there. The champion of this interpretation, and of the philosophical outlook that followed from it, was Bohr, and the interpretation has come to be known as the Copenhagen doctrine, even though later discussions (which continue to the present day) indicate that there is scope for disagreement about what Bohr really meant.[24]

Schrödinger himself was never reconciled to the probabilistic interpretation of the wave-function, and an even more persistent critic was Einstein who accepted the success of quantum mechanics as a calculational procedure, but rejected the loss of determinism as a fundamental property of the physical world. The effort it cost Bohr to refute his ingenious and subtle counter-arguments undoubtedly helped to reveal the deep distinction between Einstein's classically influenced views and Bohr's determination to form as clear a picture as possible of an essentially new set of phenomena. I shall return to this later, but first must tell of an outstanding triumph of quantum mechanics which left a legacy of still greater mystification.

Relativistic quantum theory of the electron

In 1928 Dirac formulated quantum mechanics in such a way as to satisfy in every respect the requirements of the special theory of relativity.[25] This was far more than Sommerfeld's *ad hoc* adjustment of elliptical orbits to allow for the variation of mass with velocity. Dirac's equations were wholly compatible with the four-dimensional symmetry involving both space and time, and they went much further than a minor adjustment of Schrödinger's equation. Solving the equations for a hydrogen atom was by no mean easy, but when solution was achieved it was found that all Sommerfeld's results were reproduced, so that details of the fine structure of spectral lines were correctly predicted. I shall not be concerned here with minute, but very important, imperfections which led, after 1945, to the systematic development of quantum electrodynamics.

This in turn created serious problems of interpretation for those with a classical predilection for microscopic models. For my purpose the curiosities of Dirac's equation suffice to illustrate the revolution in thought that was forced by quantum mechanics, aided by many other results of modern physics.

A feature of Dirac's equation, which caused him and others great anxiety for some years was an unavoidable symmetry of positive and negative energies.[26] In relativistic mechanics the energy of a free particle is mc^2, where m is the velocity-dependent mass $m_0/(1 - v^2/c^2)^{1/2}$, and m_0 is the rest-mass. In Dirac's formulation, any solution valid for energy mc^2 is equally valid for $-mc^2$, even though the physical meaning of a negative mass is, to put it mildly, obscure. Dirac's first thought was that the solutions for negative mass were no more than mathematical artefacts, without physical significance. It was soon found, however, by Klein and others, that to ignore those solutions led to a completely wrong result for the scattering of radiation by an electron. Dirac then suggested that the negative-energy states were normally all occupied to the limit prescribed by Pauli's exclusion principle. It was apparently for Dirac a tolerable conception that space should be occupied by an infinite density of electrons with negative energy (whatever that may mean) and carrying an infinite charge density which, according to old-fashioned ideas, must generate an infinite electric field everywhere. This extraordinary concentration of potential disasters has to be assumed to describe the ground-state of the vacuum, only deviations from which produce observable effects.

One such deviation would be the result of giving to a negative-energy electron enough energy (at least $2m_0c^2$) to excite it to a positive energy, when it would materialize as a real electron with negative charge. The vacancy would also be observable, being a positively charged departure from the uniform unobservable ground state. Dirac's first idea, that this would be observed as a proton, thus killing two birds with one stone, had many unattractive features—lack of symmetry in mass between positive and negative particles, and the expected very short lifetime of a proton in the presence of electrons any one of which could fall into the vacant state and immediately annihilate the proton and electron. The dilemma was resolved in 1932 when Carl Anderson,[27] and Blackett and Occhialini[28] confirmed Dirac's enforced conclusion by demonstrating in cloud chambers that energetic cosmic rays, on passing through a foil, could generate a pair of particles, oppositely charged and similar in mass. Figure 2 shows a more modern example of the process. With the discovery of the positron the difficulty of interpreting Dirac's equation seemed to be so unequivocally removed that it was generally accepted

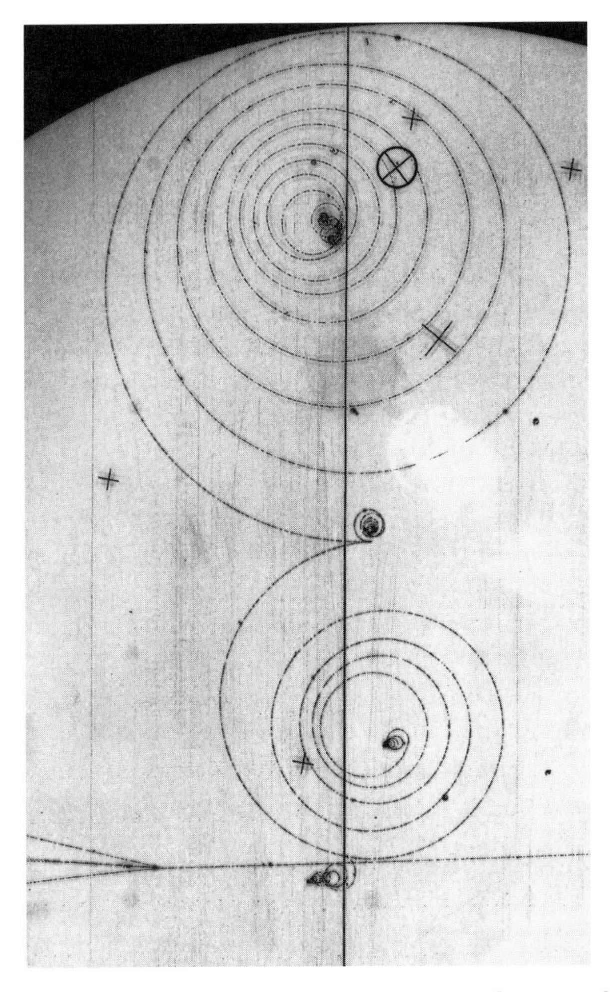

Fig. 2 Bubble-chamber photograph of the production of an electron-positron pair by an energetic neutral particle arriving from the right. A magnetic field normal to the plane of the picture bends the orbits in opposite directions, while the gradual loss of energy by collisions in the liquid diminishes their curvature. (Photograph kindly supplied by Dr T.O. White, Cavendish Laboratory).

as a true expression of the dynamics of a quantized electron. And indeed, apart from the small corrections demanded by quantum electrodynamics, this remains an article of belief.

There is, however, one further peculiar difficulty about which most textbooks are remarkably coy. Like Sommerfeld, Dirac took as his starting-point a structureless point charge. Sommerfeld's solutions were not spoilt by the assumption of electron spin, but the associated effects of spin—two directions of quantization and the magnetic properties that must be attributed to the spinning electron—were all additions to Sommerfeld's electron. Dirac's electron, on the other hand, point particle though it was, automatically exhibits all the properties that were ascribed to the spin. In other words, spin and magnetic moment are not internal properties of the electron, but necessary external dynamic properties that appear inevitably with Dirac's scheme of relativistic quantization. They are inseparable from the presence of negative-energy solutions, and presumably the difficulty of visualizing (in old-fashioned classical ways) a particle of negative energy is enough to make it equally difficult to visualize the phenomenon of spin. Most physicists, in their daily round of work, are content to attribute structure and spin to the electron, knowing that most of the time it will suffice to give them as good an explanation as they need. In the same way, they think of the electron as a little classical particle moving in a well-defined path; but they remember to patch up their model with the appropriate wave properties at such times as experience tells them they must do so.

The physical world, old and new

It is here that we have to face the fundamental switch of attitude brought about by the development of modern physics. The classical physicist, of whom we may take Maxwell and J.J. Thomson as examples, were well aware of the problems encountered in forming a consistent and intelligible model of the physical world, but they did not doubt that eventually someone would succeed. The very end of Maxwell's great *Treatise on electricity and magnetism* reads:

> ... whenever energy is transmitted from one body to another in time, there must be a medium or substance in which the energy exists after it leaves one body and before it reaches the other. ... Hence all these theories lead to the conception of a medium in which propagation takes place, and if we admit this medium as an hypothesis, I think it ought to occupy a prominent place in our investigations, and that we ought to endeavour to construct a mental representation of all the details of its action, and this has been my constant aim in this treatise.

Papers of the nineteenth century are full of attempts to model the ether, not so much perhaps in the hope of finding a definitive solution as of showing that such a thing is not incompatible with the mechanical principles so well understood in those times. It is clear from J.J. Thomson's writings that he hankered all his life for a frictionless fluid ether in which perpetual vortices would exhibit the properties of atoms. Einstein's special theory of relativity (1905) did not utterly destroy the ether, but made it intrinsically unobservable. Those who felt, not unreasonably, that if light was to be propagated as waves it was desirable to have some medium that could wave, could compromise (like the great Lorentz) by acceding to relativity while making clear that there were advantages in retaining the ether, with modified properties along the lines proposed independently by Fitzgerald and Lorentz.

In developing relativity theory Einstein had, I think, no intention of destroying the classical picture, but rather of completing it by repairing a fault. In one of his other great innovations of that time, the quantum theory of radiation, he exhibited the structure underlying Planck's hypothesis, and by showing that light had both wave and particle properties really put the cat among the pigeons.[27] Attempts to resolve in one model the modes of behaviour exemplified by diffraction and the photoelectric effect occupied prime time in the first quarter of the century, and presumably greatly helped the acceptance, after 1925, of the interpretation of quantum mechanics along the lines of the Copenhagen doctrine. This was the philosophical watershed of modern physics, and it is worthwhile tabulating some of the evidence that justified abandoning the classical world picture. As we are primarily concerned with the influence of the electron's discovery I shall frame the argument in those terms; but they are in no way exclusive.

1. *Indeterminacy.* One cannot ascribe a definite path to a particle, nor can one determining its position and momentum simultaneously beyond a well-defined limit. This is not because instruments are limited in precision, but is a necessary consequence of the mathematical structure. There have been ingenious and valiant attempts to devise theories that avoid this limitation, but (in my view) they all incur the necessity of believing in forces and effects that are no more susceptible to common-sense visualization.
2. *Wave-function collapse.* At each measurement the acquisition of new information necessitates reconsidering the Schrödinger equation with an initial wavefunction that updates one's knowledge. Those who believe in the reality of the wavefunction have to worry about how this comes about. The question has been argued over ever since 1925, but only by those who like to believe in the actuality of the world described by quantum mechanics. Those who see wavefunctions etc. as no more than concepts to enable successive observations to be

correlated are content to ignore the problem, as they are content to talk about light waves without any idea of the medium of propagation.

3. *The cluttered vacuum.* Every oscillator possesses zero-point energy of half a quantum. There is no known limit to the frequency of electromagnetic waves, and their zero-point energies add up to an infinite energy density in the vacuum. The formal treatment necessitates the idea of spontaneous fluctuations of energy in the vacuum and small adjustments to the energy levels of atoms which have been verified with extraordinary accuracy. The vacuum is also full of negative-energy electrons and indeed of analogues which also obey the Dirac equations.
4. *The structureless electron.* I have already noted the peculiar appearance of spin and magnetic moment as relativistic consequences; the point-particle (there is no experimental evidence for any structure in an electron, unlike the proton and neutron) also has the awkward classical property of possessing infinite energy, and hence infinite mass, in its external electric field. This was known before 1939, and Dirac and others had suggested there was also an infinite negative mass that very nearly compensated the electrical mass. After 1945 the technique of renormalization was developed extremely successfully to keep the balance-sheet respectable, and at the same time explain the energy-level adjustments mentioned in the last paragraph. But what it means, in a classically visualizable sense, defies the imagination.

These four headings should be enough to indicate that the classical ideal of an intuitively intelligible picture is unlikely to emerge from the present formulations. Yet these formulations are impressively precise in their predictions over an enormous range of phenomena, and one cannot imagine they can be tinkered with to improve their intuitive merits without losing their astonishing integrity. There is much to be said in favour of the anti-metaphysical, and a very ancient, point of view that reality is unknowable except in terms of what we experience and observe. The mathematical description provided by quantum mechanics is as complete as we can expect. What then, to return to basics, is this electron that was revealed 100 years ago? Since its first appearance, as a simple charged particle interacting with other particles, the description of its properties has become ever clearer while in itself it has become ever more mysterious.

To become still further embroiled, let us remember that we have really been discussing only a single electron, and that when we consider a crowd everything becomes worse. The naive observer, asked to imagine two electrons, would foresee no trouble in thinking of one here and another there; and if asked to describe them in wave-like terms would probably suggest that each has its own wavefunction to control its movement. He would be wrong—the only way to describe two electrons in wave mechanics is by introducing a single wavefunction for both, a Ψ which depends on the coordinates of both, as if it were a wave in six dimensions. One of the correct answers that comes from this treatment is that the behaviour of a composite body can be dissected into two parts—a single three-dimensional wavefunction to describe the motion of its centroid, and another multidimensional wavefunction to describe all that goes on within the body, as seen by an observer moving with the centroid. Thus a slowly moving atom may be governed by an internal wavefunction of immense complexity, but made up only of short wavelengths (because long wavelengths could not be confined to the limited space available in the atom). On the other hand, its centroid is governed by a three-dimensional, and quite simple, wavefunction whose long wavelength can by no juggling be thought to be a manifestation of the internal wave function.

A recent demonstration—a virtuoso performance—has been given by a Japanese team who used laser cooling to bring to rest and confine in a cloud a stream of neon atoms.[28] They could be released to fall under gravity through 42 cm, their de Broglie wavelength then being 7 nm, about 30 times larger than the atomic diameter. A pierced plate, made by electron beam lithography, was designed to diffract the waves so that they built up an image of the initials—NEC—of one of the research institutions involved; and that is exactly what happened. This is a convincing verification of the necessity of using many-particle wavefunctions to predict the behaviour of complex bodies, and it shows the basic flaw in any naive view of particles being described as independent classical objects.

The microscopic world, then, is entirely different from any extrapolation we may attempt from everyday experience. There are those who wish to believe that at some level between the atomic and the human scale the rules of quantum mechanics are suspended, perhaps by the destruction of wave coherence as a result of random fluctuations; or even by a modification of the rules so as to give macroscopic bodies desirable classical properties. Ingenious as these suggestions are, they are as yet without experimental support, and seem to be desperate attempts to retain belief in the reality of the physical world—in the sense of 'something out there' which can be studied by human observation and, if not perfectly visualized, at least reduced to a calculable model which would carry on even though there was no one to observe it.

The limitations of science

The great success of quantum physics has led to a species of triumphalism, the proclamation of physical science as the best way to understand the world, a panacea for all problems. Among its recent manifestations has been the attempt by many independent scientists to initiate a truly scientific theory of mind. The most optimistic are not content to describe changes in brain and body function that accompany conscious thought (a perfectly proper scientific programme)—they seek an explanation in scientific terms of how conscious thought is generated. This, I believe, is not possible. Without conscious thought there is no science—it is the pondering on observations that has generated science, and one must accept that all the enigmas that result from the development of physics are secondary to the prime experience that we conscious humans enjoy. To say this is not necessarily to deny that there is a real outside world which is responsible for making us think, and certainly not to suggest that the multitude of observations on which we can all agree are generated solely by social pressure and have no intrinsic validity. But the very incomprehensibility, in terms of everyday experience, of the formulations we find so successful should surely deter those who attempt a scientific theory of mind. These enthusiasts seem sometimes to ignore the extraordinary advances of the last hundred years, and expect to start their theorizing from a naively realistic conception of the world. Believing that the scientific method has been wholly successful they do not hesitate to put the cart before the horse, and expect science to explain the mental processes that have generated science itself. If the world had shown it conformed to the classical mechanistic ideal there might be some hope, even if we have no way of defining consciousness in scientific terms. But without any such description of consciousness, and without an intelligible picture of the basic units (particles, fields, even the vacuum) there seems to be no hope of success in this ambitious programme.

This is no way a confession of failure. No one seriously doubts that modern physics has been immensely successful, and one may go further and claim as a great success that the achievements have been taken to a point where the barriers to further progress can be seen. It is no failure to recognize that not every problem is solvable by our chosen methods, and if at some future date a totally new and better system arises that incorporates conscious thought as the most basic and essential datum, it must go beyond physical science and yet display the achievements of science as a great venture in its own terms.

References

1. J.J. Thomson, *Proc. R. Inst.*, 1897, **15**, 419. (See also chapter 1 of M. Springford, *Electron*, Cambridge University Press, 1997.)
2. J.R. Tillman and D.G. Tucker, in T.I. Williams (ed.), *A history of technology*, Vol. 7, Chapter 46, Clarendon Press, Oxford, 1978.
3. O.W. Richardson, *Phil. Trans. R. Soc.*, 1903, **A201**, 497.
4. US Patents nos 836531 and 837616, 1906.
5. G.C. Southworth, *Principles and applications of waveguide transmission*, Van Nostrand, New York, 1950.
6. J. Bardeen and W.H. Brattain, *Phys. Rev.* 1948, **74**, 230.
7. A.L. Hodgkin, *Biog. Mem. Fellows R. Soc.*, Royal Society, London, 1979, **25**, 1.
8. H. Barkhausen, *Phys. Zeitschr.*, 1919, **20**, 401.
9. L.R.O. Storey, *Phil. Trans. R. Soc.*, 1953, **A246**, 112.
10. A.P. French and P.J. Kennedy (ed.), *Niels Bohr*, Harvard University Press, 1985, p. 33.
11. A. Sommerfeld, *Sitzungber. Bayer. Akad. Wiss.*, 1915, 425 and 459.
12. G.E. Uhlenbeck and S.A. Goudsmit, *Naturwiss.*, 1925, **13**, 953.
13. K.W.H. Stevens, Chapter 13 of *Twentieth century physics*, IOP Publishing and AIP Press, 1995.
14. A.P. French and P.J. Kennedy (ed.), *Niels Bohr*, Harvard University Press, 1985, p. 50.
15. E.C. Stoner, *Phil. Mag.*, 1924, **48**, 719.
16. W. Pauli, *Z. Phys.*, 1925, **31**, 765.
17. A.P. French and P.J. Kennedy (ed.), *Niels Bohr*, Harvard University Press, 1985, p. 58.
18. L. de Broglie, *Comptes Rendues*, 1923, **177**, 507, 548, 630.
19. E. Schrödinger, *Ann. Phys.*, 1926, **79**, 361.
20. C.J. Davisson and I.H. Germer, *Nature*, 1927, **119**, 558.
21. G.P. Thomson, *Proc. R. Soc.*, 1928, **A117**, 600.
22. D.R. Hartree, *Rep. Prog. Phys.*, 1948, **11**, 113.
23. W. Heitler and F. London, *Z. Phys.*, 1927, **44**, 455.
24. J.A. Wheeler and W.H. Zurek (ed.), *Quantum theory and measurement*, Princeton University Press, 1983.
25. P.A.M. Dirac, *Proc. R. Soc.*, 1928, **A117**, 610 and **A118**, 351.
26. H. Kragh, *Dirac*, Cambridge University Press, 1990, Chapter 5.
27. C.D. Anderson, *Science*, 1932, **76**, 238.
28. P.M.S. Blackett and G.P.S. Occhialini, *Proc. R. Soc.*, 1933, **A 139**, 699.
29. B.R. Wheaton, *The tiger and the shark*, Cambridge University Press, 1983.
30. M. Morinaga, M. Yasuda, T. Kishimoto, F. Shimizu, J. Fujita, and S. Matsui, *Phys. Rev. Lett.*, 1996, **77**, 802.

SIR BRIAN PIPPARD

Born 1920, graduated in physics from the University of Cambridge, and has passed almost all his later life there, except for 4 years during the war spent developing radar. Substantial teaching experience and experimental research in low temperature physics, especially superconductivity and the electron structure of metals, led in 1971 to the Cavendish chair and headship of the department. Unable thereafter to give enough attention to research students, he wrote books on physics and then, having written all he knew, turned in an amateurish way to the history of physics in recent times. From 1990 to 1995 he was Visiting Professor of Physics at the Royal Institution.

The Coleridge experiment

RICHARD HOLMES

Introduction

Coleridge loved walking about with large Folio Volumes. He said he once walked for the whole of a hot summer's day—from Bristol to Stowey (40 miles)—'carrying Baxter's folio *On the Immortality of the Soul*—and a plucked goose—and the giblets'—for his supper. For our supper this evening we have this large Folio—the Original Minute Book of the Royal Institution (1807–9). Here is the entry for 7 December 1807.

> Mr Bernard reported that Mr Coleridge will give in the ensuing Season five courses of five Lectures each on the distinguished English Poets in illustration of the General Principles of Poetry ... from Shakespeare to the Moderns ... to begin immediately, and to give one or two lectures a week as may be convenient, for a compliment of £140, of which £60 is proposed to be paid in advance.

These twice-weekly Lectures began in February 1808.

But 4 months later we find this dramatic entry in the Royal Institution Minute Book, 13 June 1808.

> William Savage, the assistant secretary, laid before the Managers the following letter from Mr Coleridge. Dear Sir, Painful as it is to me almost to anguish, yet I find my health in such a state as to make it almost Death to me to give any further Lectures. I beg that you would acquaint the Managers that instead of expecting any remuneration, I shall, as soon as I can, repay the sum I have received. I am indeed more likely to repay it by my Executors, than myself. If I could quit my Bed-room, I would have hazarded every thing rather than not have come, but I have such violent fits of Sickness and Diarrhoea that it is literally impossible.

The subject of this chapter is what happened to Coleridge—and to Science and to Poetry—between these two entries.

Part one

The invitation to lecture had originated with his friend Humphry Davy, and it could be described as the 'Coleridge Experiment'. Davy was of course a founding figure in the history of the Royal Institution. Born in Cornwall in 1778, educated at the Bristol Pneumatic Institute, he was appointed first lecturer (aged only 23) and then Professor of Chemistry at the Royal Institution. His brilliant manner as a lecturer, and vivid demonstrations, caused such frequent traffic jams in Albemarle street (together with Byron's book-launches at John Murray's opposite) that it was made the first one-way thoroughfare in London.

Both Coleridge's and Davy's lectures were held in the Great Lecture Theatre, where we are gathered this evening. The Minute Book tells us that it was redesigned by Count Rumford in spring 1802, to became the most modern and comfortable 'in Europe'. It could seat 500 people within its 60 foot hemisphere. It was upholstered with 'Green moreen cushions', heated with copper steam pipes, and insulated with green baize from outside noise. The upper gallery had a separate entrance for late-comers and students. The steep tiers of seating were modelled on the anatomy theatre, producing a dramatic focus on the lecturer.

Founded in 1799, the Royal Institution promoted both Arts and Sciences, with no distinction between the 'two cultures'. In 1808 the Minute Book records lecture series on chemistry, botany, mechanics, Persian literature, German music, and Renaissance architecture. When the Reverend Sydney Smith lectured on moral philosophy, it was said that the laughter could be heard in the street. The Institution was quickly copied by others in London—the Surrey, the Russell, the Philosophical—and it created a new audience. Gillray's famous cartoon showed Davy producing stinks and fireworks to a mixed audience of men and women of every age and class, both Anglicans and non-conformists. This hunger for knowledge eventually led to the foundation of the University of London.

Davy had been trying to bring Coleridge to the lecture dais ever since his return from Malta in the summer of 1806. He wrote to Coleridge's West Country friend, the radical tanner Tom Poole in August 1807.

> The Managers of the Royal Institution are very anxious to engage him; and I think he might be of material service to the Public, and of benefit to his own mind, to say nothing of the benefit his Purse might receive. In the present condition of society, his opinions on matters of taste, literature, and metaphysics must have a healthy influence; and unless he soon become and actual member of the living world, he must expect to be brought to judgment for hiding his light...

Considering Coleridge's reputation at this date, this could be considered as one of Davy's most daring experiments in chemical combustibles. (Coleridge was known both as a visionary poet, and as a man suffering from opium addiction, unrequited love, and a serious writer's block.) After numerous delays Coleridge began to lecture on Friday 15 January 1808. He broke off, and started again on Friday 5 February. He broke off again—and started again on Wednesday 30 March. He eventually completed 18 or 20 lectures, but they appeared to produce nothing but a catalogue of disasters and disappointments.

The lecturer frequently postponed at the last minute

Here is an unpublished letter from the Institution archives, February 1808.

> Dear Sir, I have been confined to my Bed, with intervals of a few Hours, since Saturday last—still however anxious—a very weak word—to be present at my First (actually second) on Friday—and hourly hoping and fearing, I have delayed to acquaint you of my state.—But today my illness becoming more serious, and the very possibility of my removing from my Bed or being able even to stand in a Public Room tomorrow, becoming utterly unlikely, I am obliged to advertise you thereof—whatever expense may be incurred I shall most cheerfully pay, if indeed I live to pay it in my own person, in order to inform the Subscribers. I have sent for Dr Abernathy. I shall learn from him whether this be only an Interruption—or a final farewell. Either myself or my medical attendant will write to Mr Bernard. STC—Thursday evening.

The lecturer arrived almost speechless with opium

(Coleridge was then in the habit of consuming up to a pint of laudanum, opium dissolved in brandy, a day.) The young Thomas De Quincey recalled:

> His appearance was generally that of a person struggling with pain and overmastering illness. His lips were baked with feverish heat, and often black in colour; and, in spite of the water which he continued drinking through the whole course of his Lecture, he often seemed to labour under an almost paralytic inability to raise the upper jaw from the lower ... Most unfortunately he relied on his extempore ability to carry him through.

The lecturer was often unprepared and chaotic in appearance

Coleridge himself tells the story of how he prepared three clean white lecture shirts, but found he had slept in the first, dried his face and hands on the second, and used the third as a foot mat. Many years later a fourth shirt marked STC on the collar was found on a poor man hanging from a tree in Hyde Park.

The lecturer often digressed from his announced topics, or never touched on them at all

The *Times* journalist Henry Crabb Robinson recorded: 'Coleridge's digressions are not the worst parts of his Lectures, or rather he is always digressing—(and they are the only parts)'.

The lecturer lost his lecture notes in mysterious circumstances

Edward Jerningham:

> To continue the History of this Lecturer ... He looked sullen and told us that He previously had prepared and written down Quotations from different Authors to illustrate the present Lecture. These Quotations he had put among the Leaves of his Pocket Book which was stolen as he was coming to the Institution. This narrative was not indulgently received, and he went through his Lecture heavily ... The next day he received an Intimation from the Managers that his Lectures were no longer expected.

Finally, the lecturer was officially censured by the committee of the Royal Institution

Minute book, May 1809: 'Resolved, that the personal attack made by Mr Stephen Coleridge (*sic*) on Mr Joseph Lancaster, in a lecture delivered by him at the Royal Institution on Tuesday 3rd May 1808, was in direct violation of a known and established Rule of the RI prohibiting any personal animadversions in the Lectures there delivered.' It accordingly took more than a year to pay the balance on his fees.

The disaster was summed up in an unpublished note by Davy which is still held in the Royal Institution Archives.

> He has suffered greatly from Excessive Sensibility—the disease of Genius. His mind is a wilderness in which the Cedar and Oak which might aspire to the Skies are stunted in their growth by underwood Thorns, Briars and parasitical plants. With the most exalted Genius, enlarged views, sensitive heart and enlightened Mind, he will be the victim of want of Order, Precision and regularity. I cannot think of him without experiencing mingled feelings of admiration, regard and pity.

So this is how the 1808 lectures have gone down in history. They gained the reputation of being the most disasterous series ever delivered here. As the great Coleridge scholar, Kathleen Coburn, noted in a paper delivered 25 years ago at the Royal Institution: 'We know a little about these lectures but no full report has been found. And some of the details are best passed over.' It would seem that Davy's great Coleridge experiment was a failure.

Part two

But history may be rewritten by biography. One of the methods of biographers, in their search for truth, is to alter our conventional perceptions of historical time. They may challenge superficial chronology, and seek to place events in a new sequence or deeper context of ideas. So I want to wind back this extraordinary story, as one might wind back a video tape, by some 10 years. Let us start the story again in 1799 and look at it afresh.

A picture of Coleridge done in 1799 shows a strikingly youthful figure, who might be a hippy from the 1960s with his long hair, huge thoughtful eyes, and hungry mouth. This is a man of astonishing intellectual intensity and power. This is the man who had just written—*The Ancient Mariner*, *Kubla Khan*, and *Frost at Midnight*—some of the greatest Romantic poems in the language. (He had also written, *To a Young Ass*, which is perhaps first the animal rights poem. As a result Byron christened him 'the Laureate of the long-eared kind.')

Coleridge believed all knowledge, both scientific and artistic, could be included in poetry. He put his wonderful idea of an epic poem in a letter written to his publisher Joseph Cottle in 1796.

> I should not think of devoting less than 20 years to an Epic Poem. Ten to collect materials and warm my Mind with universal Science. I would be a tolerable Mathematician, I would thoroughly know Mechanics, Hydrostatics, Optics, and Astronomy, Botany, Metallurgy, Fossilism, Chemistry, Geology, Anatomy, Medicine—then the mind of Man—then the Minds of Men—in all Travels, Voyages and Histories. So I would spend ten years—the next five to the composition of the Poem—and the five last to the correction of it. So I would write haply not unhearing of that divine and rightly-whispering Voice, which speaks to mighty minds of predestined Garlands, starry and unwithering.

It should perhaps be pointed out that Coleridge had just missed his publisher's deadline at this date.

The 1799 picture was painted in Germany, where Coleridge was studying the latest in poetry (Schiller, Lessing), Biblical criticism (Eichhorn), Philosophy (Kant), and Life Sciences (Blumenbach) at the University of Gottingen. He came back to England breathing the heady intellectual gas of Schelling's *Naturphilosophie*. This influential movement was an extraordinary blend of early Evolution theory, Pantheism, and German Idealist philosophy. Only a quick sketch is possible here. *Naturphilosophie* regarded the hard Newtonian universe of material mechanisms—the clockwork of planetary orbits, the celestial billiards of atomism—as theoretically inadequate. Instead it proposed the concept of a fluid, dynamic, unified Nature, driven by invisible 'powers' to which chemistry and electricity provided the key. Scientific theory could be advanced through analogy, metaphor, and speculation as well as through empirical observation and controlled experiment.

Naturphilosophie produced a distinctive language of German scientific mysticism. Schelling and his circle spoke of 'vitalism', 'organicism', and 'galvanism'. All observable activity in life-forms was the result of invisible 'polarities' in Nature. There was a 'world-soul' constantly 'evolving' higher life-forms, and 'levels of consciousness' in all matter, animate or inanimate. All Nature had a tendency to move towards a 'higher state'. Several distinguished European scientists were connected with this movement: Lorenz Oken, who postulated cell theory; Johann William Ritter, who discovered ultraviolet rays; Henrik Steffens, the Scandinavian geologist who proposed a curious form of psychic evolution, making the famous observation that 'the diamond is a piece of flint that has come to its senses,' which a hard-headed British geologist was reported to have replied, 'a quartz must therefore be a diamond run mad'.

Coleridge was reading all these new European sources, as well as earlier British scientists like Erasmus Darwin (author of *The botanic garden* and *The loves of the plants*) and Joseph Priestley (whom he described in a poem as 'Patriot, Saint and Sage'). We might consider Priestley as a precursor of the brain drain, as he emigrated to Philadelphia when a patriot mob burnt down his house and laboratory in Birmingham in 1791. A crucial idea appeared in Priestley's *History of electricity* (1775):

> Hitherto philosophy has been chiefly conversant about the more sensible properties of bodies; electricity, together with chemistry and the doctrine of light and colours, seems to be giving us an inlet into their Internal Structure, on which all their sensible properties depend. By pursuing this new light, therefore, the Bounds of natural science may possibly be extended, beyond what we can now form an idea of. New worlds may open to our view, and the glory of the great Sir Isaac Newton ... be eclipsed.

Both Coleridge and Davy were full of these ideas in which new German Romantic Science challenged the eighteenth-century British empirical tradition. Indeed one can almost hear the *bang* when the two young men first met, at Dr Thomas Beddoes Pneumatic Institute, Clifton, in the summer of 1799. Here took place the famous experiments with nitrous oxide, laughing gas. Davy saw both the psychological interest of the work, as a revelation of the unconscious mind; and the medical one, as a possible form of anaesthetic. There are various accounts of their experiences, of which my favourite is by one Peter Roget, who said there were 'no words' to describe accurately what he felt, and then went on to compile Roget's *Thesaurus*.

Davy's own experimental notes provide this memorable description.

> By degrees as the pleasurable sensations increased, I lost all connection with external things; trains of vivid visible images rapidly passed through my mind and were connected with words in such a manner, as to produce perceptions perfectly novel. I existed in a world of newly connected and newly modified ideas. I theorised; I imagined that I made discoveries. When I was awakened from this semi-delirious trance by Dr Kinglake, who took the bag from my mouth, Indignation and pride were my first feelings. ... My emotions were enthusiastic and sublime; and for a minute I walked round the room perfectly regardless of what was said to me. ... With the most intense belief and prophetic manner, I exclaimed to Dr Kinglake,—'Nothing exists but Thoughts!—the Universe is composed of impressions, ideas, pleasures and pains!'

It is fascinating to compare this with Coleridge's account of writing his poem *Kubla Khan* under opium in 1797.

> The Author continued for about three hours in a profound sleep, at least of the external senses, during which time he has the most vivid confidence, that he could not have composed less than from two to three hundred lines; if that indeed can be called composition in which all the images rose up before him as *things*, with a parallel production of the correspondent expressions, without any sensation or consciousness of effort. On awaking he appeared to himself to have a distinct recollection of the whole, and taking his pen, ink, and paper, instantly and eagerly wrote down the lines that are here preserved. At this moment he was unfortunately called out by a person on business from Porlock...

Davy was also writing poetry, and was quite at home in the literary world. He corrected the proofs to the 2nd edition of *Lyrical ballads* (1800) and even parodied Wordsworth's revolutionary plain-style of verse:

As I was walking up the street
In pleasant Burny town,
In the high road I chanced to meet
My cousin Matthew Brown

My cousin was a simple man
A simple man was He
His face was of the hue of Tan
And sparkling was his—Eye.

Coleridge formed a passionate alliance with Davy in these years, endlessly discussing the links between science and poetry. When Coleridge moved to the Lakes (1800), he proposed setting up a chemical Laboratory, and bought a microscope, and said he would 'attack Chemistry like a shark'. Typically, he imagined transmitting a chemical vision of the Lake District landscape to Davy, by wrapping it up in 'a pill of opium'. He called Davy 'the founder of Philosophic Alchemy', and invited him to join a later Pantisocratic community in Italy—'what a Colony might we not make!' When, instead, Davy was appointed to lecture at the Royal Institution in 1801, he congratulated him with evident delight: 'Success my dear Davy, to Galvanism—and every other Schism—and Ism'.

When the philosopher William Godwin attacked Davy's scientific work as a form of naive materialism, Coleridge defended Davy in boisterous terms:

> Godwin talks evermore of you with lively affection.—'What a pity that such a Man should degrade his vast Talents to Chemistry'—cried he to me.—Why quoth I, how Godwin! can you thus talk of a science, of which neither you nor I understand an Iota? etc etc. And I defended Chemistry as knowingly at least as Godwin attacked it.—Affirmed that it united the opposite advantages of Immaterializing the mind without destroying the definiteness of the Ideas—nay even while it gave clearness to them.
>
> And eke that being necessarily performed with the passion of Hope, it was Poetical ... You and I, and Godwin, and Shakespeare and Milton, with what an Athanasiophagous Grin we shall march together,—We Poets!—Down with all the rest of the World!
>
> (By the word Athanasiophagous—I mean devouring Immortality, by anticipation!—Tis a sweet word!)
>
> God bless, you my dear Davy! Take my nonesense like as pinch of snuff—sneeze it off, it clears the head...

When Davy gave his First Series of Chemical Lectures at the Royal Institution in the spring of 1802, Coleridge rushed up to London to listen and 'to enlarge my stock of Metaphors'. We find these scientific metaphors everywhere in his letters. When describing Dorothy Wordsworth's extraordinary poetic sensibility—so evident in her wonderful journals—he deliberately used a laboratory image, drawn from the action of the filament of gold-leaf in an electrometer registering tiny shifts in electrical charges. 'Her manners are simple, ardent, impressive ... Her information various—her eye watchful in minutest observation of Nature—and her Taste a perfect electrometer—it bends, protrudes, and draws in, at subtlest beauties and most recondite faults.' (CL1 p. 331)

At Davy's 1802 lectures Coleridge made over 60 pages of notes, and remarked: 'Every subject in Davy's mind has the principle of Vitality. Living thoughts spring up like Turf under his feet'. One can select out a few, which convey the magical excitement of poetic suggestion, combining in Coleridge's mind with the acute pleasures of scientific precision.

Note f.4 'Experiment. Quart of Oxygen in a bladder breathed by Davy 2 minutes. Did not at all hurt the flame of the Candle. He breathed the same quantity of common air in a bladder for a less length of time—at the end of which it immediately extinguished the Candle. NB hold the end of the breathing pipe against the flame of the Candle'.

Note f.5 'Ether burns bright indeed in the atmosphere, but O! how brightly whitely vividly beautiful in Oxygen gas.'

Note f8. 'Hydrogen pistol ... electric spark detonates ... applied to a Leyden phial—BANG!'

Note f.9 'Hydrogen Gas employed for Balloons—this illustrated by Soap Bladders, filled with a mixture of Hydrogen and Oxygen Gas. Blow up soap into Bubbles by means of a BLABBER filled with mixed Gas' (In his excitement Coleridge mistranscribed the word.)

Note f.31 'Sulphuric acid pour upon hyperoxygenated muriate of Potash. NB Be sure to hold your face close to the glass. First reddens, and then destroys, vegetable Blues'.

(Then a wonderful bit of Coleridge's mischief.) Note f.31. 'If all aristocrats here, how easily Davy might poison them all—15 parts Oxygen, 85 muriatic acid Gas...!'

So when Coleridge left for Malta in 1804—seeking health, an escape from his unhappy married life, new forms of writing, and a cure to his growing opium addiction—Davy wrote him a kind of valediction which insisted on the intellectual work still to be done on the borders of science and poetry.

> In whatever part of the World you are, you will often live with me, not as a fleeting Idea but as a recollection possessed of creative energy, as an IMAGINATION winged with fire, inspiriting and rejoicing. You must not live much longer without giving to all men the proof of your Power. ... You are to be the Historian

> of the Philosophy of feeling.—Do not in any way dissipate your noble nature. Do not give up your birth-right.

This rapid sketch may suggest just a little more clearly what Coleridge and Davy had come to mean to each over the previous decade, and why Davy took the risk of inviting Coleridge to this dais. Remember especially that thought-provoking phrase: that as science was 'necessarily performed with the passion of Hope, it was Poetical.'

Part three

So having wound back the tape of biographical time, we return to our point of departure, the lectures in 1808. Here the biographer can make a second chronological adjustment—slow time down and freeze-frame it, or say put it under the microscope, for several revealing moments.

Coleridge in the summer of 1807 was hiding away in the Quantock Hills in Somerset (where 10 years before he had written *Kubla Khan*): now ill, depressed, opium addicted, and writing very little. He could not see how his literary career could develop. But he read the accounts of Davy's famous Second Bakerian Lecture at the Royal Society in November 1807, when he first isolated and named the elements of potassium and sodium using voltaic batteries to decompose potash and soda ('caustic alkalies').

The old *Naturphilosophie* excitement stirred in Coleridge. Davy had found a key to those deep powers and 'internal structures' hidden mysteriously within the material world. He wrote with something of his old enthusiasm:

> His March of Glory ... has run for the last six weeks—within which time by the aid and application of his own great discovery, of the identity of electricity and chemical attractions, he has placed all the Elements and all their inanimate combinations in the power of man; having decomposed ... and discovered as the base of the Alkalies a new metal ... Davy supposes there is only ONE POWER in the world of senses; which in particles acts as chemical attractions, in specific masses as electricity, and on matter in general, as planetary Gravitation ... when this has been proved, it will then only remain to resolve this into some Law of vital Intellect—and all human Knowledge will be Science, and Metaphysics the only Science.'

The crucial and engendering idea here for Coleridge, was that of the ONE POWER. What Davy was doing for the world of the senses, Coleridge aimed to do for the world of the Mind, Creativity, and Art. He would 'isolate, decompose and name' the one power of the imagination in his poetry lectures. He would analyse, demonstrate, and define the imagination at work, as in a laboratory.

When we look at the circumstances more closely, the apparent chaos of his lectures takes on a different dimension and context. For a start we find Davy himself had also been ill and postponed his own series. (Characteristically he had caught gaol fever, while investigating the ventilation systems at Newgate prison.) Davy's despairing note on Coleridge's 'Excessive Sensibility' was written after Coleridge's second lecture only, and did not apply to the remainder. Indeed he had composed a fine poem—partly inspired by their shared vision of a dynamic Nature:

> All speaks of change: the renovated forms
> Of long-forgotten things arise again;
> The light of suns, the breathe of angry storms
> The everlasting motions of the main.
>
> These are but engines of the Eternal will,
> The One intelligence, whose potent sway
> Has ever acted, and is acting still,
> While stars, and worlds, and systems all obey;
>
> Without whose power, the whole of mortal things
> Were dull, inert, an unharmonious band,
> Silent as are the Harp's untuned strings
> Without the touches of the Poet's hand.

This makes clear reference to Coleridge's poem *The Eolian Harp*, and suggest a continuing sublime fusion between science and poetry.

Similarly, de Quincey's account of Coleridge's opium-wrecked presence at the dais, turns out to be based on two early lectures only, before Coleridge had got into his stride. The lectures gathered strength and coherence as they went on. Coleridge quickly learned to turn his spontaneous, digressive style to dramatic advantage. His very peculiarities and risky performance manner could rivet his audience, who never knew what to expect next. This appears in a very different description from 11-year old Katharine Byerley (an indication, incidentally, of the wide appeal of his reputation across the age band).

> He came unprepared to lecture. The subject was a literary one, and the poet had either forgotten to write, or left what he had written at home. But his locks were now trimmed, and a conscious importance gleamed in his eloquent eyes. Every whisper was hushed ... and I began to think, as Coleridge went on, that the lecture had been left at home on purpose; he was so eloquent—there was such a combination of wit and poetry in his similes—such fancy, such a finish in his illustrations...

Again, from Edward Jerningham (who had mocked the story of the stolen notes) we can find an intriguing later summary of Coleridge's impact over the whole series.

> My opinion of the Lecturer is that he possess a great reach of mind; that he is a wild Enthusiast respecting the objects of his Elogium; that he is sometimes very eloquent, sometimes paradoxical, sometimes absurd. His voice has something in it particularly plaintive and interesting. ... He spoke without any assistance from a manuscript, and therefore said several things suddenly, struck off from the Anvil, some of which were entitled to high Applause ... He too often wove himself into the Texture of his Lecture ... He was in some respects, I told him one day, like Peter Abelard.

(This unexpected reference to the great medieval lecturer, Peter Abelard, at the Sorbonne gives one pause for reflection. Abelard was renowned for his personal style of lecturing, and the great following he achieved among the younger students.)

Moreover, though it used to be thought that almost nothing of the 1808 lectures had survived, scholars have now recovered substantial fragments: from Coleridge's recently published notebooks, from the Egerton Mss in the British Library, and from a set of drafts unearthed in the New York Public Library. From these we have time to consider four snapshots, which show Coleridge working towards his great theory of the imagination and steadily using scientific vocabulary and metaphors to explore it. They also show the special, popularizing gift he had for vivid homely anecdotes (which we might consider as the equivalent to Davy's use of desktop demonstrations with bladders and bell jars).

1. Coleridge proposed that the critical language used to discuss all the arts must be philosophically based, and almost scientifically sharpened. Take the concept of beauty.

> In a Boat on the Lake of Keswick, at the time that recent Rains had filled all the Waterfalls, I was looking at the celebrated Cataract of Lodore, then in all its force and greatness—a Lady of no mean Rank observed, that it was sublimely beautiful, & indeed absolutely—PRETTY.
> I have never mentioned this without occasioning a laugh...

This casual confusion of terms was inadequate. Beauty must be part of a philosophic system, a strict hierarchy of forms and perceptions. (The epistemology behind this was in fact Kantian, but the explanation was drawn from Davy's calorific experiments, melting ice by friction.) 'If our language have any defect, it is in want of Terms expressing KIND, as distinguished from degree—thus we have Cold and Heat—but no words that instantly give the idea of Heat independent of comparison—if we say to persons who have never studied even the Elements of Chemistry, there is heat in Ice, we should appear to talk paradoxes.'

2. Coleridge suggested that the way the human imagination responds to all works of art was far more subtle than the traditional mechanistic theories of admiration and judgement. Imagination was an active process, like an an electrical current pulsing between objective and subjective polarities. The mind does not stand passively outside its experience, registering and recording, but enters dynamically into what it sees, reads, or hears. Coleridge used the example of stage illusion (a favourite of Dr Johnson and other eighteenth-century critics) and illustrated it characteristically by a tender anecdote of his son Hartley looking at the picture of a seascape.

> As Sir George Beaumont was showing me a very fine Engraving from Rubens, representing a storm at sea (without any Vessel or Boat introduced), my little Boy (then about 5 years old) came dancing and singing into the room, and all at once (if I may dare use so low a phrase) tumbled in upon the Print. He instantly started, stood silent and motionless, with the strongest expression of Wonder and then of Grief in his eyes and countenance, and at length said—'And where is the Ship? But that is sunk!—and the men all drownded!'—still keeping his eye fixed on the Print.
>
> Now what Pictures are to little Children, Stage Illusion is to Men, provided they retain any part of a Child's sensibility. Except that in the latter instance, this Suspension of the Act of Comparison, which permits this sort of Negative Belief, is somewhat more assisted by the will.

A proper commentary on this remarkable passage would take us very far into the new Romantic theory of imagination. It proclaims the child-like part of the creative sensibility, ever fresh and spontaneous, which both the scientist and the poet must retain. It enacts the emotional energy—the passion of Hope—which accompanies the imaginative impulse. It employs the metaphor of electrical polarity in the striking idea of 'Negative Belief'—an image which recurs in a central passage in Coleridge's *Biographia Literaria* (1817)—'that willing suspension of Disbelief for the moment that constitutes poetic faith'. And arguably it prepares the way for John Keats's brilliant definition of 'Negative Capability' as the crucial power of the poet to enter into other 'modes of being' outside himself. But these unformed hints must suffice tonight.

3. Coleridge himself rose to a brilliant invocation of the poet's imaginative power, as the capacity to project multiple images on the human

mind, and fuse them into a single impulse of feeling and vision. Great writers have 'that power and energy of what a living poet has grandly and appropriately described as: "To flash upon that Inward Eye, which is the Bliss of Solitude" and to make everything present by a Series of Images.—This an absolute Essential of Poetry ...' Again the metaphors behind this argument are both electrical and chemical (the notion of 'fusion' and dynamic 'energy'), with the concept of a single power or principle lying within them: 'imagination, the power of modifying one image or feeling by the precedent or following ones ... to produce that ultimate end of Human Thought, and human Feeling,—Unity, and thereby the reduction of the Spirit to its Principle and Fountain ...' This fountain of inexhaustible energy was ultimately the same both in the physical world and the imaginative one.

It found its greatest literary source in Shakespeare's poetry.

> ...A sort of fusion to force many into one ... Even as Nature, the greatest of Poets, acts upon us when we open our eyes upon an extended prospect.—Thus the flight of (Shakespeare's) Adonis from the enamoured Goddess in the dusk of the Evening:
>
> Look! how a bright star shooteth from the Sky
> So glides he in the night from Venus' Eye
>
> How many Images and Feelings are here brought together without effort and without discord: the Beauty of Adonis—the rapidity of his Flight—the yearning yet Hopelessness of the enamoured gazer (Venus)—and a shadowy, Ideal character thrown over the whole.

Here the creation of poetry seems to share in the dynamic process of the continuous creation of the physical universe itself. The one is in harmony with the other. That is exactly what Coleridge would later argue, with more technical philosophic language, in his theory of Primary and Secondary Imagination in the *Biographia Literaria*.

4. Finally, I would like to glance at that lecture for which Coleridge was censured. Surprisingly, it turns out to be a 'Supernumerary' one, which he actually volunteered to make up for his earlier absences and placate the committee. The subject at first appears completely peripheral to the topic of the imagination: a contemporary controversy over educational methods. Two opposing pedagogic systems were espoused by Dr Andrew Bell (an Anglican) and Dr Joseph Lancaster (a Quaker). The central issue in dispute was that of the use of punishments in teaching children. Dr Lancaster—an early Pavlovian—promulgated an astonishing arsenal of correctives: the use of pillories in the classroom, the

shackling of legs with wooden logs, trussing dunces up in a sack, making them walk backwards through the school corridors, and suspending them in 'punishment baskets' from the ceiling. It was this system that Coleridge attacked with such passionate vehemence.

Robert Southey later recalled the scene.

> When Mr Coleridge in a Lecture at the Royal Institution, upon the New System of Education, came to this part of the subject (punishments), he read Mr Lancaster's account of these precious inventions verbatim from his own book, and throwing the book down with a mixture of contempt and indignation, exclaimed, 'No boy who has been subject to punishments like these will stand in fear of Newgate, or feel any horror at the thought of a Slave Ship!'

It fact this was probably one of Coleridge's finest lectures. He drew upon his own childhood experiences, and he effortlessly turned the subject back to his central theme. He argued that 'true Education comes from Love and Imagination.'

Crabb Robinson reported with enthusiasm.

> The extraordinary Lecture on Education was most excellent, delivered with great animation and Extorting praise from those whose prejudices he was mercilessly attacking. And he kept his audience on the rack of pleasure and offence two whole hours and 10 minutes, and few went away during the lecture ... The cardinal rules of early Education: 1. to work by love and so generate love; 2. to habituate the mind to intellectual accuracy and truth; 3. to excite (imaginative) power.

He concluded: 'Little is taught by contest or dispute, everything by sympathy and love.'

Part four

Having wound-back time, and slowed down time, the biographer can now fast-forward time in the few moments left to us.

In view of subsequent events, the 1808 series should be considered as a triumph snatched from the jaws of disaster, and one of the eccentric glories of the Royal Institution. Against all the odds, Davy had succeeded in launching Coleridge's career as Public Lecturer. It saved Coleridge from despair, it saw him through his opium addiction, and set him writing again. Between 1811 and 1813 Coleridge gave over 120 lectures in London and Bristol, and his great reputation as the foremost English Romantic critic began. It eventually produced his two volumes of *Shakespeare criticism*, which is still studied in schools and colleges, and frequently shapes the thinking behind many modern stage productions. It also formed the basis of his aesthetic theories in the *Biographia Literaria*, the most influential book of Romantic critical philosophy ever published.

The interplay between science and poetry continued the fill Coleridge's later notebooks, and shape his broadest speculations for the rest of his life. It is typical that in one lecture he compares the poetic cosmology of Milton's *Paradise Lost* with Erasmus Darwin's scientific one, and gives a vividly coherent account of the Big Bang Theory. He then raises an objection which might still give us pause for thought: the Big Bang Theory is not beautiful enough! He also grappled with Evolution—what he called the Orang-outang Theory—and gave an Idealist interpretation of it in *The theory of life* (1816). 'I define Life as the Tendency to Individuation; and the degrees or intensities of Life, to consist in a progressive realisation of this tendency.' Again he raises a philosophical problem which still haunts us: Does the mechanism of Evolution necessarily exclude the notion of purpose? He remained friendly with a wide circle of scientists, and it is no coincidence that his last amanuensis at Highgate, J.H. Green, later became President of the Royal College of Surgeons.

It is heartening to remember the great tribute that Coleridge paid to Davy and Science, in an essay published in his philosophical collection *The Friend* in 1818.

> This is, in truth, the first charm of Chemistry, and the secret of the almost universal interest excited by its discoveries ... the propounding and the solving of an Enigma. It is the sense of a principle of connection given by the Mind, and sanctioned by the correspondency of Nature. Hence the strong hold which, in all ages, Chemistry has had on the Imagination. If in Shakespeare we find Nature idealized into Poetry, through the creative power of a profound yet observant meditation; so through the meditative observation of a Davy, a Wollaston, or a Hatchett ... we find Poetry, as it were, substantiated and realised in Nature.

So what, in sum, have I tried to demonstrate?

Perhaps three very simple ideas. First, that there was once such a thing as Romantic Science, and a 'porous interface' between the two cultures which proved immensely creative both for Davy and for Coleridge. It may be, with recent growth in popular science writing (one thinks of Stephen Hawking, of Richard Dawkins, of Stephen Jay Gould) that we are entering such a magic period again.

Second, that in a proper perspective, Davy's Coleridge experiment here at the Royal Institution can be seen as an historic success, and a tribute to the values that the Institution has always stood for.

And third, that the methods of biography—a form that is often derided as the 'higher gossip'—can actually bring back to life not only the most significant figures in our national history, but also the most significant challenges—intellectual and spiritual challenges—that impelled them. And which impel us still.

Biography, like science, is for the living. It too, in Coleridge's memorable words, is 'necessarily performed with the passion of Hope'.

Note on sources

Apart from material in the Royal Institution archives, and the published works of Sir Humphry Davy and ST Coleridge, the author would like to acknowledge the following sources. Leslie Stephen, 'Coleridge', (a Royal Instution Lecture) published in *Hours in a library*, 1879). Kathleen Coburn, 'Coleridge: a bridge between science and poetry', published in *Proceedings of the Royal Institution of Great Britain*, Vol. 46, 1973; David Knight, *Humphry Davy: science and power*, 1992; and Trevor H. Levere, *Poetry realized in nature: Samuel Taylor Coleridge and early nineteenth-century science*, 1981.

RICHARD HOLMES

Born 1945 in London. Graduated from Churchill College, Cambridge, travelled haphazardly in France, and contributed for twenty years as historical features writer to *The Times*. He is known mainly for his literary biographies: *Shelley: The Pursuit* (1977); *Footsteps* (1985); *Coleridge: Early Visions* (1989); Dr Johnson and Mr Savage (1993). He has edited a new edition of Coleridge's Selected Poems (1996) and is just completing a second volume of his biography, Coleridge: Darker Reflections. He has lectured widely in defence of the biographical method, for the British Council in France, Portugal, Germany and Australia; at the Institute of Psycho-Analysis; and at the Royal College of Surgeons. He is a Fellow of the Royal Society of Literature, and a Fellow of the British Academy, and was made an OBE in 1992.

THE ROYAL INSTITUTION

The Royal Institution of Great Britain was founded in 1799 by Benjamin Thompson, Count Rumford. It has occupied the same premises for nearly 200 years and, in that time, a truly astounding series of scientific discoveries has been made within its walls. Rumford himself was an early and effective exponent of energy conservation. Thomas Young established the wave theory of light; Humphry Davy isolated the first alkali and alkaline earth metals, and invented the miners' lamp; Tyndall explained the flow of glaciers and was the first to measure the absorption and radiation of heat by gases and vapours; Dewar liquefied hydrogen and gave the world the vacuum flask; all who wished to learn the new science of X-ray crystallography that W.H. Bragg and his son had discovered came to the Royal Institution, while W.L. Bragg, a generation later, promoted the application of the same science to the unravelling of the structure of proteins. In the recent past the research concentrated on photochemistry under the leadership of Professor Sir George (now Lord) Porter, while the current focus of the research work is the exploration of the properties of complex materials.

Towering over all else is the work of Michael Faraday, the London bookbinder who became one of the world's greatest scientists. Faraday's discovery of electromagnetic induction laid the foundation of today's electrical industries. His magnetic laboratory, where many of his most important discoveries were made, was restored in 1972 to the form it was known to have had in 1854. A museum, adjacent to the laboratory, houses a unique collection of original apparatus arranged to illustrate the more important aspects of Faraday's immense contribution to the advancement of science in his fifty years at the Royal Institution.

Why the Royal Institution Is Unique

It provides the only forum in London where non-specialists may meet the leading scientists of our time and hear their latest discoveries explained in everyday language.

It is the only Society that is actively engaged in research, and provides lectures covering all aspects of science and technology, with membership open to all.

It houses the only independent research laboratory in London's West End (and one of the few in Britain)—the Davy Faraday Research Laboratory.

What the Royal Institution Does for Young Scientists

The Royal Institution has an extensive programme of scientific activities designed to inform and inspire young people. This programme includes lectures for primary and secondary school children, sixth form conferences, Computational Science Seminars for sixth-formers and Mathematics Masterclasses for 12–13 year-old children.

What the Royal Institution Offers to its Members

Programmes, each term, of activities including summaries of the Discourses; synopses of the Christmas Lectures and annual Record.

Evening Discourses and an associated exhibition to which guests may be invited.

An annual volume of the *Proceedings of the Royal Institution of Great Britain* containing accounts of Discourses.

Christmas Lectures to which children may be introduced.

Meetings such as the RI Discussion Evenings; Seminars of the Royal Institution Centre for the History of Science and Technology, and other specialist research discussions.

Use of the Libraries and borrowing of the books. The Library is open from 9 a.m. to 9 p.m. on weekdays.

Use of the Conversation Room for social purposes.

Access to the Faraday Laboratory and Museum for themselves and guests. Invitations to debates on matters of current concern, evening parties and lectures marking special scientific occasions.

Royal Institution publications at privileged rates.

Group visits to various scientific, historical, and other institutions of interest.

Evening Discourses

The Evening Discourses have been given regularly since 1826. They cover all aspects of science and technology (with regular ventures into the arts) in a form suitable for the interested layman, and many scientists use them to keep in touch with fields other than their own. An exhibition, on a subject relating to the Discourse, is arranged each evening, and light refreshments are available after the lecture.

Christmas Lectures

Faraday introduced a series of six Christmas Lectures for children in 1826. These are still given annually, but today they reach a much wider audience through television. Titles have included: 'The Languages of Animals' by David Attenborough, 'The Natural History of a Sunbeam' by Sir George Porter, 'The Planets' by Carl Sagan and 'Exploring Music' by Charles Taylor.

The Library

The Royal Institution library reflects the functions and the activities of the RI. The subject coverage is science, its history, its role in society including education, and its interaction with religion, literature, and the arts. The emphasis is on the popular science books, the history of science, and the research monographs of interest to the research group in the Davy Faraday Research Laboratories.

It is probably the only library of its kind specializing in the public understanding of science, that is science for the non-specialist. It also has a junior section.

Schools' Lectures

Extending the policy of bringing science to children, the Royal Institution provides lectures throughout the year for school children of various ages, ranging from primary to sixth-form groups. These lectures, attended by thousands, play a vital part in stimulating an interest in science by means of demonstrations, many of which could not be performed in schools.

Seminars, Masterclasses, and Primary Schools' Lectures

In addition to educational activities within the Royal Institution, there is an expanding external programme of activities which are organized at venues throughout the UK. These include a range of seminars and master classes in the areas of mathematics, technology and, most recently, computational science. Lectures aimed at the 8–10 year-old age group are also an increasing component of our external activities.

Teachers' Workshops

Lectures to younger children are commonly accompanied by workshops for teachers which aim to explain, illustrate, and amplify the scientific principles demonstrated by the lecture.

Membership of the Royal Institution

Member

The Royal Institution welcomes all who are interested in science, no special scientific qualification being required. By becoming a Member of the Royal Institution an individual not only derives a great deal of personal benefit and enjoyment but also the satisfaction of helping to support the unique contribution made to our society by the Royal Institution.

Family Associate Subscriber

A Member may nominate one member of his or her family residing at the same address, and not being under the age of 16 (there is no upper age limit), to be a Family Associate Subscriber. Family Associate Subscribers can attend the Evening Discourses and other lectures, and use the Libraries.

Associate Subscriber

Any person between the ages of 16 and 27 may become an Associate Subscriber. Associate Subscribers can attend the Evening Discourses and other lectures, and use the Libraries.

Junior Associate

Any person between the ages of 11 and 15 may become a Junior Associate. Junior Associates can attend the Christmas Lectures and other functions, and use the Libraries. There are also visits organized during Easter and Summer vacations.

Corporate Subscriber

Companies, firms and other bodies are invited to support the work of the Royal Institution by becoming Corporate Subscribers; such organizations make a very valuable contribution to the income of the Institution and so endorse its value to the community. Two representatives may attend the Evening Discourses and other lectures, and may use the Libraries.

College Corporate Subscriber

Senior educational establishments may become College Corporate Subscribers; this entitles two representatives to attend the Evening Discourses and other lectures, and to use the Libraries.

School Subscriber

Schools and Colleges of Education may become School Subscribers; this entitles two members of staff to attend the Evening Discourses and other lectures, and to use the Libraries.

Membership forms can be obtained from: The Membership Secretary, The Royal Institution, 21 Albemarle Street, London W1X 4BS. Telephone: 0171 409 2992. Fax: 0171 629 3569.

Discourses

Reproductive Fallacies — *27 January 1995*
Jack Cohen, PhD

Science and Political Power — *3 February 1995*
The Rt Hon Tony Benn, MP

Risks, Costs, Choice and Rationality — *10 February 1995*
Sir Colin Berry, PhD, FRCPath, FRCP, FFPM

Laser Treatment of Myopia—Is It Really The Bottom Line? — *17 February 1995*
David S. Gartry, FRCS, FRCOphth, DO, FBCO

Nuclear Power Plant Safety—What's the Problem? — *24 February 1995*
John Collier FRS, FEng

Will Computers Ever Permanently Dethrone the Human Chess Champions? — *3 March 1995*
Raymond Keene, OBE

Sticking Up for Adhesives — *10 March 1995*
Anthony J. Kinloch, PhD, DSc(Eng), FIM, FRSC, CChem, CEng

The Art of Scientific Demonstrations (The Woodhull Lecture) — *17 March 1995*
David Jones, PhD

Complex Systems: a Physicist's Viewpoint — *5 May 1995*
Giorgio Parisi

The Chemistry of the Interstellar Medium as Revealed by Spectroscopy — *12 May 1995*
William Klemperer, PhD

The Origin of Life (The Woodhull Lecture) — *19 May 1995*
John Maynard Smith, FRS

A Strategy for Nature Conservation in England — *26 May 1995*
The Earl of Cranbrook, PhD, DSc, CBiol, DL

Safety's Debt to Davy and Faraday *13 October 1995*
James McQuaid, PhD, DSc, FEng, FIMinE

Advertising Washes Whiter? *20 October 1995*
Winston Fletcher, MA, FIPA

Aphrodisiacs, Psychedelics and the Elusive Magic Bullet *27 October 1995*
John Mann, PhD, DSc, FRSC

An Imaging Renaissance—New Opportunities for Medicine *3 November 1995*
Ian Isherwood, MD, FRCP, FRCR

Incremental Decisions in a Complex World *10 November 1995*
Chris Elliott, MA, PhD, CEng, FRAeS

Science and Fine Art *17 November 1995*
David Bomford, MSc, FIIC

Exploring the Universe with the Hubble Space Telescope *24 November 1995*
Alexander Boksenberg, PhD, FRS

Visual Art and the Visual Brain (The Woodhull Lecture) *1 December 1995*
Semir Zeki, FRS

Faraday in the Pits, Faraday at Sea *26 January 1996*
Frank James, MSc, PhD, DIC, FRAS

Tick, Tick, Tick Pulsating Star: How We Wonder What You Are *2 February 1996*
S. Jocelyn Bell Burnell, PhD

Pondering on Pisa *9 February 1996*
John B. Burland, DSc, FEng, FICE, MIStructE

Pollution: a Local, National and World Problem *16 February 1996*
The Lord Lewis, Kt, FRS

The Colossus of Bletchley Park *23 February 1995*
Anthony E. Sale, FBCS

To Divide or Not to Divide? *1 March 1996*
Paul Nurse, FRS

Lasers—Making Light Work *8 March 1996*
Colin Webb, FRS

There or Thereabouts—the Story of Measurement, Past, Present and Future *15 March 1996*
Andrew Wallard, PhD

On the Air, the Constituents of the Atmosphere, their Separation and Uses (The BOC Lecture) *3 May 1996*
Michael E. Garrett

Sheathing the Two-Edged Sword: One Hundred Years of Radioactivity *10 May 1996*
Peter Christmas, MA, DPhil

Human Use and Abuse of Alcohol, a Historical Perspective *17 May 1996*
Bert L. Vallee, MD

The Crucible of Creation: the Burgess Shale and the Rise of Animals (The Woodhull Lecture) *24 May 1996*
Simon Conway Morris, FRS

Meteorites: Messengers from the Past *18 October 1996*
Monica Grady, PhD

Asthma and Allergy: Disorders of Civilisation *25 October 1996*
Stephen T. Holgate

Tower Bridge: Enduring Monument or Victorian Folly *1 November 1996*
Roy Aylott, CEng, FICE, FIHT

Insects and Chemical Signals: A Volatile Situation *8 November 1996*
John Pickett

Evidence and Speculation in Astronomy *15 November 1996*
Sir Martin Rees, FRS

Metals Combined—or, Chaos Vanquished *22 November 1996*
Robert W. Cahn, FRS

Disappearing Species: Problems *and* Opportunities *29 November 1996*
Norman Myers, PhD

Television Beyond the Millennium *6 December 1996*
Will Wyatt

Magellen Looks at Venus *24 January 1997*
D. P. McKenzie, FRS

The Search for Extraterrestrial Life—And the Future of Life on Earth *31 January 1997*
Sir Arnold Wolfendale, FRS

Molecular Information Processing: Will It Happen? *7 February 1997*
Peter Day, FRS

One Hundred Years of the Electron *14 February 1997*
Sir Brian Pippard, FRS

Self-Assembly: Nature's Way To Do It *21 February 1997*
Kuniaki Nagayana, PhD

J. J. Thomson, Discoverer of the Electron: The Man in His Setting *28 February 1997*
David Thomson

An Arts/Science Interface: Mediaeval Manuscripts, Pigments and Spectroscopy *7 March 1997*
Robin J. H. Clark, FRS

The Coleridge Experiment *14 March 1997*
Richard Holmes, OBE, FRSSL

Carbon Nanotubes, the Tiniest Man-made Tubes *9 May 1997*
Sumio Iijima, PhD

The Biology of Nitric Oxide *16 May 1997*
Salvador Moncada, FRCP, FRS

The Epidemic of Mad Cow Disease (BSE) in the United Kingdom *23 May 1997*
Roy M. Anderson, FRS

To See a World in a Grain of Sand *30 May 1997*
Sir John Meurig Thomas, FRS

Electricity, Magnetism and the Body *17 October 1997*
Anthony T. Barker, PhD, FIEE, FIPEM

Risky Business *24 October 1997*
Ben Selinger, FTS, FRACI

Vegetation and Climate: Virtually Worlds in Themselves *31 October 1997*
P.A. Spicer, PhD

MRI: A Window into the Human Body *7 November 1997*
Laurie Hall, FRS(Can)

Environmental Campaigners and Scientists—Friends or Foes *14 November 1997*
Lord Melchett

Fuel Cells for the Future *21 November 1997*
Andrew Hamnett, DPhil

Gardening as Ecology *28 November 1997*
Stefan Buczacki, DPhil

Killers in the Brain: New Discoveries in Neurodegenerative Diseases *5 December 1997*
Nancy Rothwell, PhD, DSc

The Human Singing Voice *30 January 1998*
David M. Howard, PhD, FIOA

The Nino and its Significance *6 February 1998*
Sir Crispin Tickell, GCMG, KCVO

Mendeleev's Palette *13 February 1998*
Anthony K. Cheetham, FRS

Destructive Rust, Creative Rust: From the Pantheon to the Millennium Dome *20 February 1998*
Jack Harris, MBE, FEng, FRS

The Science of Murphy's Law *27 February 1998*
Robert Matthews, MA, CPhys, MInstP, FRAS

FLIMsy Constructions *6 March 1998*
David Phillips, BSc, PhD, FRSC

Every Drop to Drink *13 March 1998*
K. J. Ives, CBE, DSc, FEng

God, Time, and Cosmology *8 May 1998*
F. Russell Stannard, PhD, FInstP, Cphys

The Royal Institution as a Mayfair Landholder *15 May 1998*
H. J. V. Tyrrell, MA, Dsc, FRSC

Solar Fuel from Chemistry *22 May 1998*
Harry B. Gray

A Philosopher's Tree: Michael Fara*day* To*day* at the Royal Institution *29 May 1998*
Peter Day, FRS